Selected Paper from 6th International Conference on Renewable Energy Sources (ICoRES 2019)

Selected Paper from 6th International Conference on Renewable Energy Sources (ICoRES 2019)

Editor

Marcin Jewiarz

MDPI • Basel • Beijing • Wuhan • Barcelona • Belgrade • Manchester • Tokyo • Cluj • Tianjin

Editor
Marcin Jewiarz
University of Agriculture in Krakow
Poland

Editorial Office
MDPI
St. Alban-Anlage 66
4052 Basel, Switzerland

This is a reprint of articles from the Special Issue published online in the open access journal *Energies* (ISSN 1996-1073) (available at: https://www.mdpi.com/journal/energies/special_issues/ICoRES2019).

For citation purposes, cite each article independently as indicated on the article page online and as indicated below:

LastName, A.A.; LastName, B.B.; LastName, C.C. Article Title. *Journal Name* **Year**, *Article Number*, Page Range.

ISBN 978-3-03943-422-0 (Hbk)
ISBN 978-3-03943-423-7 (PDF)

© 2020 by the authors. Articles in this book are Open Access and distributed under the Creative Commons Attribution (CC BY) license, which allows users to download, copy and build upon published articles, as long as the author and publisher are properly credited, which ensures maximum dissemination and a wider impact of our publications.

The book as a whole is distributed by MDPI under the terms and conditions of the Creative Commons license CC BY-NC-ND.

Contents

About the Editor vii

Preface to "Selected Paper from 6th International Conference on Renewable Energy Sources (ICoRES 2019)" ix

Daniel Chludziński and Michał Duda
A New Concept and a Test of a Bubble Pump System for Passive Heat Transport from Solar Collectors †
Reprinted from: *Energies* **2020**, *13*, 1185, doi:10.3390/en13051185 1

Weronika Kruszelnicka
A New Model for Environmental Assessment of the Comminution Process in the Chain of Biomass Energy Processing †
Reprinted from: *Energies* **2020**, *13*, 330, doi:10.3390/en13020330 17

Joanna Hałacz, Aldona Skotnicka-Siepsiak and Maciej Neugebauer
Assessment of Reducing Pollutant Emissions in Selected Heating and Ventilation Systems in Single-Family Houses †
Reprinted from: *Energies* **2020**, *13*, 1224, doi:10.3390/en13051224 39

Sławomir Francik, Adrian Knapczyk, Artur Knapczyk and Renata Francik
Decision Support System for the Production of Miscanthus and Willow Briquettes
Reprinted from: *Energies* **2020**, *13*, 1364, doi:10.3390/en13061364 59

Jakub Styks, Marek Wróbel, Jarosław Frączek and Adrian Knapczyk
Effect of Compaction Pressure and Moisture Content on Quality Parameters of Perennial Biomass Pellets
Reprinted from: *Energies* **2020**, *13*, 1859, doi:10.3390/en13081859 83

Radovan Nosek, Maw Maw Tun and Dagmar Juchelkova
Energy Utilization of Spent Coffee Grounds in the Form of Pellets
Reprinted from: *Energies* **2020**, *13*, 1235, doi:10.3390/en13051235 103

Elżbieta Hałaj, Leszek Pająk and Bartosz Papiernik
Finite Element Modeling of Geothermal Source of Heat Pump in Long-Term Operation †
Reprinted from: *Energies* **2020**, *13*, 1341, doi:10.3390/en13061341 111

Marek Wróbel, Marcin Jewiarz, Krzysztof Mudryk and Adrian Knapczyk
Influence of Raw Material Drying Temperature on the Scots Pine (*Pinus sylvestris* L.) Biomass Agglomeration Process—A Preliminary Study
Reprinted from: *Energies* **2020**, *13*, 1809, doi:10.3390/en13071809 129

Marcin Jewiarz, Krzysztof Mudryk, Marek Wróbel, Jarosław Frączek and Krzysztof Dziedzic
Parameters Affecting RDF-Based Pellet Quality
Reprinted from: *Energies* **2020**, *13*, 910, doi:10.3390/en13040910 147

Jakub Sikora, Marcin Niemiec, Anna Szeląg-Sikora, Zofia Gródek-Szostak, Maciej Kuboń and Monika Komorowska
The Effect of the Addition of a Fat Emulsifier on the Amount and Quality of the Obtained Biogas
Reprinted from: *Energies* **2020**, *13*, 1825, doi:10.3390/en13071825 165

Krzysztof Pilarski, Agnieszka A. Pilarska, Piotr Boniecki, Gniewko Niedbała, Karol Durczak, Kamil Witaszek, Natalia Mioduszewska and Ireneusz Kowalik
The Efficiency of Industrial and Laboratory Anaerobic Digesters of Organic Substrates: The Use of the Biochemical Methane Potential Correction Coefficient
Reprinted from: *Energies* **2020**, *13*, 1280, doi:10.3390/en13051280 . . . . **177**

Beata Brzychczyk, Tomasz Hebda, Jakub Fitas and Jan Giełżecki
The Follow-up Photobioreactor Illumination System for the Cultivation of Photosynthetic Microorganisms
Reprinted from: *Energies* **2020**, *13*, 1143, doi:10.3390/en13051143 . . . . **191**

Jakub Sikora, Marcin Niemiec, Anna Szeląg-Sikora, Zofia Gródek-Szostak, Maciej Kuboń and Monika Komorowska
The Impact of a Controlled-Release Fertilizer on Greenhouse Gas Emissions and the Efficiency of the Production of Chinese Cabbage
Reprinted from: *Energies* **2020**, *13*, 2063, doi:10.3390/en13082063 . . . . **201**

Przemysław Motyl, Marcin Wikło, Julita Bukalska, Bartosz Piechnik and Rafał Kalbarczyk
A New Design for Wood Stoves Based on Numerical Analysis and Experimental Research [†]
Reprinted from: *Energies* **2020**, *13*, 1028, doi:10.3390/en13051028 . . . . **215**

About the Editor

Marcin Jewiarz is Assistant Professor in the Department of Mechanical Engineering and Agrophysics at the University of Agriculture in Krakow. His research focuses on solid biofuel production and combustion and biochar production. He has supervised more than 12 M.Sc. candidates and is the author of more than 30 articles indexed in Web of Science. Since 2015, he has taken part in the organizing committee of the International Conference on Renewable Energy Sources held in Krynica, Poland.

Preface to "Selected Paper from 6th International Conference on Renewable Energy Sources (ICoRES 2019)"

Thank you for reaching for this book. It is a summary of the research presented at the 6th International Conference on Renewable Energy Sources (ICORES19), which took place in Krynica, Poland, in June 2019. This event is the most recognizable scientific meeting connected to RES in Poland. From the very beginning, this conference has been a unique occasion for gathering Polish and international researchers' perspectives on renewable energy sources and balancing them against governmental policy considerations. Accordingly, the conference has also offered panels to discuss best practices and solutions with local entrepreneurs and federal government bodies. The meeting attracts not only scientists but also industry representatives, as well as local and federal government personnel. We are open to new and fresh ideas concerning renewable energy, which is why so many scientists from Central and Eastern Europe visit Krynica to discuss the "Green Future" of this region. In 2019, the conference was organized by the University of Agriculture in Krakow, in cooperation with the AGH University of Science and Technology (Krakow), the State Agrarian and Engineering University in Podilya, the University of Žilina, the International Commission of Agricultural and Biosystems Engineering (CIGR) and the Polish Society of Agricultural Engineering. Honorary auspices were made by the Ministry of Science and Higher Education of the Republic of Poland, the rector of the University of Agriculture in Krakow, the rector of the AGH University of Science and Technology and the rector of the State Agrarian and Engineering University in Podilya.

Marcin Jewiarz
Editor

Article

A New Concept and a Test of a Bubble Pump System for Passive Heat Transport from Solar Collectors †

Daniel Chludziński * and Michał Duda

Department of Electrical Engineering, Power Engineering, Electronics and Automation, University of Warmia and Mazury in Olsztyn, Oczapowskiego 11, 10-736 Olsztyn, Poland; michal.duda@uwm.edu.pl
* Correspondence: daniel.chludzinski@uwm.edu.pl
† This paper is an extended version of article presented at the 6th Scientific Conference Renewable Energy Systems, Engineering, Technology, Innovation ICoRES, Krynica, Poland, 12–14 June 2019, to be published in E3S Web of Conferences.

Received: 30 January 2020; Accepted: 3 March 2020; Published: 5 March 2020

Abstract: Heat is usually transported by hydraulic circuits where the working medium is circulated by an electric pump. Heat can also be transferred by natural convection in passive systems. Passive systems where heat is transported downward have also been described and studied in the literature. These types of devices are referred to as reverse thermosiphons. However, systems of the type are not widely applied in practice due to the problems associated with the selection of the optimal working medium. In water-based thermosiphons, negative pressure is produced when water temperature falls below 100 °C, and non-condensable gas can enter the system. These problems are not encountered in systems where the working medium has a low boiling point. However, liquids with a low boiling point can be explosive, expensive, and harmful to the environment. The solution proposed in this paper combines the advantages of water and a liquid with a low boiling point. The described system relies on water as the heat transfer medium and small amounts of a substance with a low boiling point. The developed model was tested under laboratory conditions to validate the effectiveness of a passive system where heat is transported downward with the involvement of two working media. The system's operating parameters are also described.

Keywords: thermosiphon; anti-gravity; bubble pump; solar collector; passive heat transport

1. Introduction

Rapid technological development spurs the search for new solutions to improve the performance and reliability of equipment. The demand for energy is directly linked with economic growth, and most of the energy consumed world-wide is derived from fossil fuels [1].

Thermal energy plays a considerable role in the economy, which is why heat transfer processes deserve special attention in a critical approach. Novel solutions should be characterized by the lowest possible energy consumption, sound, and reliable design, low cost, and low environmental impact.

Conventional heat exchange systems rely on hydraulic circuits with natural (thermosiphon) and forced circulation (electrical pump). From the point of view of energy conservation, passive systems (thermosiphons) are preferred because they do not require an external power source (electricity). They are also characterized by simple design and high reliability. Despite many advantages, passive heat exchange systems have a major drawback: heat is transferred in one direction only—from the bottom to the top of the tank.

Passive circulation systems that transport heat downward from the hot water tank have been described in the literature [2–6]. Two-phase thermosiphons are one of such solutions. In two-phase thermosiphons, the temperature of the working medium reaches boiling point, and heat energy is transmitted by vapor. The main drawback of this solution is that the condensate has to be returned to the heating zone against the force of gravity. Condensate is lifted by the difference in vapor pressure between heat supply and heat extraction zones. Some systems feature porous materials to lift the condensate by capillary force. However, in such solutions, heat can be transferred only across small distances (several meters) [7].

A two-phase thermosiphon with a liquid heat exchanger supports passive downward transport of heat across greater distances [8,9]. In this solution, condensate does not have to be lifted. These types of thermosiphons have been widely studied [10–14]. Most of them rely on a single working medium, usually water or coolant. In water-based thermosiphons, negative pressure is produced when the temperature of the heat source falls below 100 °C, which creates practical problems. Vast amounts of coolant are needed to fill the entire system, which increases cost and exerts a negative impact on the environment. These types of systems are not widely used due to the problems associated with the selection of the optimal working medium.

This article presents the results of an experiment where two working media were applied to induce passive downward transfer of heat in a system with a low-temperature heat source. Unlike thermosiphons that operate cyclically [15–17], the proposed system is characterized by continuous operation. The liquid medium is water, and the second medium is a substance with a low boiling point [18,19].

Reverse thermosiphons can be applied to transfer heat from solar collectors. Renewable energy plays an increasingly important role in the development of many countries, and the results of this study were used to assess the applicability of reverse thermosiphons in the context of renewable energy generation.

2. A Review of Passive Heat Transport Systems

2.1. Reverse Thermosiphons

Passive heat exchange systems generally rely on three physical phenomena: phase change, capillary flow, and gravity. Depending on the structure of a passive heat exchange system, these phenomena can occur simultaneously, separately, or in different combinations (Figure 1) [20].

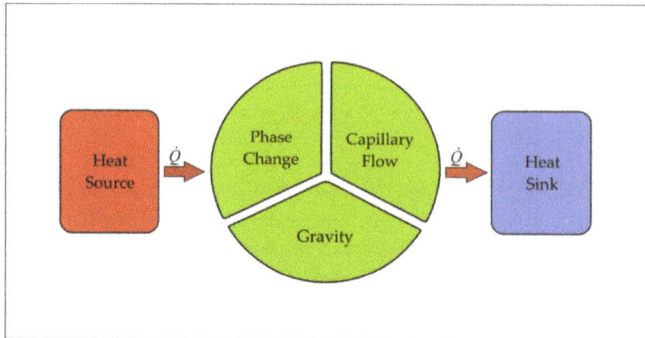

Figure 1. General diagram of a passive heat exchanger.

A reverse thermosiphon (RT) enables passive heat transfer in a direction opposite to natural convection (downward). One of the first devices of the type was developed by Tamburini in 1977. The proposed solution was used to cool electronic systems [21]. In reversed thermosiphons, the condensate has to be returned to the evaporation zone, which poses certain problems. Two methods for lifting the condensate have been proposed in the literature:

- anti-gravity thermosiphon with a vapor lift pump,
- anti-gravity thermosiphon with a passive pumping module [22].

Depending on the applied solution, heat is transferred by:

- a gaseous medium (vapor),
- a liquid medium.

2.2. Reverse Thermosiphons with a Vapor Lift Pump and One Working Medium

Systems where heat is transported downward rely on a vaport lift pump, which is also known as a bubble pump. These systems use only one working medium. The main advantage of systems with a vapor lift pump is their relatively simple structure and the absence of moving elements.

A heat transfer device with a vapor lift pump is presented in Figure 2. The solution proposed by Movick (Figure 2a) has a significant drawback because the lifting pipe and the return pipe remain in direct contact, which decreases the system's efficiency. The remaining devices in Figure 2 have separate lifting and return pipes. Devices developed specifically for use with solar collectors are presented in Figure 2b–d,g. General-purpose heat transfer devices without additional pumps are presented in Figure 2a,e,f.

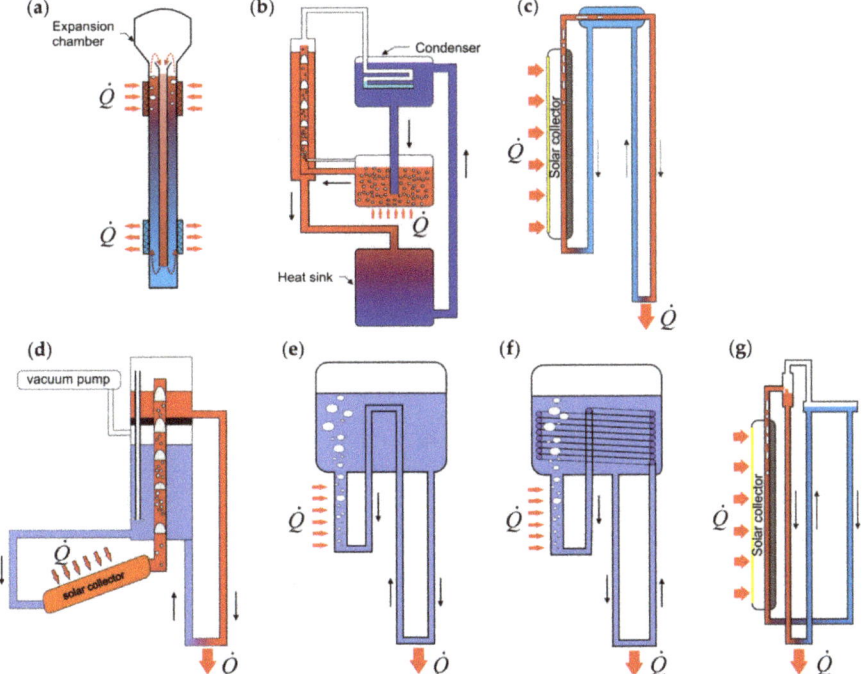

Figure 2. Diagrams of devices with a bubble pump for transporting heat downward (The figure is based on patents): (**a**) US Patent 1976 [23]; (**b**) US Patent 1981 [24]; (**c**) US Patent 1985 [25]; (**d**) US Patent 1994 [26]; (**e**) US Patent 2007 [27]; (**f**) US Patent 2008 [28]; (**g**) US Patent 2010 [29].

Reverse thermosiphons rely on a vapor lift pump, also known as a bubble pump, to transfer heat downward from the tank. These devices are filled with one working medium. Their main advantage is a relatively simple structure and the absence of moving elements. A diagram of a solar heater with a vapor lift pump is presented in Figure 3 [30].

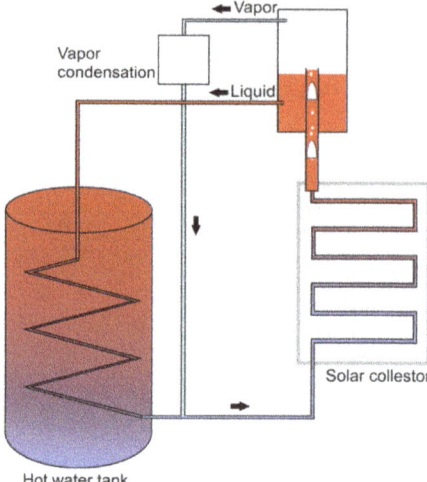

Figure 3. Diagram of a solar heater with a vapor lift pump (The figure is based [30]).

In this system, heat is transferred by a liquid medium. The heated medium is lifted by gas bubbles to the separator, and it returns to the hot water tank and the flat panel by force of gravity. The two-phase (liquid-vapor) reservoir is referred to as the bubble pump.

A system with a vapor lift pump and a solar collector was analyzed by Han-shik et al. [31]. This system was filled with water as the working medium for transporting heat from vacuum tube collectors. The tested installation had the height of 1 m and 5 m. The device was activated when temperature inside the collector reached 90–100 °C. The experiment was conducted in a real-world setting, and the operation of the tested device was considerably influenced by variations in solar radiation. According to the authors, the developed device can replace an electric circulator pump. The discussed system is presented in Figure 4.

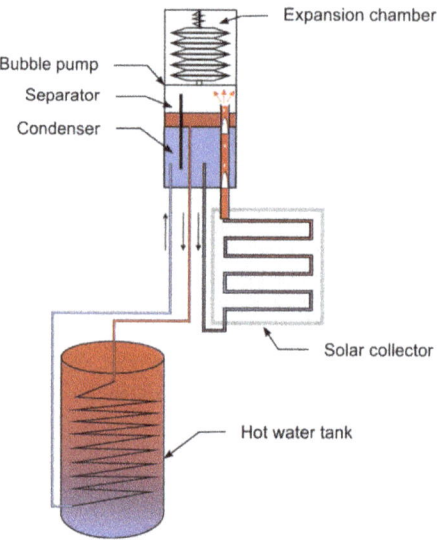

Figure 4. Diagram of a heat exchanger with a vapor lift pump (The figure is based [31]).

Similar systems that are filled with one working medium and use a vapor lift pump to transfer heat downward have been analyzed by Ito et al. [32], Yandri et al. [33], and Klugmann et al. [34].

3. The Operating Principle of a Reverse Thermosiphon with a Vapor Lift Pump

3.1. Heat Transfer System with One Working Medium

Two-phase flow takes place in reverse thermosiphons where heat is transported by force of gravity. Two-phase flow involves a continuous carrier phase, such as liquid or gas, and a dispersed carrier phase in the form of a solid, liquid, or gas.

Vapor bubbles present in liquid intensify heat transfer in systems where heat is transported by natural convection as well as in systems where heat is transported downward. The operating principle of a reverse thermosiphon with a vapor lift pump and one working medium is presented in Figure 5.

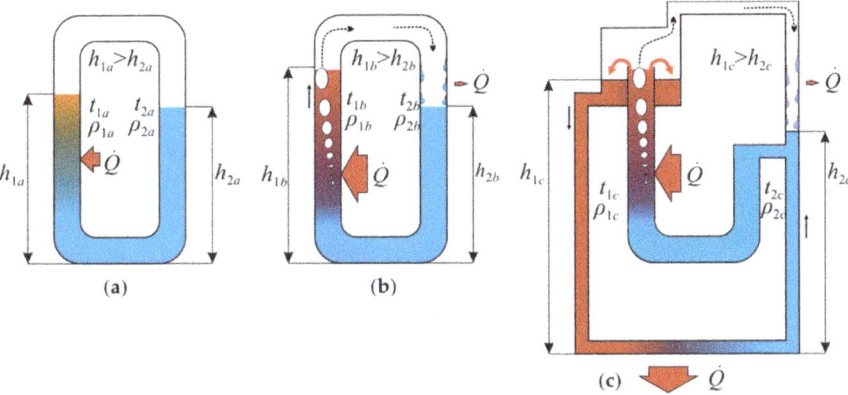

Figure 5. The operating principle of a reverse thermosiphon with one working medium: (**a**) difference in the height of the liquid column in heat pipes resulting from differences in temperature; (**b**) boiling liquid increases the difference in the height of the liquid column in heat pipes; (**c**) circulation in a reverse thermosiphon with one working medium; h_{1a}, h_{2a}, h_{1b}, h_{2b}, h_{1c}, h_{2c}—height of the liquid column in heat pipes; t_{1a}, t_{2a}, t_{1b}, t_{2b}, t_{1c}, t_{2c}—liquid temperature; ρ_{1a}, ρ_{2a}, ρ_{1b}, ρ_{2b}, ρ_{1c}, ρ_{2c}—liquid density.

In the system presented in Figure 5a, if the working medium that partly fills heat pipes has a different temperature ($t_{1a} > t_{2a}$), the liquid inside pipes with a higher temperature has lower density ($\rho_{1a} < \rho_{2a}$). To equalize hydrostatic pressure in the system, the liquid column with a higher temperature has to be taller than the liquid column with a lower temperature ($h_{1a} > h_{2a}$) (Figure 5a). When the liquid reaches boiling point, the produced vapor bubbles increase the difference in the height of liquid columns in heat pipes ($h_{1b} > h_{2b}$) (Figure 5b). The difference in the height of liquid columns induced by vapor bubbles is greater than the difference observed in a system where the working medium does not reach boiling point (Figure 5a).

Reverse thermosiphons with one working medium rely on the flow of boiling liquid (Figure 5c). In a closed system that is partly filled with liquid, heat is transported to one heat pipe in the heating zone. Vapor bubbles lift the liquid, which leads to two-phase single-component flow. Boiling liquid is transported to a vessel positioned above the heating zone (separator), and it is driven downward by force of gravity. Vapor is condensed in the upper part of the heat pipe containing cold liquid. The resulting condensate is mixed with cold liquid and is recirculated to the heating zone. Hydrostatic pressure is higher in the heat pipe containing hot liquid than in the pipe containing cold liquid. As a result, the height of the hot liquid column significantly exceeds the height of the cold liquid column ($h_{1c} > h_{2c}$). The difference in the height of liquid columns is sufficient to compensate for the loss of hydrostatic pressure resulting from the lower density of the liquid with a higher temperature ($\rho_{1c} < \rho_{2c}$).

The resulting difference in hydrostatic pressure is the driving force for liquid circulation inside the system. The heat transfer medium flows from the heat supply zone to the heat extraction zone below. The described system is a reverse thermosiphon where heat is transferred downward (unlike in systems where heat is transferred by natural convection). The system's efficiency is determined by the efficiency of the vapor lift pump. The efficiency of lifting the liquid with the vapor lift pump is not very high, compared with other methods, but this solution has a simple structure, and it is devoid of moving components. The efficiency of a vapor lift pump is affected by the structure of two-phase flow. There are five main flow structures: bubble flow, slug flow, froth flow, annular flow, and mist flow. The basic types of two-phase flow in a vertical pipe are presented in Figure 6. The type of flow is determined mainly by the proportions of vapor and liquid. Slug flow is the optimal type of flow in the analyzed system [35–37].

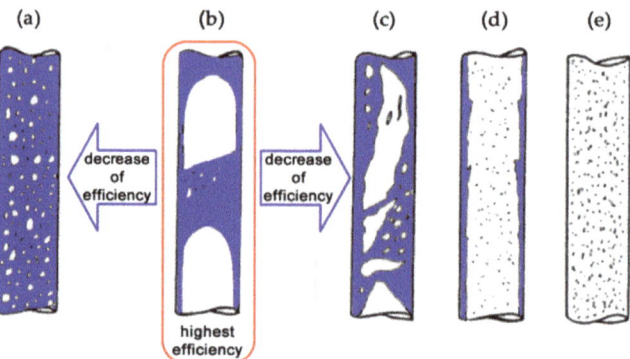

Figure 6. Two-phase flow structures: (**a**) bubble flow; (**b**) slug flow; (**c**) froth flow; (**d**) annular flow; (**e**) mist flow.

3.2. Heat Transfer System with Two Working Media

The reviewed reverse thermosiphons with a vapor lift pump rely on one working medium. Such systems are burdened by the following technical problems:

- In water-based thermosiphons, negative pressure is produced when water temperature falls below 100 °C. The above compromises the system's tightness, and the application of systems that maintain negative pressure is not cost-effective.
- Coolants can be used to maintain negative pressure in a system when water temperature falls below 100 °C, but large amounts of coolant are needed to fill the entire system (such as solar collectors), which increases cost and exerts a negative impact on the environment.

A circuit diagram of a reverse thermosiphon with two working media is presented in Figure 7. Working media in a reverse thermosiphon should have the following characteristics:

- they should not be mutually soluble in the liquid phase,
- the pumping medium should have a lower boiling point than the liquid heat transfer medium, and
- the pumping medium should have lower density than the liquid heat transfer medium.

The proposed thermosiphon with two working media is filled with a liquid working medium and a small amount of a pumping medium (with a low boiling point) that does not exceed 0.3 kg.

A reverse thermosiphon with two working media has five basic components: a vapor lift pump, an evaporator, a separator, a condenser, and a thermal receiver with a heat exchanger. All components are connected to form a closed liquid system. The operating principle of a reverse thermosiphon with two working media is presented in Figure 7. The heat flux reaches the evaporator which is partly filled with the liquid heat transfer medium. A thin layer of the pumping medium (6–12 mm) floats on the liquid medium. The liquid medium is heated, which raises the temperature of the pumping medium. When the vaporized pumping medium reaches the required pressure, it is transferred to the vapor lift pump via a dedicated channel. As a result, two-phase two-component flow (liquid heat transfer medium—vaporized pumping medium) takes place in the vapor lift pump, and the liquid medium is pumped to the separator. The liquid medium and the vaporized pumping medium are separated in the separator. Gravity drives the liquid medium to the thermal receiver. The liquid medium is cooled and recirculated to the evaporator. The vaporized pumping medium flows to the condenser, and the condensate returns to the evaporator by force of gravity.

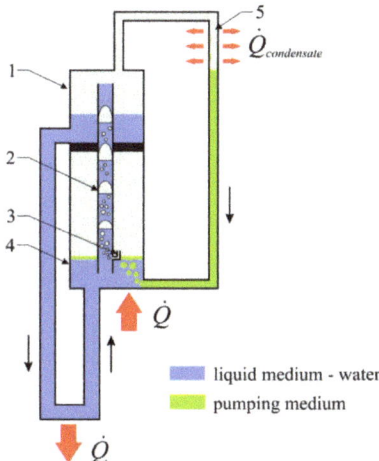

Figure 7. Circuit diagram of a reverse thermosiphon with two working media: 1—separator, 2—vapor lift pump, 3—steam inlet, 4—evaporator, 5—thermal receiver, 6—condenser.

A solar collector can be used as a heat source in the proposed thermosiphon. In this case, the thermosiphon has to be positioned above the solar collector (Figure 4). Heat is transferred from the evaporator to the collector by natural convection.

4. Materials and Methods

A diagram of a laboratory stand for testing a reverse thermosiphon with two working media is presented in Figure 8.

The vapor lift pump (3) is the main component of a reverse thermosiphon with two working media. The vapor lift pump (3) is positioned inside a vessel divided into two zones: the separator in the top zone (1) and the evaporator in the bottom zone (2). The electric heater (5) and the pumping medium return pipe (11) are connected to the bottom zone of the evaporator (2). The vapor pipe (14) is connected to the top zone of the separator (1), and the downcomer pipe (6) is connected to the bottom zone of the separator. The other end of the downcomer pipe (6) is connected to the heat exchanger (8) in the water hot water tank (7). The hot water tank (7) has a volume of 25 liters, it is thermally insulated and equipped with an electric stirrer to prevent liquid stratification inside the tank. The condenser (12) is positioned above the vessel containing the evaporator and the separator. The vapor produced by the pumping medium is transported by the pumping medium pipe (14) to the upper part of the coil pipe. The condensate produced in the coil pipe returns to the evaporator (2) via the pumping medium return pipe (11). The condensate returns to the evaporator by force of gravity. The pumping medium has lower density than the heat transfer medium; therefore, the condenser (12) has to be positioned above the vessel containing the separator (1) and the evaporator (2). The riser pipe (9) and the liquid medium return pipe (10) are connected to the bottom part of the condenser. The vapor lift pump, the evaporator, and the separator were made of transparent material (Plexiglas) to facilitate observations of two-phase (liquid-gas) flow and to control liquid levels in each segment of the model.

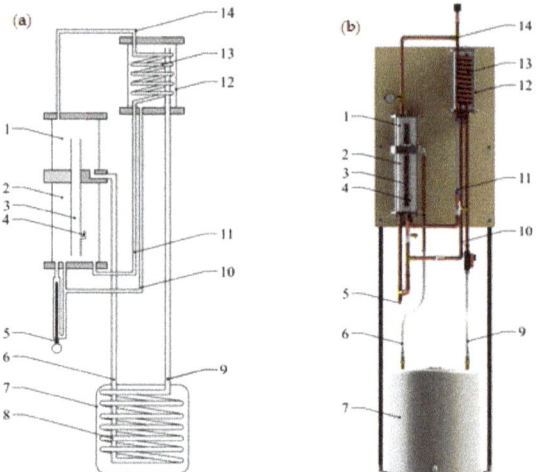

Figure 8. Diagram of the laboratory stand for testing a reverse thermosiphon with two working media: (**a**) circuit diagram, (**b**) laboratory model; 1—separator; 2—evaporator, 3—vapor lift pump, 4—vapor inlet, 5—electric heater, 6—downcomer pipe, 7—hot water tank, 8—coiled pipe inside the hot water tank, 9—riser pipe, 10—liquid medium return pipe, 11—pumping medium return pipe, 12—condenser, 13—condenser coil, 14—pumping medium pipe.

A series of experiments were carried out to evaluate the applicability of two working media for transporting heat in a direction opposite to natural convection in a thermosiphon with a vapor lift pump. The laboratory test stand is presented in Figure 9.

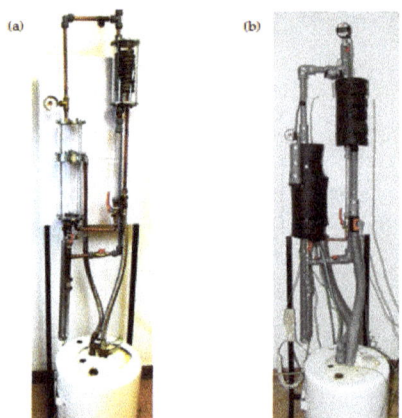

Figure 9. Laboratory test stand: (**a**) without thermal insulation; (**b**) with thermal insulation.

Water (6 kg) was the heat transfer medium, and pentane (0.1 kg) was the pumping medium. These substances are mutually insoluble in the liquid phase. Pentane is less dense than water [38], which is why it is found in the top part of the thermosiphon. Heat was supplied by a 1500 W electric heater. The experiments were conducted at three power settings: 600, 900, and 1200 W. Water temperature inside the hot water tank was 20 °C at the beginning of each experiment. The experiment was terminated when water temperature inside the hot water tank exceeded 40 °C. The system was thermally insulated to minimize heat loss (Figure 9b).

The following physical parameters were measured during the experiment:

- liquid temperature,
- vapor temperature,
- volumetric flow rate of the heat transfer medium, and
- electric heater power.

The location of measuring sensors is presented in Figure 10.

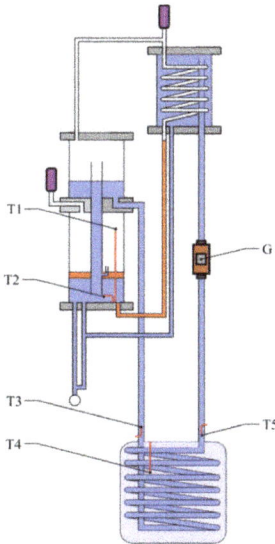

Figure 10. Location of measuring sensors: T1–T5—temperature sensors; G—flow sensor.

All physical parameters were measured and registered at a frequency of 1 Hz.

5. Results

5.1. Flow Rate of the Heat Transfer Medium

The flow rate of the heat transfer medium in the reverse thermosiphon was measured during the experiment. Instantaneous flow rates of the liquid medium (dotted line) at 600 W are presented in Figure 11. Data fluctuations were smoothed with a moving average trend line (continuous line). A total of 60 data points (time period) were averaged (data were averaged for 60 measurements, i.e., over a period of one minute).

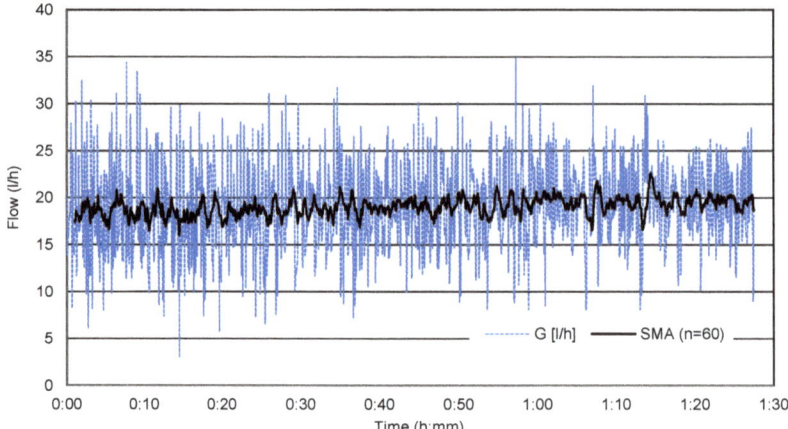

Figure 11. Instantaneous flow rate of the liquid heat transfer medium in the laboratory model at 600 W.

The flow rate of the working medium at every power setting is presented in Figure 12. Only moving average trend lines for 60 measurements were presented to preserve the diagram's readability.

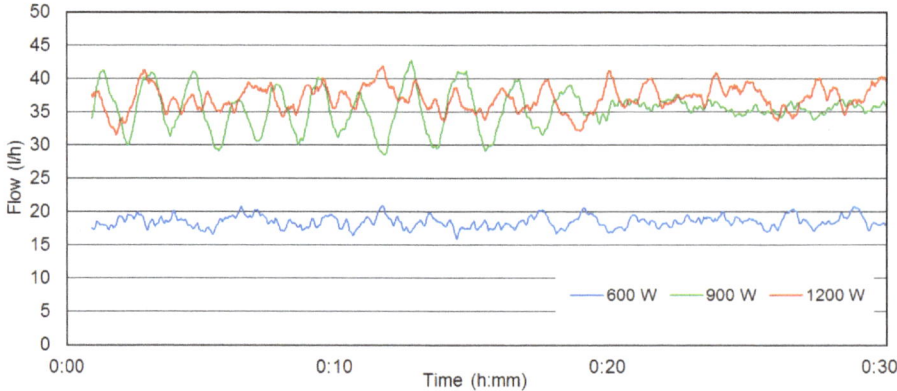

Figure 12. Flow rate of the liquid heat transfer medium represented by moving average trend lines (time period—60) at 600, 900 and 1200 W.

Medium flow was unstable in all experiments. The average flow rate was determined at 15 ÷ 20 L/h at 600 W and 30 ÷ 40 L/h at 900 and 1200 W. The flow rate of the liquid heat transfer medium did not increase when the heater's power output was raised from 900 to 1200 W.

5.2. Temperatures in the System

During the experiment, the temperature was measured in the following five points of the system:

- vapor temperature of the pumping medium in the evaporator (T1),
- liquid medium temperature in the evaporator (T2),
- liquid medium temperature at the hot water tank inlet (T3),
- water temperature in the hot water tank (T4), and
- liquid medium temperature at the hot water tank outlet (T5).

Temperature values for the experiments conducted with the following three power settings, 600, 900, and 1200 W, are presented in Figures 13–15.

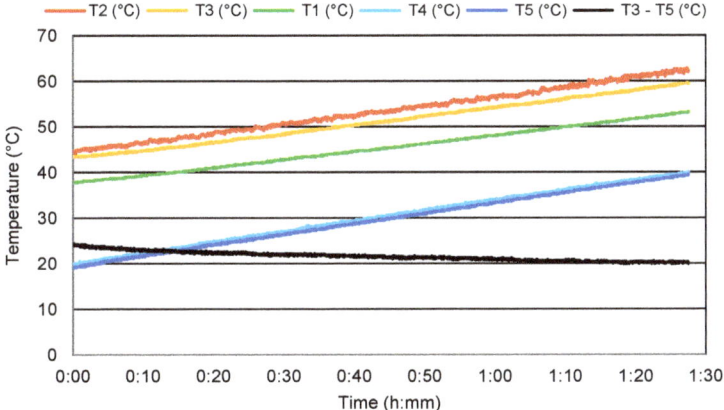

Figure 13. Liquid and vapor temperatures—heat source power 600 W.

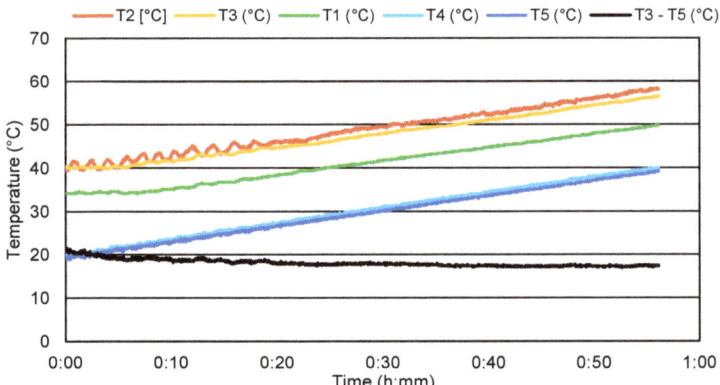

Figure 14. Liquid and vapor temperatures—heat source power 900 W.

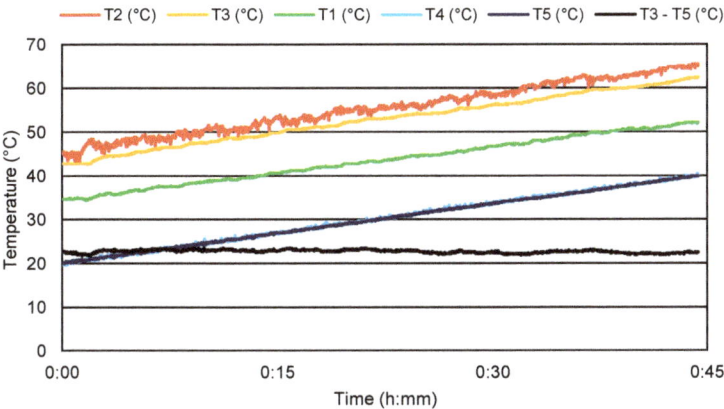

Figure 15. Liquid and vapor temperatures—heat source power 1200 W.

During the experiments, the temperature difference (T3–T5) for the entering liquid and the liquid leaving the hot water tank got changed. The lowest temperature difference occurred with power 900 W and ranged from 17 to 21 °C. After increasing the heat source power to 1200 W, the temperature difference increased and came to 22–24 °C. However, at the lowest heat source power of 600 W, the temperature difference at the beginning of the experiment came to 23 °C and kept decreasing until it reached 20 °C at the end of the experiment.

5.3. Heat Transfer Efficiency

A reverse thermosiphon is a device that transports heat, and heat transfer efficiency is one of the most important parameters in the system. A system's heat transfer efficiency η_{HT} (%) is defined as the ratio of the energy accumulated in the hot water tank (Q_{WT}) to the energy supplied to the system (Q_{in}), and it is expressed by Equation (1):

$$\eta_{HT} = \frac{Q_{WT}}{Q_{in}} \times 100\% \qquad (1)$$

A similar approach was used by Roonprasang et al. [39] to determine the efficiency of experimental systems transporting heat in a direction opposite to natural convection. Thermal energy accumulated in the hot water tank was calculated with the use of Equation (2):

$$Q_{WT} = m_w \times c_{p,w} \times \Delta T \qquad (2)$$

where m_w is the mass of water, $c_{p,w}$ specific heat capacity of water and ΔT is temperature difference (T3–T5, Figure 10).

Thermal energy supplied to the system by the electric heater was calculated with Equation (3):

$$Q_{in} = P \times \tau \qquad (3)$$

where P is power consumed by the electric heater and τ experiment time.

The heat transfer efficiency of the reverse thermosiphon at different power settings is presented in Figure 16.

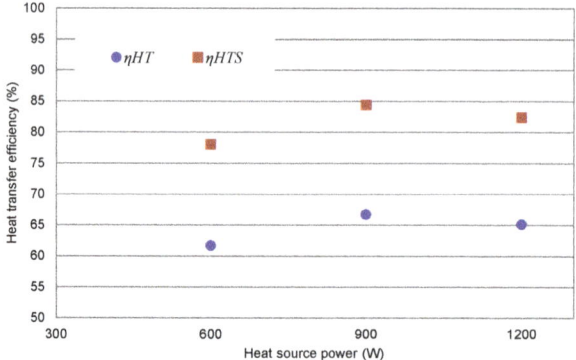

Figure 16. Heat transfer efficiency in the laboratory model.

Heat transfer efficiency was lowest when the electric heater was set to 600 W (η_{HT} = 62%). The analyzed parameter reached η_{HT} = 67% at 900 W and η_{HT} = 65% at 1200 W. These results indicate that heat transfer efficiency in the laboratory model was highest when the heat source was set to 900 W. Heat transfer efficiency was not bound by a linear correlation with the output of the electric heater. When power was increased above 900 W, the proportion of the gaseous phase in two-phase flow increased rapidly, which changed flow structure from slug flow to froth flow. The froth flow regime in two-phase flow does not guarantee maximum flow of the heat transfer medium in a system and, consequently, decreases heat transfer efficiency. The flow structures observed at 600 W and 1200 W are compared in Figure 17.

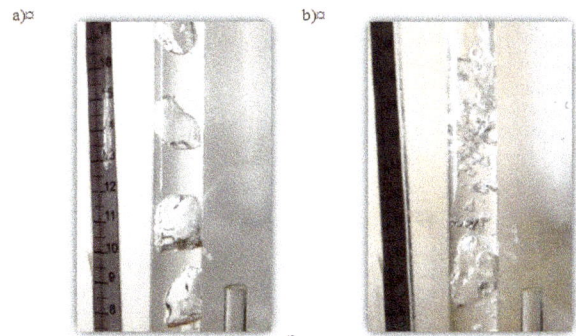

Figure 17. Structure of two-phase flow in a vapor lift pump: (**a**) slug flow at 600 W; (**b**) froth flow at 1200 W.

In a vapor lift pump, the liquid medium has the highest flow rate during two-phase slug flow [35].

The experiment was conducted with a constant heat flux reaching the device and variable temperature of the hot water tank. If the water tank has a constant temperature, heat can be exchanged more efficiently because it is not accumulated in the system. The following components influence a system's heat capacity:

- liquid heat transfer medium (water), 6 kg,
- copper pipes, 2.5 kg, and
- steel components in reverse thermosiphon vessels (evaporator, separator, condenser), 3.4 kg.

The system's heat capacity Q_S was taken into account in Formula (1) to calculate heat transfer efficiency with the Formula (4):

$$\eta_{HTS} = \frac{Q_{HWT} + Q_S}{Q_{in}} \times 100\% \qquad (4)$$

Heat transfer efficiency that accounts for the system's heat capacity η_{HTS} (red squares) is presented in Figure 16. The lowest heat transfer efficiency was 78%. Heat transfer efficiency was highest (84%) when the heater was set to 900 W.

6. Conclusions

The presented laboratory experiment confirmed that two working media can be applied in a reverse thermosiphon with a vapor lift pump. The heat flux reaching the system had constant value. The vapor produced by the pumping medium guaranteed the system's continuous operation.

When the system's heat capacity was taken into account, heat transfer efficiency was highest (84%) when the electric heater was set to 900 W, and it was lowest (78%) at 600 W. The supplied heat and the obtained heat were taken into account in the heat efficiency analysis. Heat is a system parameter that is not easily scaled or converted to expected efficiency in differently scaled systems, which is why the parameters of a vapor lift pump should be determined at the design stage.

The flow rate of the liquid heat transfer medium was highest (40 L/h) at 900 W. Two-phase slug flow was observed in the heat lift pipe. When power was increased above 1000 W, flow structure changed from slug to froth flow, which disrupted liquid circulation in the system. The dimensions of the vapor lift pipe should be adapted to the heat source to guarantee two-phase slug flow.

In the studied laboratory model the minimal temperature difference of liquid heat transfer medium (water) entering and leaving the hot water tank occurred with the heat source power 900 W and was 18 °C.

A downward heat transfer system with a vapor lift pump and two working media (water as the heat transfer medium and a small amount of a pumping medium with a low boiling point) is a viable solution with many practical applications.

In the next stage of the research, the described system will be tested in a real-world setting with a solar collector as the heat source. The proposed solution can replace pumps and control elements in systems powered by solar energy.

The proposed solution can be applied in other system where the temperature of the heat source is below 100 °C and heat has to be transported downward.

7. Patents

A device for gravitational heat transfer in a direction opposite to natural convection; PL227757, (2018).

Author Contributions: Conceptualization, D.C.; methodology, D.C. and M.D.; formal analysis, D.C. and M.D.; investigation, D.C.; writing—original draft preparation, D.C.; writing—review and editing, D.C. and M.D. All authors have read and agreed to the published version of the manuscript.

Funding: This research received no external funding.

Conflicts of Interest: The authors declare no conflict of interest.

References

1. Kołodziej, B.; Matyka, M. *Renewable Energy Sources*; State Agricultural and Forestry Publishing House: Poznań, Poland, 2012; pp. 11–40.
2. Charles, C.; Roberts, J. A review of heat pipe liquid delivery concepts. *J. Heat Recovery Syst.* **1981**, *1*, 261–266. [CrossRef]
3. Dobriansky, Y. Concepts of self-acting circulation loops for downward heat transfer (reverse thermosiphons). *Energy Convers. Manag.* **2011**, *52*, 414–425. [CrossRef]
4. Khandekar, S.; Groll, M. Roadmap to realistic modelling of closed loop pulsating heat pipes. In Proceedings of the 9th International Heat Pipe Symposium, Kuala Lumpur, Malaysia, 17–20 November 2008.
5. Chludziński, D.; Duda, M.; Sołowiej, P.; Hałacz, J.; Lnage, A. The concept of a passive heat transport system from solar collectors. In Proceedings of the Renewable Energy Sources-Engineering, Technology, Innovation, Krynica, Poland, 12–14 June 2019.
6. Dobrinsky, Y.; Wojcik, R. State of the art review of conventional and anti-gravity thermosyphons: Focus on two working fluids. *Int. J. Therm. Sci.* **2019**, *136*, 491–508. [CrossRef]
7. Faghri, A. *Heat Pipe Science and Technology*; Taylor & Francis: Boca Raton, FL, USA, 1995.
8. Ozsoy, A.; Corumlu, V. Thermal performance of a thermosyphon heat pipe evacuated tube solar collector using silver-water nanofluid for commercial applications. *Renew. Energy* **2018**, *122*, 26–34. [CrossRef]
9. Kassai, M.; Simonson, C.J. Experimental effectiveness investigation of liquid-to-air membrane energy exchangers under low heat capacity rates conditions. *Exp. Heat Transf.* **2016**, *29*, 445–455. [CrossRef]
10. Dobriansky, Y. Simple devices for heat transfer downward. In Proceedings of the VII International Conference on Solar Energy at High Latitudes, Tummavuoren Kirjapaino Oy, Espoo-Otaniemi, Finland, 9–11 June 1997.
11. Dobriansky, Y.; Yohanis, Y. Cyclical reverse thermosiphon. *Arch. Thermodyn.* **2010**, *31*, 3–32. [CrossRef]
12. Filippeschi, S. On periodic two-phase thermosyphons operating against gravity. *Int. J. Therm. Sci.* **2006**, *45*, 124–137. [CrossRef]
13. Koito, Y.; Ahamed, M.S.; Harada, S.H.; Imura, H.I. Operational characteristics of a top-heat-type long heat transport loop through a heat exchanger. *Appl. Therm. Eng.* **2008**, *29*, 259–264. [CrossRef]
14. Koito, Y.; Ikemizu, Y.; Tomimura, T.; Mochizuki, M. A vapor-pressure-driven heat pipe for sideward long-distance heat transport. *Front. Heat Pipes* **2010**, *1*, 1–7. [CrossRef]
15. Chludzinski, D.; Dobrianski, J.; Duda, M.; Piechocki, J.; Samsel, M.; Wojcik, R. Method and Device for Self-Acting Heat Transfer in a Direction Reverse to Natural Convection. U.S. Patent US20130105117A1, 15 July 2011. Available online: http://www.google.com/patents/US20130105117 (accessed on 10 May 2019).

16. Duda, M. Reverse Thermosiphon with Cyclical Operation with the Separation of the Function of Pumping Heat Carrier and Heat Transfer. Ph.D. Thesis, University of Warmia and Mazury in Olsztyn, Olsztyn, Poland, 2012.
17. Dobrinski, J.; Duda, M. Experimental solar installation with a self-acting circulation pump powered by local heat. *Tech. Sci.* **2010**, *13*, 181–188.
18. Kalogirou, S.A.; Agathokleous, R.; Barone, G.; Buonomano, A.; Forzano, C.; Palombo, A. Development and validation of a new TRNSYS Type for thermosiphon flat-plate solar thermal collectors: Energy and economic optimization for hot water production in different climates. *Renew. Energy* **2019**, *136*, 632–644. [CrossRef]
19. Kassai, M.; Simonson, C.J. Performance investigation of liquid-to-air membrane energy exchanger under low solution/air heat capacity rates ratio conditions. *Build. Serv. Eng. Res. Technol.* **2015**, *36*, 535–545. [CrossRef]
20. Mameli, M. Pulsating Heat Pipes Numeriicall Modelliing and Experiimenttall Assessment. Ph.D. Thesis, Universita Degli Studi Di Bergamo, Bergamo, Italia, 2012.
21. Filippeschi, S.; Latrofa, E.; Salvadori, G. On the possibility of evaporator drastic scale reduction in a periodically operating two-phase thermosyphon. In Proceedings of the 3rd International Symposium on Two-Phase Flow Modelling and Experimentation, Pisa, Italy, 22–24 September 2004.
22. Groll, M.; Rösler, S. Operation principles and performance of heat pipes and closed two-phase thermosyphons. *J. NonEquilib. Thermodyn.* **1992**, *17*, 91–151.
23. Movick, N.O. Method and Apparatus for Thermally Circulating a Liquid. U.S. Patent US3951204A, 22 July 1974. Available online: http://www.google.com/patents/US3951204 (accessed on 22 March 2019).
24. Brekke, C.E. Solar Heating System. U.S. Patent US4270521A, 15 August 1979. Available online: https://www.google.nl/patents/US4270521 (accessed on 22 March 2019).
25. Sorensen, W. Heat Actuated System for Circulating Heat Transfer Fluids. U.S. Patent 4,552,208, 12 November 1985.
26. Sorensen, W.B. Solar Energy Generator. U.S. Patent US5351488A, 31 January 1994. Available online: https://www.google.nl/patents/US5351488 (accessed on 26 March 2019).
27. Ippoushi, S.; Uehara, N.; Yamada, A.; Ogushi, T.; Yamakage, H. Vapor-Lift Pump Heat Transport Apparatus. U.S. Patent US7201215B2, 10 April 2007. Available online: https://www.google.com/patents/US7201215 (accessed on 20 March 2019).
28. Ippoushi, S.; Uehara, N.; Yamada, A.; Ogushi, T. Pump-Free Water-Cooling System. U.S. Patent US7380584B2, 7 October 2004. Available online: http://www.google.nl/patents/US7380584 (accessed on 20 March 2019).
29. Van Houten, A. Adaptive Self Pumping Solar Hot Water Heating System with Overheat Protection. U.S. Patent US7798140B2, 8 April 2010. Available online: https://www.google.nl/patents/US7798140 (accessed on 26 March 2019).
30. Zhang, Q.; Stewart, S.W.; Brownson, J.R.S.; Witmer, L.T. Geyser Pump Solar Water Heater System Modeling Design Optimization. Ph.D. Thesis, Pennsylvania State University, Centre County, PA, USA.
31. Chung, H.-S.; Woo, J.-S.; Shin, Y.-H.; Kim, J.-H.; Jeong, H.-M. Experimental assessment of two-phase bubble pump for solar water heating. *J. Cent. South Univ.* **2012**, *19*, 1590–1599. [CrossRef]
32. Ito, S.; Tateishi, K.; Miura, N. Studies of a thermosyphon system with a heat source near the top and heat sink at the bottom. In Proceedings of the ISES World Congress, Beijing, China, 18–21 September 2007; pp. 930–934. [CrossRef]
33. Yandri, E.; Miura, N.; Kawashima, T.; Fujisawa, T.; Yoshinaga, M. Seasonal Outdoor Performance Evaluation and Analysis of Thermosyphon with Heat Source on the Top and Heat Sink at The Bottom. In Proceedings of the ISES-AP—3rd International Solar Energy Society Conference—Asia Pacific Region, Sydney, Australia, 25–28 November 2008.
34. Klugmann, M.; Dąbrowski, P.; Mikielewicz, D. Flow Boiling in Minigap in the Reversed Two-Phase Thermosiphon Loop. *Energies* **2019**, *12*, 3368. [CrossRef]
35. Hanafizadeh, P.; Karimi, A.; Saidi, M.H. Effect of step geometry on the performance of the airlift pump. *Int. J. Fluid Mech. Res.* **2011**, *38*, 387–408. [CrossRef]
36. Abu-Mulaweh, H.I.; Mueller, D.W.; Wegmann, B.; Speith, K.; Boehne, B. Design of a Bubble Pump Cooling System Demonstration Unit. *Int. J. Therm. Environ. Eng.* **2011**, *2*, 1–8. [CrossRef]
37. Ligus, G.; Zając, D.; Masiukiewicz, M.; Anweiler, S. A New Method of Selecting the Airlift Pump Optimum Efficiency at Low Submergence Ratios with the Use of Image Analysis. *Energy* **2019**, *12*, 735. [CrossRef]

38. Birdi, K.S. *Handbook of Surface and Colloid Chemistry*, 3rd ed.; Taylor & Francis Group: Boca Raton, FL, USA, 2009; p. 27.
39. Roonprasang, N.; Namprakai, P.; Pratinthong, N. Experimental studies of a new solar water heater system using a solar water pump. *Energy* **2008**, *33*, 639–646. [CrossRef]

© 2020 by the authors. Licensee MDPI, Basel, Switzerland. This article is an open access article distributed under the terms and conditions of the Creative Commons Attribution (CC BY) license (http://creativecommons.org/licenses/by/4.0/).

Article

A New Model for Environmental Assessment of the Comminution Process in the Chain of Biomass Energy Processing [†]

Weronika Kruszelnicka

Department of Technical Systems Engineering, Faculty of Mechanical Engineering, University of Science and Technology in Bydgoszcz, 85-796 Bydgoszcz, Poland; weronika.kruszelnicka@utp.edu.pl

† This paper is an extended version of article presented at the 6th Scientific Conference Renewable Energy Systems, Engineering, Technology, Innovation ICoRES 2019, to be published in E3S Web of Conferences.

Received: 9 November 2019; Accepted: 8 January 2020; Published: 9 January 2020

Abstract: Acquiring energy contained in biomass requires its prior appropriate preparation. These treatments require some energy inputs, which significantly affect the reduction of the energy and the environmental balance in the entire life cycle of the biomass energy processing chain. In connection with the above, the aim of this work is to develop a methodology for the environmental assessment of biomass grinding in the processing chain for energy purposes. The research problem is formulated as follows: Is it possible to provide an assessment model that takes into account the environmental inputs and benefits of the grinding process of biomass intended for further energy use (for example, combustion)? How do the control variables of the grinding machine affect the environmental process evaluation? In response to these research problems, an original, carbon dioxide emission assessment index of the biomass grinding process was developed. The model was verified by assessing the process of rice and maize grinding on a real object—a five-disc mill—with various speed settings of the grinding disc. It was found that the carbon dioxide emission assessment model developed provides the possibility of comparing grinding processes and identifying the grinding process with a better CO_2 emission balance, where its values depend on the control parameters of the mill.

Keywords: disc-mill; efficiency; carbon dioxide; rice; corn; grinding

1. Introduction

Grinding processes are one of the most commonly used preparatory processes for energy carriers (fossil and alternative) intended for combustion and co-combustion [1,2] and biofuel production [3–5]. These biomass forms may include plant-based lignocellulose waste [6,7], sewage waste [8,9], and animal-based meat processing waste [10]. To enhance the utilization of these biomass forms for direct combustion [11,12] or as a biorefinery feedstock [1,7], size reduction approaches typically constitute a significant pre-treatment step that must be undertaken [13]. This is because size reduction operations serve to enhance accessibility to the stored carbon present in biomass. Such size reduction is predominantly carried out on cylindrical mills, drum mills, ball mills, hammer mills, and disc mills [13,14]. Hammer and disc mills yield the best results in terms of the grinding product quality, energy consumption, and efficiency [1,13]. The main goals of grinding granular biomass, as well as other energy materials (e.g., coal, wood), include the reduction of dimensions, the release of substances contained in the material structure, and an increase in the material-specific surface area so that the energy contained in its structure can be released faster [15–18]. In order to maximize the potential of ground raw materials, energy consumption during grinding (processing) should be as low as possible [19]. Unfortunately, mills and grinders currently in use are still characterized by high energy consumption and low efficiency [15,20]. Undertaking research, creating structural and technological solutions aimed at improving the energy and environmental efficiency of grinding systems, is significant. The key element is the implementation of eco-innovative systems, starting with assessment at the design stage with the use of a multi-criteria analysis, as one example [21,22]. Criteria for the selection of grinding technology and the assessment of its operation should clearly indicate solutions that meet the assumptions of sustainable development [23–25].

Assessment criteria that refer to the concept of environmental efficiency can be interpreted as the efficiency of innovative and environmental actions and the consequences of the environmental impact of machines, devices, and engineering equipment [26]. The concept of a product assessment based on the so-called eco-efficiency and related indexes is becoming more and more popular [26,27]. The major assumption of eco-efficiency is to work out a technological solution with the best cost-to-benefit ratio while maintaining a reduction of the environmental impacts [27,28]. Huppes and Ishikawa [29] proposed four indexes to be used for the assessment of a technology's (Table S1) environmental productivity (Table S1, Equation (1)), environmental intensity of production (Table S1, Equation (2)), environmental improvement cost (Table S1, Equation (3)) and environmental cost-effectiveness (Table S1, Equation (4)) which is a combination of the relations between the environmental indexes and the economic results. LCA (life cycle assessment) is a method for assessing environmental impacts [30]. LCC (life cycle costing) is used for assessing economic efficiency [31] and PSIA product impact social assessment is a method used for assessing social impacts [32]. LCA and LCC methods for assessing the grinding process constitute just one of many other elements to be used to assess technological biomass processing in its entirety, either for energy or food production purposes [30,33–36]. MFA (material flow analysis) is another assessment method that focuses on the flow of material between particular stages of the technological process [37]. Knowledge of the quantitative demand for materials in the manufacturing process is also used by the MAIA (material intensity analysis), which is based primarily on determining MIPS (material input per service unit index) (Table S1, Equation (5)), which defines the amount of natural resources needed to manufacture a product or provide a service [38]. X factor, defined as the ratio of the eco-efficiency of the product under assessment to that of the reference product, described in [39], can also be used for eco-efficiency assessment (Table S1, Equation (6)). X factor makes it possible to determine how far the manufactured product is from the performance level of the reference product. It allows one to compare different manufacturing variants of the same product on the basis of the ratio of the product being assessed to the reference product's ecoefficiency level (Table S1, Equation (6)). Tahara et al. [40] have proposed the ICEICE (integrated CO_2 efficiency index for company evaluation) including indirect and direct CO_2 emissions in combination with economic effects and total CO_2 efficiency, direct CO_2 efficiency, and indirect CO_2 efficiency calculated on the basis of indexes (Table S1, Equations (7)–(9). Bennet et al. [41] have developed an optimization tool—Eco Compass—which can be used to support decision making processes in business to meet the demand according to sustainable development assumptions. Kasner [42] has proposed an integrated life cycle index (Table S1, Equation (10)) to determine efficiency in obtaining benefits from technological costs involved in the entire life cycle, which includes manufacturing costs, operational costs, and post-use management costs. This index takes advantage of data regarding the environmental impact obtained with the use of LCA [43]. The indexes and criteria presented herein can successfully be adapted for evaluation of the grinding process through appropriate modeling. Eco-efficiency models developed for technology make it possible to compare the technologies under consideration (products) in a scalar form and select a technology that is best in terms of productivity, the economy, and the environment [44].

Only a few works address the environmental assessment of the grinding process exclusively in terms of ecology [45–49]. In many works, specific energy consumption and fragmentation degree have been indicated as two basic criteria for the assessment of this process [50–53]. Flizikowski et al. [45] proposed an environmental efficiency index to assess the grinding process defined as a ratio of CO_2 emissions produced by grinding to the biomass energy use for outlays in the form of emissions of equivalent CO_2 in the process of grinding (Table S2, Equation (1)). Mroziński et al. [46] showed an index of environmental non-destructivity defined as a ratio of equivalent CO_2 emissions to electric energy consistent with this equivalent (Table S2, Equation (2)). Kruszelnicka et al. [48] discussed a material energy efficiency index, which refers to indirect emissions involved in the process of energy of different materials (Table S2, Equation (3)). In other studies [47,49], Kruszelnicka proposed a sustainable emissivity index for the environmental assessment of grinding, which also refers to indirect emissions involved in grinding (Table S2, Equation (4)), making it possible to optimize the grinder disc process parameter settings.

In view of the above, the aim of the work is to develop a methodology for the carbon dioxide emission assessment of biomass grinding in the processing chain for energy purposes. The research problem is formulated as follows: Is it possible to create an assessment model that takes into account the environmental inputs and benefits of the grinding process of biomass intended for further energy use (for example combustion)? How do the control variables (the angular velocity of the cutting edges of working elements) of the grinding machine affect the carbon dioxide emissions of the process [54]? To obtain an answer to the research problem, a model was developed for the carbon dioxide emission assessment of the grinding process, including the relations between the benefits from energy biomass grinding and the environment-related costs used for grinding. Verification of the model was carried out on a real object—a five-disc mill. Dependences between the angular speed settings of the grinding discs and the values of the proposed assessment model were identified to indicate parameters that provide the assessment index with the best values. A mathematical model of the proposed sustainable CO_2 emissions indicator, the test stands and the experiment conditions are described in Section 2. Section 3 contains the results and a discussion. The discussion has been summarized and the most important conclusions are included in Section 4.

2. Materials and Methods

2.1. Model of a Sustainable CO_2 Emissions Index

The proposed model of a sustainable CO_2 emissions index is based on the assumption of environmental assessment models of the grinding process (Table S2, Equations (1) and (2)) presented in [45,46]. Equivalent CO_2 emissions in the grinding process are closely related to its energy consumption [45–49]. Considering a system whose aim is to reduce the process, product, machine, and environmental impacts (e.g., by eliminating CO_2) while maintaining its higher efficiency and providing a high-quality product as well as energy efficiency, the integrated energy purpose of product (for example, combustion) sustainable CO_2 emissions index can be expressed as [54,55]

$$W_{ZCO_2} = \Delta B_{CO_2}/N_{CO_2} \qquad (1)$$

where

ΔB_{CO_2}—environmental benefits—change in CO_2 emissions,
N_{CO_2}—environmental costs of grinding—energy consumption.

The one factor influencing the positive change in carbon dioxide balance can be described as CO_2 emissions, for example, from the burnt ground energy biomass X_{BCO_2}, because biomass is a renewable fuel, which during the growth phase, absorbs an amount of carbon dioxide equal to the emissions from its energy use [56]. Additionally, in industrial emission monitoring systems, CO_2 emission indexes from biomass are treated as zero [57]. These should be reduced by equivalent CO_2 emissions related to energy consumption in the grinding process X_{RCO_2} in accordance with the equation on environmental benefits [54]:

$$\Delta B_{CO_2} = X_{BCO_2} - X_{RCO_2} \qquad (2)$$

where

X_{BCO_2}—CO_2 emissions from the energy use (for example, combustion) of ground biomass, kgCO$_2$eq,

X_{RCO_2}—the amount of CO_2 emissions associated with the use of electricity in the grinding process, $kgCO_2eq$.

The results of previous research show that the energy properties of biomass (especially its digestibility, exergy) change depending on the degree of fineness (post-grinding particle size) [46,55,58]; therefore, it was accepted that emissions from the combustion of ground biomass X_{BCO_2} are expressed as the sum of emissions from the combustion of biomass divided into size classes, according to the equation [54]:

$$X_{BCO_2} = m_B \cdot \sum (J_{ACO_2i} \cdot W_{Bi} \cdot q_i) \qquad (3)$$

where

m_B—mass of ground biomass, kg,
J_{ACO_2i}—unit CO_2 emissions for the i-th size class of biomass, $kg \cdot kWh^{-1}$,
W_{Bi}—calorific value of the i-th size class of biomass, $kWh \cdot kg^{-1}$,
q_i—mass share of the i-th size class of biomass.

The emissions involved in electricity consumption during grinding are described as follows [54]:

$$X_{RCO_2} = E_{cM} \cdot J_{KCO_2} \qquad (4)$$

where

E_{cM}—total energy consumption of the grinding machine during the grinding of a mass of biomass m_B, kWh,
J_{KCO_2}—emissions of carbon dioxide for the production of electric energy from coal, $kgCO_2eq \cdot kWh^{-1}$.

In this case, environmental costs were assumed as total energy consumption E_{cM} for biomass grinding. Taking into account the above and Dependencies (3) and (4), the sustainable CO_2 emissions index will take the form [54]:

$$W_{ZCO_2} = (X_{BCO_2} - X_{RCO_2})/E_{cM} = (m_B \cdot \sum (J_{ACO_2i} \cdot W_{Bi} \cdot q_i) - E_{cM} \cdot J_{KCO_2})/E_{cM} \qquad (5)$$

Knowing that the mass of the ground biomass m_B is related to grinding efficiency Q:

$$Q = m_B/t \Rightarrow m_B = Q \cdot t \qquad (6)$$

Q—grinding efficiency, $kg \cdot h^{-1}$,
t—time, h,

and that the energy used in the grinding process E_{cM} is equal to

$$E_{cM} = \sum P_i \cdot t = P_c \cdot t \qquad (7)$$

where

P_i—power on the i-th grinding element,
P_c—total power of the grinder.

Then Equation (5) assumes the following form:

$$W_{ZCO_2} = (Q \cdot t \cdot \sum (J_{ACO_2i} \cdot W_{Bi} \cdot q_i) - P_c \cdot t \cdot J_{KCO_2})/P_c \cdot t \qquad (8)$$

which after being reduced by time t, gives

$$W_{ZCO_2} = (Q \cdot \sum (J_{ACO_2i} \cdot W_{Bi} \cdot q_i) - P_c \cdot J_{KCO_2})/P_c \qquad (9)$$

In this way, Equation (9), including the most important characteristics of the grinding process, that is, efficiency, the power of the grinding unit, and the product degree of fineness, is provided. Both the power and the product degree of fineness depend on the angular speeds of rotating elements. In the case of multi-disc grinders, this concerns the angular speeds of the grinding discs. Thus, it can be supposed that the values of the sustainable emissivity index will also depend on the angular speeds of the discs, which have been verified in this study.

For the machine idle gear, when $E_{cM} > 0$ and $m_B = 0$, the quantity of emissions from groundmass is equal to 0, which yields:

$$W_{ZCO_2(min)} = (-P_c \cdot J_{KCO_2})/P_c = -J_{KCO_2} \tag{10}$$

Thus, the lowest value of emissions that can be assumed by the sustainable indicator is emissions equal to unit emissions from electrical energy produced from coal (Figure 1).

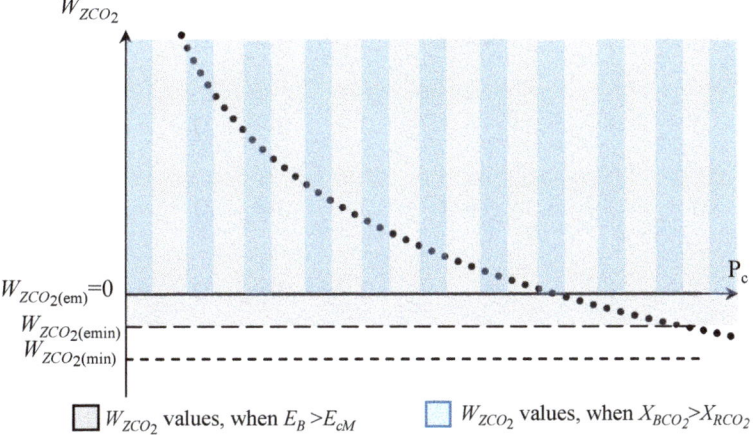

Figure 1. Graphic interpretation of the sustainable CO_2 emissions index.

The level of emissions of CO_2eq from fossil fuels to be sustained by emissions from alternative fuels occurs at $W_{ZCO_2} = 0$ (Figure 1), that is, when

$$Q \cdot \sum (J_{ACO_2i} \cdot W_{Bi} \cdot q_i) - P_c \cdot J_{KCO_2} = 0 \tag{11}$$

$$Q \cdot \sum (J_{ACO_2i} \cdot W_{Bi} \cdot q_i) = P_c \cdot J_{KCO_2} \tag{12}$$

In order to reach energy sustainability (with an assumption that the mass to be ground will be used for the production of electrical energy to power the grinder), the value of electric energy produced from the combustion of biomass E_B should be at least equal to that used in the grinding process:

$$E_{cM} = E_B \tag{13}$$

The level of energy obtained from combusted biomass is

$$E_B = Q \cdot t \cdot k_k \sum W_{Bi} \cdot q_i \tag{14}$$

where:

W_{Bmin}—minimal calorific value from the set of values of the i-th biomass fractions W_{Bi},
$k_k = 0.4$—co-generation coefficient [59].

Substituting Equations (7) and (14) for Equation (13) yields:

$$P_c = Q \cdot k_k \sum W_{Bi} \cdot q_i \tag{15}$$

In an associated production of electric energy from heat energy obtained from biomass combustion, unit emissions (per 1 kWh$_e$ of generated electric energy) J_{ACO_2ie} is expressed in the following way:

$$J_{ACO_2ie} = J_{ACO_2i}/k_k \tag{16}$$

hence, the value of the sustainable CO_2 emissions index for sustainable $W_{ZCO_2(e)}$ is

$$W_{ZCO_2(e)} = (Q \cdot k_k \sum W_{Bi} \cdot q_i \cdot J_{ACO_2ie} - Q \cdot k_k \cdot J_{KCO_2} \sum W_{Bi} \cdot q_i)/Q \cdot k_k \sum W_{Bi} \cdot q_i \tag{17}$$

after reduction it is

$$W_{ZCO_2(e)} = \sum W_{Bi} \cdot q_i \cdot J_{ACO_2i} / \sum W_{Bi} \cdot q_i - J_{KCO_2} \tag{18}$$

and considering Equation (16):

$$W_{ZCO_2(e)} = (\sum W_{Bi} \cdot q_i \cdot J_{ACO_2i}/k_k)/\sum W_{Bi} \cdot q_i - J_{KCO_2} \tag{19}$$

The minimal value of the sustainable CO_2 emissions index occurs for energy sustainability when 100% ($q_i = 1$) of the grinding product (combusted biomass) is the fraction with the lowest unit CO_2 emissions, thus yielding:

$$W_{ZCO_2(emin)} = (W_{B(JACO_2min)} \cdot q_i \cdot J_{ACO_2min}/k_k)/W_{B(JACO_2min)} \cdot q_i - J_{KCO_2} \tag{20}$$

Considering that $q_i = 1$, Equation (20) takes the form:

$$W_{ZCO_2(emin)} = W_{B(JACO_2min)} \cdot J_{ACO_2min}/k_k \cdot W_{B(JACO_2min)} - J_{KCO_2} \tag{21}$$

and being reduced:

$$W_{ZCO_2(emin)} = J_{ACO_2min}/k_k - J_{KCO_2} \tag{22}$$

Equation (22) implies that

- if $J_{ACO_2min}/k_k > J_{KCO_2}$ then $W_{ZCO_2\,(e)} > 0$,
- if $J_{ACO_2min}/k_k < J_{KCO_2}$ then $W_{ZCO_2\,(e)} < 0$.

2.2. Conditions of Experimental Model Verification

Verification of the model of the sustainable CO_2 emissions index involved carrying out an experiment with the use of a real object. Tests were performed in the following order:

1. Choice and preparation of the test stand;
2. Choice and preparation of the biomass to be tested;
3. Determination of grinding conditions and the tests program;
4. Comminution of the biomass while monitoring functional characteristics of grinding
5. Determination of calorific values and emission values for given biomass fractions after comminution;
6. Calculation of the value of the sustainable CO_2 emissions index;
7. Analysis of the relations between the emissions index and the angular speed of the working elements cutting edges.

2.2.1. Test Stand

A mathematical model of the sustainable CO_2 emissions index for the grinding process in the biomass processing chain has been developed to be implemented in a self-regulating control grinding system. The tests were conducted for a five-disc mill drive. The test stand consisted of a grinding unit (of a five-disc grinder) equipped with modules to control and monitor the functional characteristics of the mill control unit and the feed system. Figure 2 shows a view of the test stand.

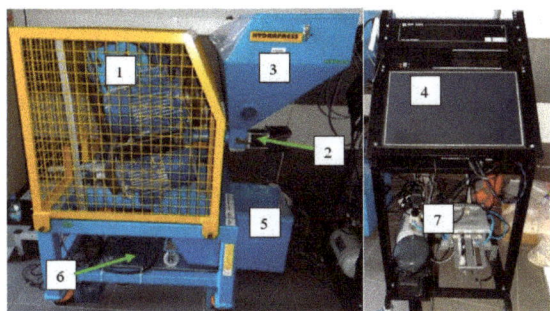

Figure 2. Test stand for monitoring the process of multi-disc grinding, 1—five-disc mill, 2—slide feeder, 3—hopper, 4—control panel, 5—product reception basket, 6—scales for the product, 7—chamber for measuring the size of the product after grinding.

The main elements of the grinder include the housing, body, working chamber, hopper, product reception basket, and grinding unit. The grinding unit includes five discs powered by five electric motors which make it possible to control and monitor the grinding characteristics independently for each disc. Table 1 provides the most important structural characteristics of the grinder discs.

Table 1. Structural characteristics of a five-disc grinder working discs RWT-KZ5.

Parameter	Disc 1	Disc 2	Disc 3	Disc 4	Disc 5
Disc diameter D_n (mm)	274	274	274	274	274
Number of holes l_n (psc.)	14	22	27	33	39
Diameter of holes d_n (mm)	30	23	21	17.5	17.5
Radius of hole arrangement in a row 1 (mm)	85	82.4	79.5	79.5	82
Radius of hole arrangement in a row 2 (mm)	101.5	107.4	95.5	99.5	102
Radius of hole arrangement in a row 3 (mm)	-	-	110.5	114.5	117

2.2.2. Comminuted Biomass

The study involved samples of corn and rice grains that were not suitable for food or animal feed uses. Corn is widely used in the energy sector, especially in biogas production. In Canada, it has also been used for direct combustion in special stoves [60,61]. Rice, as the plant with the greatest cultivation area, has great potential to be used for energy production. It is most commonly used in the form of briquettes [62]. Both comminuted grains can be a good substitute for coal, considering their relatively high heating values [61,63–65].

Before grinding, the moisture of the grains was assessed on a MAC 210/NP moisture balance (RADWAG, Radom, Poland) on the basis of the sample mass difference before and after drying, in accordance with the weighing method given in ISO norm 1446 [66,67]. The moisture of the rice grains was 13.47% ± 0.02%, while for corn it was 12.68% ± 0.02%, and the average dimension D_{80} of the grains was determined on the basis of granulometric analysis carried out in accordance with ISO norm 13322-2:2006 [68], by means of a CAMSIZER device (Retsch Technology GmbH, Haan, Germany), where the results were 2.14 ± 0.02 mm and 8.15 ± 0.02 mm, respectively. Before the grinding process, grains were subjected to prior initial drying. The rice used in the study was deprived of its husk, while the corn was separated from the cobs. Before being burned, pellets were made from ground material that was divided into appropriate dimensional fractions.

2.2.3. Comminution Process Conditions

Emissivity assessment was carried out for a five-disc mill, for the comminution of two grainy materials: rice and corn. In each case, 1 kg of the material was used; both power intake and torques that occurred on particular discs were recorded. Data were recorded every 5 s. Grinding efficiency was determined according to Equation (6). The biomass granulometric content after grinding was determined by means of a CAMSIZER device (Retsch Technology GmbH, Haan, Germany), according to ISO norm 13322-2:2006 [68]. The material was delivered by means of a slide feeder with a fixed input equal to 112 kg·h^{-1}. The angular speeds of the discs were accepted as process variables in order to find out their impact on the sustainable CO_2 emissions index. Summary $S\Delta\omega$ change in the angular speed on all the discs, starting with the minimal level, that is, from 20 rad·s^{-1}, was accepted as the variable of the analysis of the relations between the emissions index values and the angular speed.

The experiment plan included five different test programs (PB) and angular speed settings depending on the manner of the speed growth $\Delta\omega$ (Table 2), accepted according to [49]. Angular speed settings were repeatedly changed in the range from 20 to 100 rad·s^{-1} with a certain gradient $\Delta\omega$ (Table 2). Next, the values of the proposed sustainable CO_2 emissions index were determined for each configuration.

Table 2. Settings of the five-disc mill control parameters.

Test Program	Configuration No.	$S\Delta\omega$ rad·s^{-1}	$\Delta\omega$ rad·s^{-1}	ω_1 rad·s^{-1}	ω_2 rad·s^{-1}	ω_3 rad·s^{-1}	ω_4 rad·s^{-1}	ω_5 rad·s^{-1}
I	1	50	5	20	25	30	35	40
	2	100	10	20	30	40	50	60
	3	150	15	20	35	50	65	80
	4	200	20	20	40	60	80	100
II	1	200	20	100	80	60	40	20
	2	150	15	80	65	50	35	20
	3	100	10	60	50	40	30	20
	4	50	5	40	35	30	25	20
III	1	40	20	20	40	20	40	20
	2	85	20	45	25	45	25	45
	3	225	25	75	50	75	50	75
	4	360	20	100	80	100	80	100
IV	1	40	20	20	40	20	40	20
	2	80	40	20	60	20	60	20
	3	120	60	20	80	20	80	20
	4	160	80	20	100	20	100	20
V	1	240	80	100	20	100	20	100
	2	280	60	100	40	100	40	100
	3	320	40	100	60	100	60	100
	4	360	20	100	80	100	80	100

2.2.4. Determination of Calorific and Emissions Values for Given Biomass Fractions after Comminution

Rice and corn grains were comminuted and divided into dimensional fractions with the use of steel sieves (Table 3). The division was made by means of a sieve shaker Analysette 3 PRO (Fritsch GmbH,

Idar-Oberstein, Germany) in accordance with DIN norm 66165 [69]. After the division into fractions, a pellet maker was used to prepare pellets. Next, 1.5 kg samples were combusted in a pellet Greń furnace of the EG-Pellet type (GREN sp. z o.o., Pszczyna, Poland). CO_2 emissions measurements were performed by means of a dedicated exhaust fumes analyzer ULTRAMAT 23 (Siemens AG, Nürnberg, Germany) for approximately 10 min for each sample. The analyzer measures the content of carbon dioxide (CO_2), carbon monoxide (CO), nitrogen oxide (NO), sulfur dioxide (SO_2), and dioxide (O_2) in the exhaust emissions.

Table 3. Dimensions of rice and corn fractions used for pellets.

	Fraction Dimension (μm)			
Rice	0–630	630–1250	1250–2000	>2000
Corn	0–500	500–1250	1250–2000	>2000

Calorific values of the pellets prepared from the ground biomass with dimensions presented in Table 3 were determined in earlier tests whose results can be found in the authors' other publications [55].

2.2.5. Analytical Methods

Tools for statistical analysis available in MS Excel (Microsoft, Redmond, Washington, USA) and Statistica (TIBCO Software Inc., Palo Alto, CA, USA) were used to analyze the results. Basic descriptive statistics of sustainable emissivity were determined. Relations between angular speeds of the grinder discs and emissivity were examined using correlation analysis with Spearman's method and the analysis of regression and the adequacy of the proposed models. The significance level was accepted as $p < 0.05$.

3. Results and Discussion

3.1. Input Variables of the Sustainable CO_2 Emissions Model

The following assumptions and limitations were adopted for this study:

- CO_2 emissions involved in the production of electric energy from coal is 0.812 $kgCO_2 \cdot kWh^{-1}$ [70].
- In the emissions analysis of this paper, only emissions from the grinding and energy-use processes were considered. Emissions relating to the pelletization process were excluded. The analysis was limited to energy use in the form of pellet combustion. No other methods were considered, such as gasification, fermentation, or digestion.

3.1.1. Power Consumption

The power input on each disc of the five-disc grinder was recorded during the experiment. Figure 3 shows the grinder's total power input for each configuration of the five testing programs (Table 2).

An increase in the angular speed of the multi-disc grinder was caused by power consumption growth (Figure 3). It can also be noted that the highest power input was found for high-speed disc settings (setting no. 4 from PB III and setting no. 4 from PB V) and lowest for low angular speeds (setting no. 1 from PB III and setting no. 1 from PB IV). The power input for all the settings was higher for corn grinding. The power requirement during corn grinding is higher than that during rice grinding, and this results from the strength properties of both types of grains, which largely depend on their physical properties, e.g., humidity and internal structure [71]. The forces used to destroy the corn grain should be higher than those used to permanently deform the rice grain, which translates into a higher resistance to the movement of the working elements of the crusher (cutting discs) and results in an increased power requirement in the case of corn grinding [47]. The above statements result from the different internal structures of the grains examined—differences in the structure of the endosperm and tegument (the ground rice was deprived of its husk) [72–74].

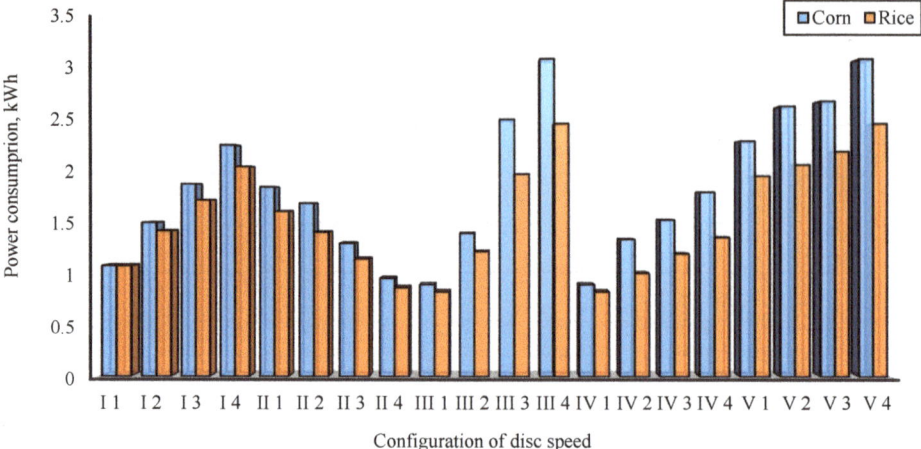

Figure 3. Total power input of a five-disc grinder for the analyzed angular speed settings of working discs.

3.1.2. Grinding Efficiency

Figure 4 shows the results of monitoring the grinding process efficiency for each configuration of the disc angular speed setting according to five research programs (Table 2). Grinding efficiency increased along with an increase in the disc angular speeds (Figure 4). It can also be seen that the highest efficiency was found for the disc high angular speed settings (setting no. 4 from PB III and setting no. 4 from PB V).

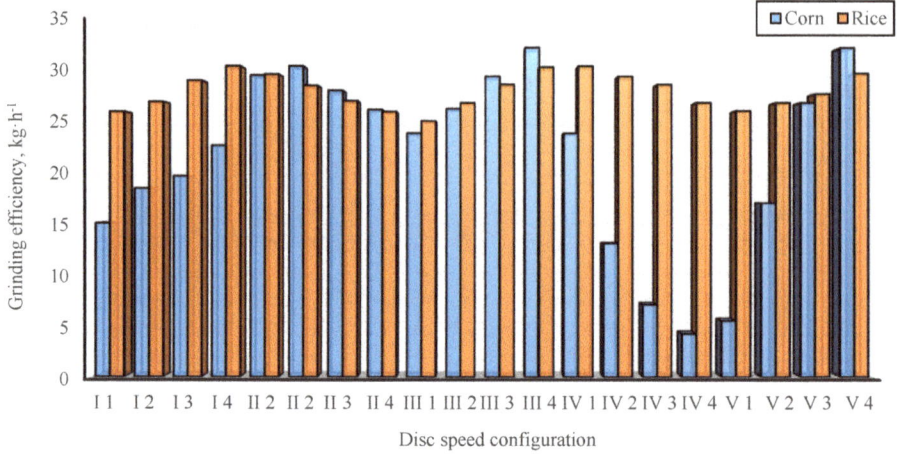

Figure 4. Grinding efficiency for the analyzed settings of the disc angular speeds.

3.1.3. Granulometric Content of Biomaterial after Grinding

Knowledge of the percentage share of the accepted dimensional fractions of biomass is needed to determine the values of the sustainable CO_2 emissions index. For this purpose, a granulometric analysis of the ground material was performed with the use of a CAMSIZER device (Retsch Technology GmbH, Haan, Germany). Table 4 shows the results of this analysis for particular settings of the disc angular speed for rice and corn. The results show that the fraction percentage share was different for each disc angular speed setting configuration.

Table 4. Granulometric content of biomaterial after grinding for the analyzed disc angular speed settings.

Config.		Fraction Percentage Share (%)							
		Rice				Corn			
		0–630 μm	630–1250 μm	1250–2000 μm	>2000 μm	0–500 μm	500–1250 μm	1250–2000 μm	>2000 μm
I	1	9.4	29.3	56.4	4.9	2.4	32.2	42.3	23.1
	2	3.4	13.8	67.8	15.0	2.9	38.2	40.1	18.8
	3	6.9	21.0	62.4	9.7	3.0	47.5	38.3	11.2
	4	5.8	18.4	63.5	12.3	2.9	44.7	39.4	13.0
II	1	5.8	18.4	63.5	12.3	1.8	26.9	42.0	29.3
	2	2.9	12.7	66.6	17.8	2.2	30.9	41.0	25.9
	3	2.3	10.1	66.8	20.8	2.0	30.8	37.0	30.2
	4	0.9	6.4	67.1	25.6	2.4	32.2	36.3	29.1
III	1	1.7	9.0	66.4	22.9	1.9	27.9	35.8	34.4
	2	2.8	11.7	65.7	19.8	2.1	31.1	37.8	29.0
	3	8.9	22.5	60.3	8.3	2.0	31.3	42.3	24.4
	4	13.3	35.0	48.7	3.0	2.3	39.2	44.4	14.1
IV	1	1.7	9.0	66.4	22.9	1.9	27.9	35.8	34.4
	2	2.4	13.7	67.7	16.2	2.4	33.2	38.0	26.4
	3	6.3	23.4	62.1	8.2	2.4	36.2	48.7	14.7
	4	8.4	28.1	58.1	5.4	2.2	33.0	48.1	16.7
V	1	7.0	27.3	60.5	5.2	1.9	38.1	42.5	17.5
	2	9.4	27.2	58.2	5.2	2.1	44.1	42.7	11.1
	3	12.4	30.8	53.7	3.1	2.2	42.2	44.4	11.2
	4	13.3	35.0	48.7	3.0	2.2	33.0	48.1	16.7

3.1.4. Unit Emissions Index J_{ACO2i} for the i-th Dimensional Fraction

A unit CO_2 emissions index for a given biomass fraction was determined on the basis of the analysis of the CO_2 content in exhaust emissions. The values of the unit J_{ACO2i} index are presented in Table 5. It can be noticed that CO_2 emissions increased along with an increase in the size of the particles of the biomass, which was used for the preparation of pellets.

Table 5. Unit index J_{ACO_2i} of CO_2 emissions of the i-th fraction.

Rice		Corn	
Fraction Dimension (μm)	Unit Emissions Index (kg·kWh^{-1})	Fraction Dimension (μm)	Unit Emissions Index (kg·kWh^{-1})
0–630	0.0453	0–500	0.04859
630–1250	0.0356	60–1250	0.05432
1250–2000	0.0745	1250–2000	0.059701
>2000	0.11428	>2000	0.068181

3.1.5. Calorific Values

Calorific values for given dimensional fractions of biomass were accepted according to [55]. Table 6 shows the calorific values calculated into kWh·kg^{-1}.

Table 6. Calorific values of selected fractions of rice and corn calculated based on [55].

Rice		Corn	
Fraction Dimension (μm)	Calorific Value (kWh·kg^{-1})	Fraction Dimension (μm)	Calorific Value (kWh·kg^{-1})
0–630	3.97	0–500	3.91
630–1250	3.93	60–1250	4.05
1250–2000	3.89	1250–2000	4.02
>2000	3.85	>2000	3.96

3.2. Carbon Dioxide Emissions Assessment of Grinding by Means of the Sustainable CO_2 Emissions Index

The grinding processes carried out in particular research programs were subjected to emissions assessment. For this purpose, the values of the sustainable CO_2 emissions index were determined for each case on the basis of both Equation (9) and the values of the variables obtained (Figures 3 and 4, and Tables 2–6). Figure 5 shows the results of the sustainable CO_2 emissions index calculation for the tested multi-disc grinder disc angular speed configurations and the two ground materials—rice and corn.

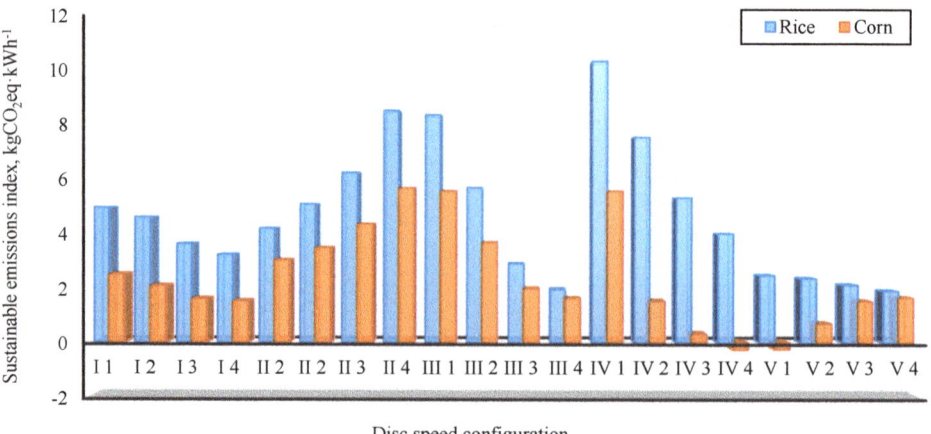

Figure 5. Results of the sustainable CO_2 emissions index for each disc speed configuration.

When analyzing the variability of the sustainable CO_2 emissions index, it should be noted that the desired condition is to reduce the energy consumption for grinding E_{cM} and increase the environmental benefits ΔB_{CO_2}, i.e., replacing CO_2 emissions from coal with CO_2 emissions from biomass combustion. In this case, the value of the proposed index should increase. When comparing the grinding processes, the more favorable in terms of emissions will be the one for which the sustainable CO_2 emissions index assumes higher values [54].

Based on the results, it was found that the best, in terms of emissivity, both for rice and corn grinding, were settings VI1 of the angular speed of the grinder discs (Figure 5). The least advantageous for rice grinding were settings V4, and for corn, settings IV4 (Figure 5). It was noticed that better emissivity results were obtained for the configuration in which the disc angular speed was low (for example, I1, II4, III1, IV1). The values of the sustainable emissions index were higher for rice grinding than for corn grinding in all the disc angular speed configurations, which was caused primarily by lower energy consumption during rice grinding and higher grinding efficiency. The values of the emissions index provided sufficient grounds to state that the emissions from electric energy used in the rice comminution process were sustainable or in fact higher than the emissions from combustion of the comminuted biomass for all the analyzed disc angular speed settings because $W_{ZCO_2} > W_{ZCO_2(em)} = 0$ (Figures 5 and 6). In the case of corn grinding, two configurations (IV4 and V1) did not provide emissions sustainability ($W_{ZCO_2} > W_{ZCO_2(em)} = 0$).

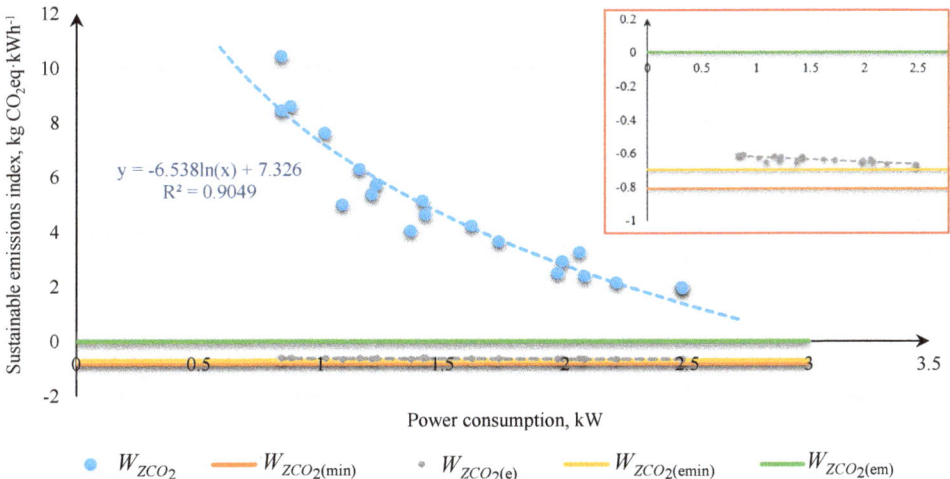

Figure 6. Sustainable CO_2 emissions index for rice grinding on a multi-disc grinder in a function of power input.

The minimal value of the sustainable emissions index CO_2 $W_{ZCO_2(min)}$, determined on the basis of Equation (10), was -0.812 kgCO_2eq·kWh^{-1} with the assumption that the electric energy used in the process of grinding was produced from coal (Figures 6 and 7). In the case of electric energy produced from natural gas, this would be -0.201 kgCO_2eq·kWh^{-1}, from biogas: -0.196 kgCO_2eq·kWh^{-1}, from diesel oil: -0.276 kgCO_2eq·kWh^{-1} [75].

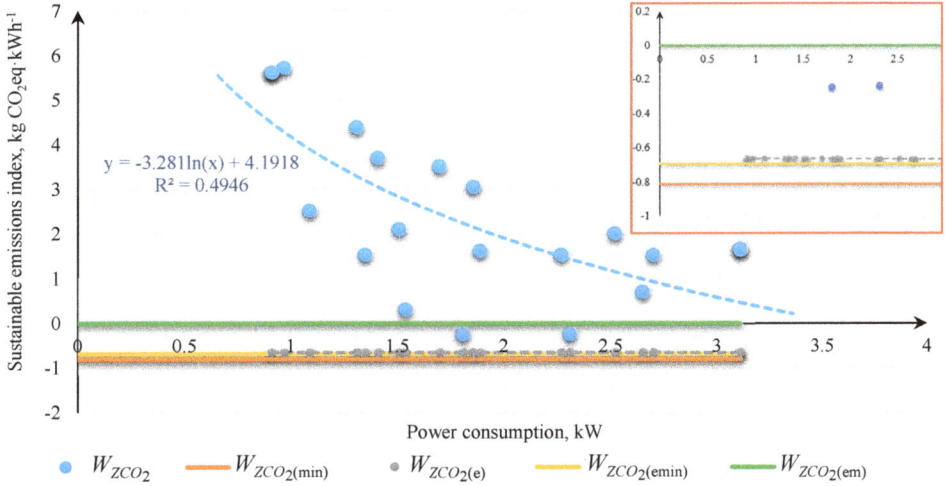

Figure 7. Sustainable CO_2 emissions index for corn grinding on a multi-disc grinder in a function of power consumption.

The dependences shown in Figures 6 and 7 indicate that there is no obvious correlation between power consumption and the sustainable CO_2 emissions index when grinding corn, mainly due to the physico-mechanical properties of the corn grains that affect the grinding process. The influence of the working unit on the materials being ground and the mass flow phenomena in the grinding chamber are also crucial. The results obtained clearly show that the grinding process is different for both materials, which is influenced by differences in the physical and mechanical properties of both grains. In the case of corn, among other things, a greater unevenness of the grinding process (transient idling phases) was observed as a result of the material deposited in the grinding chamber as a consequence of the phenomenon of particle agglomeration, which was not observed in the case of rice. The balanced CO_2 emission rate depends, apart from power consumption, on yield and the share of fractions of different particle sizes. The lack of an obvious relationship between power consumption and a balanced CO_2 emission factor in maize grinding is due to the less clear relationship between power consumption (angular speeds of the grinding discs) and the yield and grain size composition of the grinding product than for rice. This, in turn, is also dictated by the difference in grain properties such as size. The ratio between grain size and disc gap is of particular importance for the yield and granulometric composition of the product. Rice grains are smaller so that they are crushed and drained through the inter-disc slots more quickly than larger maize grains, which must collide with the cutting edges of the shredder more times than rice grains before they leave the shredding chamber. The above-mentioned aspects make less regularity in the shredding of maize, resulting in a moderate relationship between power consumption and sustainable CO_2 emissions.

The minimal value of the emissions index for energy sustainability $W_{ZCO_2(emin)}$ for rice grinding was -0.699 kgCO$_2$eq·kWh^{-1} (Figure 6), whereas for corn grinding, it was -0.690 kgCO$_2$eq·kWh^{-1} (Figure 7). The value of the emissions index for energy sustainability $W_{ZCO_2(e)}$ for the analyzed configurations of the disc angular speed settings are marked with a grey line (Figures 6 and 7). The data show that the level of energy sustainability was exceeded for all the tested disc angular speed settings (Figures 6 and 7), that is, the energy produced from biomass combustion was higher than the energy used in the process of grinding (obtainment of the comminuted product).

3.3. Analysis of the Relations between the Values of the Sustainable CO_2 Emissions Index and the Angular Speed of the Working Elements of the Cutting Edges

To determine the relations between the values of the sustainable CO_2 emissions index and the variable speeds of the working discs, a statistical analysis of the variables was carried out. Table 7 presents the most important statistics that describe the variables and the results of the sustainable CO_2 emissions index. Based on the skewness and kurtosis values, it was found that the distribution of the values of variables and the results of the sustainable CO_2 emissions index differs from the normal distribution. Therefore, Spearman's coefficient was used in the correlation analysis to describe monotonic relations between the variables.

Table 7. Results of the basic statistical analysis for the sustainable CO_2 emissions index distribution.

		R_p	M	S	K	V	$\bar{x}$	s	Min.	Max.
W_{ZCO_2}	Corn	5.96	1.83	0.54	−0.53	76.31	2.42	1.84	−0.24	5.72
	Rice	8.51	4.14	0.8	−0.08	50.73	4.82	2.45	1.93	10.44
$S\Delta\omega$		320	260	0.6	−0.72	63.11	165.5	104.44	40	360
ω_1		80	20	0.32	−1.82	66.54	54.00	35.93	20	100
ω_2		80	40	0.54	−0.72	44.03	51.75	22.78	20	100
ω_3		80	25	0.51	−1.26	57.83	54.00	31.23	20	100
ω_4		80	40	0.54	−0.72	44.03	51.75	22.78	20	100
ω_5		80	20	0.32	−1.82	66.54	54.00	35.93	20	100

$\bar{x}$—mean value, s—standard deviation, M—median, Max—maximal value, Min—minima value, R_p—range, V—variability coefficient, S—skewness, K—kurtosis.

Correlation analysis using Spearman's method revealed that the values of the sustainable CO_2 emissions index are negatively correlated with speed increase $S\Delta\omega$ (Table 8). For rice grinding, negative correlations between the sustainable CO_2 emissions index and the first, third, and fifth disc angular speeds were found as well as the summary increase in angular speeds $S\Delta\omega$. The results show that along with an increase in the multi-disc grinder angular speeds the values of the sustainable CO_2 emissions index decrease. This results, among others, from the dependence of power consumption and efficiency on angular speeds (of the rotating elements) as demonstrated in [25,47,76,77], which show that power consumption and efficiency increase along with an increase in grinder disc angular speeds, which, in the case of the sustainable CO_2 emissions index, means a decrease in its value.

Table 8. Analysis of sustainable CO_2 emissions index correlation with independent variables.

		ω_1	ω_2	ω_3	ω_4	ω_5	$S\Delta\omega$
Corn	Rho Spearman's	−0.107	−0.258	−0.307	−0.557	−0.480	−0.617 *
	p-value	0.653	0.272	0.188	0.011	0.032	0.004
Rice	Rho Spearman's	−0.609 *	−0.232	−0.855 *	−0.434	−0.853 *	−0.946 *
	p-value	0.004	0.325	0.000	0.056	0.000	0.000

* Significant correlations: Rho Spearman's >0.6, p-value <0.05; ω_1, ω_2, ω_3, ω_4, ω_5—disc angular speeds, rad·s^{-1}, $S\Delta\omega$—summary increase in angular speeds, rad·s^{-1}.

Multiple regression analysis was carried out for the analyzed variables of the rice grinding process using the backward stepwise method. A simple linear regression analysis was carried out for corn due to its correlation with $S\Delta\omega$ only. Table 9 shows the results of the regression analysis. The only model with significant coefficients for rice was a linear model of variable $S\Delta\omega$, which demonstrated a 74.6% variability of the sustainable emissions index. With regard to corn grinding, the linear model of variable $S\Delta\omega$ explained merely 32.7% of the sustainable emissions index.

Table 9. Results of regression analysis for the sustainable CO_2 emissions index.

		Coefficients	t-stat.	p-Value *	F	Significance	R^2
Corn	Constant	4.09	6.16	8.08×10^{-6}	8.75	0.008	0.3271
	$S\Delta\omega$	−0.01	−2.96	0.008			
Rice	Constant	8.17	15.13	1.11×10^{-11}	53.00	9.13×10^{-7}	0.7464
	$S\Delta\omega$	−0.02	−7.28	9.13×10^{-7}			

$S\Delta\omega$—summary angular speed increase, * results p > 0.5 were accepted as significant.

Bearing in mind that linear models account for less than 80% of the variability of the sustainable emissions, the adequacy of nonlinear models was checked. The non-linear model which best described the changes in the sustainable emissions index of rice grinding on a five-disc grinder depending on the total increase in angular velocity on the discs $S\Delta\omega$ was an exponential model and its coefficient of determination was $R^2 = 0.889$ (Figure 8). For corn grinding, the best model was a logarithmic one (Figure 9). The dependencies provided indicate that the sustainable emissions index of grinding decreases along with an increase in the grinder disc angular speed.

The dependencies presented in Figures 8 and 9 can be used to determine the CO_2 emissions level predicted for the analyzed five-disc grinder, for the other materials and for emissivity control of the grinding process—establishing optimal values for the settings of the grinding process parameters (angular velocities), providing a reduction in CO_2 emissions from alternative fuels while decreasing energy use.

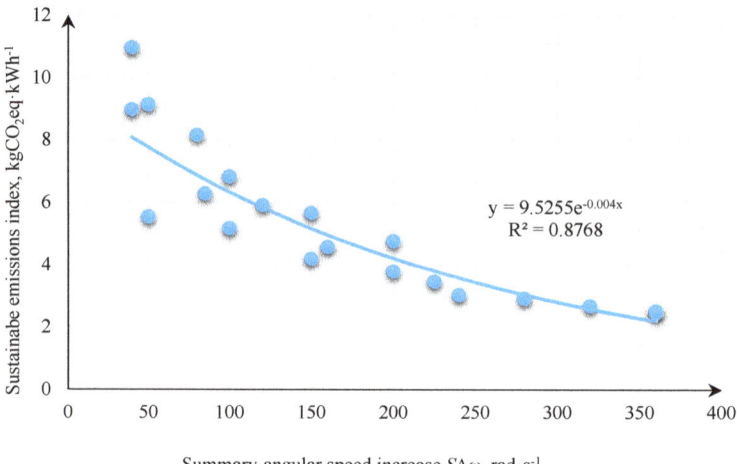

Figure 8. Sustainable CO_2 emissions index for rice grinding in a function of summary speed increase $S\Delta\omega$.

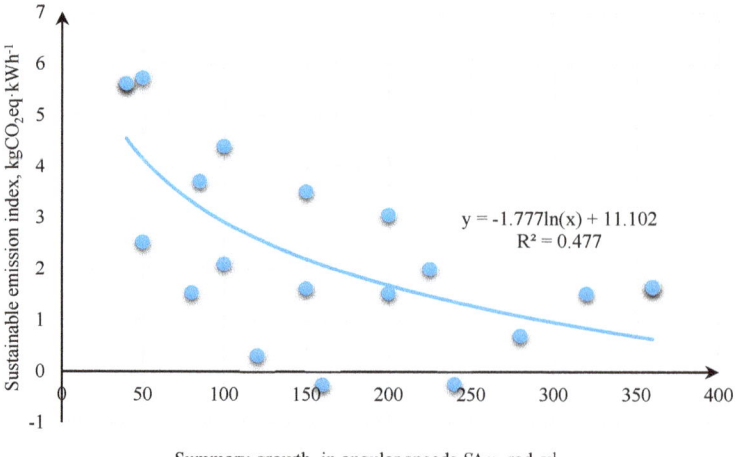

Figure 9. Sustainable CO_2 emissions index for corn grinding in a function of summary speed increase $S\Delta\omega$.

The sustainable CO_2 emissions index, in accordance with Equation (9), depends on power consumption. The power consumption, in turn, depends on the speed of the angular discs (in this case, on the total speed increase) and the moments associated with this, as well as the moments and movement resistance from the grinding of the material. Since angular speeds and power consumption are interrelated, as is the case with the relationship between power consumption and the sustainable CO_2 emissions index, the absence of a clear relationship between the summary velocity gain in the case of corn grinding (Figure 9) is due to its physico-mechanical properties and a different grinding process than in the case of rice, as described in Section 3.2.

The index presented in this paper can be designated for other types of materials besides grain and wood biomass. It can be used to compare the emissions parameters during the processing of materials for energy purposes and to indicate the parameters of the grinding process that will ensure the highest possible increase in environmental benefits in the form of a change in CO_2 emissions per unit of grinding machine power. If other materials, e.g., lignocellulose biomass, are ground, it can be expected that the results would be different from those presented in this paper, which is of course related both to the different specificity of the grinding process itself and to the fact that the physico-mechanical properties of the cellulose materials differ significantly from those of grains used in this study.

4. Conclusions

The aim of the work—to develop a mathematical model of the sustainable CO_2 emissions index for the carbon dioxide emissions assessment of the biomass grinding process—was achieved. The methodology for evaluating the emissivity of grinding made it possible to provide a measurable assessment of the process with a focus on energy consumption and emissivity. An analysis of the sustainable CO_2 emissions index allows the authors to state that

- the sustainable CO_2 emissions index of rice and corn grinding on a five-disc mill, depending on the disc angular speed increase $S\Delta\omega$, can be described with high accuracy using a nonlinear model (Figures 8 and 9);
- it was observed that the sustainable CO_2 emissions index decreases with an increase in the value of $S\Delta\omega$;
- from the point of view of emissivity, it is better to grind at lower disc angular speeds;
- higher values of the emissions index were obtained for rice, and from the point of view of emissivity, rice is better than corn in energy applications.

The values of the sustainable CO_2 emissions index will vary depending on the source of energy fueling the grinding processes. The minimal value of the sustainable emissions index $W_{ZCO_2(min)}$ was -0.812 kgCO_2eq·kWh^{-1} with the assumption that the electric energy used in the process of grinding was produced from coal (Figures 6 and 7). In the case of electric energy produced from natural gas, it would be -0.201 kgCO_2eq·kWh^{-1}, from biogas: -0.196 kgCO_2eq·kWh^{-1}, from diesel oil: -0.276 kgCO_2eq·kWh^{-1}. The best emissions balance would occur if the electricity fueling the grinding process came from a biogas plant. Different index values would also occur if biomass energy use from non-combustion methods were considered.

The index presented in this paper can be designated for other types of materials besides grain and wood biomass. It can be used to compare the emissions parameters during the processing of materials for energy purposes and to indicate the parameters of the grinding process, which would ensure the highest possible increase in environmental benefits in the form of a change in CO_2 emissions per unit of grinding machine power.

The model proposed in this study is dedicated strictly for industrial applications for real-time assessment of CO_2 emissions. The model implemented in the monitoring end control system of biomass pretreatment processes can, in a short time (using the input of real-time data from industrial sensors, counters, and meters), calculate the balance of emissions. This offers the possibility of indicating the best process parameters, for example, disc speed, which makes it possible to reduce the emissions of carbon dioxide from fossil fuels. In general, the proposed model can be part of an industrial control system because it is easy to calculate its values, depending on real-time data from sensors.

Supplementary Materials: The following are available online at http://www.mdpi.com/1996-1073/13/2/330/s1, Table S1: Eco efficiency indexes of technology presented in the literature, Table S2: Environmental assessment indexes for grinding according to the literature.

Author Contributions: The work was prepared entirely by the author Weronika Kruszelnicka. All authors have read and agreed to the published version of the manuscript.

Funding: Scientific work was financed by the budget resource for science in 2017–2021 as a research project under the "Diamentowy Grant" program.

Conflicts of Interest: The author declare no conflict of interest. The funders had no role in the design of the study; in the collection, analyses, or interpretation of the data, in the writing of the manuscript, or in the decision to publish the results.

References

1. Miao, Z.; Grift, T.E.; Hansen, A.C.; Ting, K.C. Energy requirement for comminution of biomass in relation to particle physical properties. *Ind. Crop. Prod.* **2011**, *33*, 504–513. [CrossRef]
2. Tomporowski, A.; Flizikowski, J. Motion characteristics of a multi-disc grinder of biomass grain. *Przem. Chem.* **2013**, *92*, 498–503.
3. Abo El-Khair, B.E.-S.; Manal, G.M.; Shimaa, R.H. Complementary production of biofuels by the green alga Chlorella vulgaris. *Int. J. Renew. Energy Res.* **2015**, *5*, 936–943.
4. Adapa, P.K.; Tabil, L.G.; Schoenau, G.J. Factors affecting the quality of biomass pellet for biofuel and energy analysis of pelleting process. *Int. J. Agric. Biol. Eng.* **2013**, *6*, 1–12.
5. Aguado, R.; Cuevas, M.; Pérez-Villarejo, L.; Martínez-Cartas, M.L.; Sánchez, S. Upgrading almond-tree pruning as a biofuel via wet torrefaction. *Renew. Energy* **2020**, *145*, 2091–2100. [CrossRef]
6. Chandra, R.; Takeuchi, H.; Hasegawa, T. Methane production from lignocellulosic agricultural crop wastes: A review in context to second generation of biofuel production. *Renew. Sustain. Energy Rev.* **2012**, *16*, 1462–1476. [CrossRef]
7. Che Kamarludin, S.N.; Jainal, M.S.; Azizan, A.; Safaai, N.S.M.; Daud, A.R.M. Mechanical pretreatment of lignocellulosic biomass for biofuel production. *Appl. Mech. Mater.* **2014**, *625*, 838–841. [CrossRef]
8. Jiang, L.; Liang, J.; Yuan, X.; Li, H.; Li, C.; Xiao, Z.; Huang, H.; Wang, H.; Zeng, G. Co-pelletization of sewage sludge and biomass: The density and hardness of pellet. *Bioresour. Technol.* **2014**, *166*, 435–443. [CrossRef] [PubMed]
9. Poudel, J.; Ohm, T.-I.; Lee, S.-H.; Oh, S.C. A study on torrefaction of sewage sludge to enhance solid fuel qualities. *Waste Manag.* **2015**, *40*, 112–118. [CrossRef]
10. Ware, A.; Power, N. Biogas from cattle slaughterhouse waste: Energy recovery towards an energy self-sufficient industry in Ireland. *Renew. Energy* **2016**, *97*, 541–549. [CrossRef]
11. Abelha, P.; Vilela, C.M.; Nanou, P.; Carbo, M.; Janssen, A.; Leiser, S. Combustion improvements of upgraded biomass by washing and torrefaction. *Fuel* **2019**, *253*, 1018–1033. [CrossRef]
12. Basu, P. Chapter 11—Biomass Combustion and Cofiring. In *Biomass Gasification, Pyrolysis and Torrefaction*, 3rd ed.; Basu, P., Ed.; Academic Press: Cambridge, MA, USA, 2018; pp. 393–413. ISBN 978-0-12-812992-0.
13. Eisenlauer, M.; Teipel, U. Comminution of biogenic materials. *Can. J. Chem. Eng.* **2017**, *95*, 1236–1244. [CrossRef]
14. Hu, J.; Chen, Y.; Ni, D. Effect of superfine grinding on quality and antioxidant property of fine green tea powders. *LWT Food Sci. Technol.* **2012**, *45*, 8–12. [CrossRef]
15. Salman, A.D.; Ghadiri, M.; Hounslow, M. *Particle Breakage*; Elsevier: Amsterdam, The Netherlands, 2007; ISBN 978-0-08-055346-7.
16. Mannheim, V. Examination of Thermic Treatment and Biogas Processes by Lca. *Ann. Fac. Eng. Hunedoara Int. J. Eng.* **2014**, *12*, 225–234.
17. Mannheim, V.; Siménfalvi, Z. *Determining a Priority Order between Thermic Utilization Processes for Organic Industrial Waste with LCA.*; WIT Press: Southampton, UK, 2012; pp. 153–166.
18. Johnson, N.W. Review of Existing Eco-efficient Comminution Devices. Unpublished work. 2013. Available online: http://www.ceecthefuture.org/wp-content/uploads/2013/02/Bills-review-of-Eco-eff-devices.pdf (accessed on 5 November 2019).
19. Kruszelnicka, W.; Flizikowski, J.; Tomporowski, A. Auto-monitoring system of grainy biomass comminution technology. *IOP Conf. Ser. Mater. Sci. Eng.* **2018**, *393*, 012076. [CrossRef]
20. Wołosiewicz-Głąb, M.; Foszcz, D.; Saramak, D.; Gawenda, T.; Krawczykowski, D. Analysis of a grinding efficiency in the electromagnetic mill for variable process and feed parameters. In *E3S Web of Conferences*; EDP Sciences: Les Ulis, France, 2017; Volume 18, p. 01012.
21. Jachimowski, R.; Szczepanski, E.; Klodawski, M.; Markowska, K.; Dabrowski, J. Selection of a Container Storage Strategy at the Rail-road Intermodal Terminal as a Function of Minimization of the Energy Expenditure of Transshipment Devices and CO2 Emissions. *Rocz. Ochr. Sr.* **2018**, *20*, 965–988.
22. Tomporowski, A.; Flizikowski, J.; Kasner, R.; Kruszelnicka, W. Environmental Control of Wind Power Technology. *Rocz. Ochr. Środowiska* **2017**, *19*, 694–714.
23. Kruszelnicka, W.; Tomporowski, A.; Flizikowski, J.; Kasner, R.; Cyganiuk, J. Basis of Biomass Grinders Sustainable Designing. *Syst. Saf. Hum. Tech. Facil. Environ.* **2019**, *1*, 542–549. [CrossRef]

24. Marczuk, A.; Misztal, W.; Savinykh, P.; Turubanov, N.; Isupov, A.; Zyryanov, D. Improving Efficiency of Horizontal Ribbon Mixer by Optimizing Its Constructional and Operational Parameters. *Eksploat. Niezawodn.* **2019**, *21*, 220–225. [CrossRef]
25. Tomporowski, A.; Flizikowski, J.; Kruszelnicka, W. A new concept of roller-plate mills. *Przem. Chem.* **2017**, *96*, 1750–1755.
26. Szymańska, E. Efektywność przedsiębiorstwa-definiowanie i pomiar. *Rocz. Nauk Rol.* **2010**, *97*, 152–164.
27. Kleiber, M.; Czaplicka-Kolarz, K. *Ekoefektywność Technologii: Praca Zbiorowa*; Wydawnictwo Naukowe Instytutu Technologii Eksploatacji—Państwowego Instytutu Badawczego: Radom, Poland, 2011; ISBN 978-83-7789-050-9.
28. Aoe, T. Eco-efficiency and ecodesign in electrical and electronic products. *J. Clean. Prod.* **2007**, *15*, 1406–1414. [CrossRef]
29. Huppes, G.; Ishikawa, M. Eco-efficiency and Its xsTerminology. *J. Ind. Ecol.* **2005**, *9*, 43–46. [CrossRef]
30. Schneider, L.; Finkbeiner, M. *Life Cycle Assessment of EU Oilseed Crushing and Vegetable Oil Refining*; Technische Universitat Berlin: Berlin, Germany, 2013; p. 59.
31. Homagain, K.; Shahi, C.; Luckai, N.; Sharma, M. Life cycle cost and economic assessment of biochar-based bioenergy production and biochar land application in Northwestern Ontario, Canada. *For. Ecosyst.* **2016**, *3*, 21. [CrossRef]
32. Fontes, J. *Handbook for Product Social Impact Assessment*; PRe Sustainability: Amersfoort, The Netherlands, 2016.
33. Pierobon, F.; Eastin, I.L.; Ganguly, I. Life cycle assessment of residual lignocellulosic biomass-based jet fuel with activated carbon and lignosulfonate as co-products. *Biotechnol. Biofuels* **2018**, *11*, 139. [CrossRef]
34. Liu, W.; Wang, J.; Richard, T.L.; Hartley, D.S.; Spatari, S.; Volk, T.A. Economic and life cycle assessments of biomass utilization for bioenergy products. *Biofuels Bioprod. Biorefining* **2017**, *11*, 633–647. [CrossRef]
35. Johnson, L.; Lippke, B.; Oneil, E. Modeling biomass collection and woods processing life-cycle analysis. *Prod. J.* **2012**, *62*, 258–272. [CrossRef]
36. Fantozzi, F.; Buratti, C. Life cycle assessment of biomass chains: Wood pellet from short rotation coppice using data measured on a real plant. *Biomass Bioenergy* **2010**, *34*, 1796–1804. [CrossRef]
37. Brunner, P.H.; Rechberger, H. Practical Handbook of Material Flow Analysis. In *Advanced Methods in Resource and Waste Management*; CRC/Lewis: Boca Raton, FL, USA, 2004; ISBN 978-1-56670-604-9.
38. Liedtke, C.; Bienge, K.; Wiesen, K.; Teubler, J.; Greiff, K.; Lettenmeier, M.; Rohn, H. Resource Use in the Production and Consumption System—The MIPS Approach. *Resources* **2014**, *3*, 544–574. [CrossRef]
39. Ściążko, M.; Czaplicka-Kolarz, K. *Model Ekologicznego i Ekonomicznego Prognozowania Wydobycia i Użytkowania Czystego Węgla. T. 2, Ekoefektywność Technologii Czystego Spalania Węgla*; Główny Instytu Górnictwa: Katowice, Poland, 2004.
40. Tahara, K.; Sagisaka, M.; Ozawa, T.; Yamaguchi, K.; Inaba, A. Comparison of "CO2 efficiency" between company and industry. *J. Clean. Prod.* **2005**, *13*, 1301–1308. [CrossRef]
41. Bennett, M.; James, P.; Klinkers, L. *Sustainable Measures—Evaluation and Reporting of Environmental and Social Performance*; Greenleaf Publishing Ltd.: Sheffield, UK, 1999.
42. Kasner, R. *Ocena Korzyści i Nakładów Cyklu Życia Elektrowni Wiatrowej*; Politechnika Poznańska: Poznań, Poland, 2016; p. 186.
43. Tomporowski, A.; Flizikowski, J.; Opielak, M.; Kasner, R.; Kruszelnicka, W. Assessment of Energy Use and Elimination of Co2 Emissions in the Life Cycle of an Offshore Wind Power Plant Farm. *Pol. Marit. Res.* **2017**, *24*, 93–101. [CrossRef]
44. Burchart-Korol, D.; Kruczek, M.; Czaplicka-Kolarz, K. Wykorzystanie ekoefektywności w ocenie poziomu ekoinnowacyjności. In *Innowacje w Zarządzaniu i Inżynierii Produkcji*; Oficyna Wydawnicza Polskiego Towarzystwa Zarządzania Produkcją: Opole, Poland, 2013.
45. Flizikowski, J.; Wełnowski, J.; Dudziak, P. Analysis end eco-technological evaluation of micro-grain supersonic milling. Part I. Models and indicators. *Inżynieria I Apar. Chem.* **2017**, *56*, 161–163.
46. Mroziński, A.; Flizikowski, J.; Macko, M. *Inżynieria Rozdrabniania Biomasy*; Grafpol: Wrocław, Poland, 2016; ISBN 978-83-64423-35-2.
47. Kruszelnicka, W. Analiza Procesu Wielotarczowego Rozdrabniania Biomasy w Ujęciu Energochłonności i Emisji CO2. Ph.D. Thesis, Uniwersytet Technologiczno-Przyrodniczy im. Jana i Jędrzeja Śniadeckich w Bydgoszczy, Bydgoszcz, Poland, 2019.

48. Kruszelnicka, W.; Bałdowska-Witos, P.; Flizikowski, J.; Tomporowski, A.; Kasner, R. Energy balance of wood biomass utilization. Part I—Characteristics of the problem and research methodology. *TEKA Kom. Mot. Energ. Roln* **2017**, *17*, 55–60.
49. Kruszelnicka, W.; Bałdowska-Witos, P.; Kasner, R.; Flizikowski, J.; Tomporowski, A.; Rudnicki, J. Evaluation of emissivity and environmental safety of biomass grinders drive. *Przem. Chem.* **2019**, *98*, 1494–1498.
50. Fuerstenau, D.W.; Abouzeid, A.-Z.M. The energy efficiency of ball milling in comminution. *Int. J. Miner. Process.* **2002**, *67*, 161–185. [CrossRef]
51. Kawatra, S.K. *Advances in Comminution*; Society for Mining, Metallurgy, and Exploration: Littleton, CO, USA, 2006; ISBN 978-0-87335-246-8.
52. Tomporowski, A.; Flizikowski, J.; Wełnowski, J.; Najzarek, Z.; Topoliński, T.; Kruszelnicka, W.; Piasecka, I.; Śmigiel, S. Regeneration of rubber waste using an intelligent grinding system. *Przem. Chem.* **2018**, *97*, 1659–1665.
53. Naimi, L.J.; Sokhansanj, S.; Bi, X.; Lim, C.J.; Womac, A.R.; Lau, A.K.; Melin, S. Development of Size Reduction Equations for Calculating Energy Input for Grinding Lignocellulosic Particles. *Appl. Eng. Agric.* **2013**, *29*, 93–100. [CrossRef]
54. Kruszelnicka, W. New model for ecological assessment of comminution process in energy biomass processing chain. In Proceedings of the Renewable Energy Sources-Engineering, Technology, Innovation, Krynica, Poland, 12–14 June 2019.
55. Kruszelnicka, W. Study of Physical Properties of Rice and Corn Used for Energy Purposes. In *Renewable Energy Sources: Engineering, Technology, Innovation*; Wróbel, M., Jewiarz, M., Szlęk, A., Eds.; Springer International Publishing: Manhattan, NY, USA, 2020; pp. 149–162.
56. Szyszlak-Bargłowicz, J.; Zajac, G.; Slowik, T. Research on Emissions from Combustion of Pellets in Agro Biomass Low Power Boiler. *Rocz. Ochr. Sr.* **2017**, *19*, 715–730.
57. Kowalczyk-Juśko, A. Redukcja emisji zanieczyszczeń dzięki zastąpieniu węgla biomasą spartiny preriowej. *Probl. Inżynierii Rol.* **2010**, *18*, 69–77.
58. Kruszelnicka, W.; Bałdowska-Witos, P.; Flizikowski, J.; Tomporowski, A. Investigation of grain biomass properties as an alternative fuel. *Int. Sci. J. Trans Motauto World* **2018**, *3*, 45–48.
59. Dańko, R.; Szymała, K.; Holtzer, M.; Holtzer, G. Skojarzone wytwarzanie energii elektrycznej i ciepła w systemie kogeneracji. *Arch. Foundry Eng.* **2012**, *12*, 6.
60. Buffington, D.E. Burning Shalled Corn—A Renewable Fuel Source. *Agric. Biol. Eng.* **2005**, *H*, 78.
61. Lizotte, P.-L.; Savoie, P.; De Champlain, A. Ash Content and Calorific Energy of Corn Stover Components in Eastern Canada. *Energies* **2015**, *8*, 4827–4838. [CrossRef]
62. Pode, R. Potential applications of rice husk ash waste from rice husk biomass power plant. *Renew. Sustain. Energy Rev.* **2016**, *53*, 1468–1485. [CrossRef]
63. Moe, P.W.; Tyell, H.F.; Hooven, N.W. Physical Form and Energy Value of Corn Grain. *J. Dairy Sci.* **1973**, *56*, 1298–1304. [CrossRef]
64. Panzeri, D.; Cesari, V.; Toschi, I.; Pilu, R. Seed Calorific Value in Different Maize Genotypes. *Energy Sour. Part A Recovery Util. Environ. Eff.* **2011**, *33*, 1700–1705. [CrossRef]
65. Shen, J.; Zhu, S.; Liu, X.; Zhang, H.; Tan, J. Measurement of Heating Value of Rice Husk by Using Oxygen Bomb Calorimeter with Benzoic Acid as Combustion Adjuvant. *Energy Procedia* **2012**, *17*, 208–213. [CrossRef]
66. Determination of Moisture Content. Available online: http://www.ico.org/projects/Good-Hygiene-Practices/cnt/cnt_sp/sec_3/docs_3.2/Determine%20m%20c.pdf (accessed on 21 October 2019).
67. International Organization for Standardization. *Green Coffee—Determination of Water Content—Basic Reference Method*; International Organization for Standardization: Geneva, Switzerland, 2001; ISO 1446:2001.
68. International Organization for Standardization. *Particle Size Analysis—Image Analysis Methods—Part 2: Dynamic Image Analysis Methods*; International Organization for Standardization: Geneva, Switzerland, 2006; ISO 13322-2:2006.
69. International Organization for Standardization. *DIN DIN 66 165-1 and 2: Sieve Analysis, Fundamentals (1) and Procedure (2)*; International Organization for Standardization: Geneva, Switzerland, 2016.
70. CO2 Emission Factors [Wskaźniki Emisji CO2]. Available online: https://prawomiejscowe.pl/api/file/GetZipxAttachment/155/1147170/preview (accessed on 22 October 2019).
71. Tumuluru, J.S.; Tabil, L.G.; Song, Y.; Iroba, K.L.; Meda, V. Grinding energy and physical properties of chopped and hammer-milled barley, wheat, oat, and canola straws. *Biomass Bioenergy* **2014**, *60*, 58–67. [CrossRef]

72. Warechowska, M. Some physical properties of cereal grain and energy consumption of grinding. *Agric. Eng.* **2014**, *1*, 239–249.
73. Greffeuille, V.; Mabille, F.; Rousset, M.; Oury, F.-X.; Abecassis, J.; Lullien-Pellerin, V. Mechanical properties of outer layers from near-isogenic lines of common wheat differing in hardness. *J. Cereal Sci.* **2007**, *45*, 227–235. [CrossRef]
74. Greffeuille, V.; Abecassis, J.; Barouh, N.; Villeneuve, P.; Mabille, F.; Bar L'Helgouac'h, C.; Lullien-Pellerin, V. Analysis of the milling reduction of bread wheat farina: Physical and biochemical characterisation. *J. Cereal Sci.* **2007**, *45*, 97–105. [CrossRef]
75. Rogowska, D.; Lubowicz, J. Analiza możliwości obniżenia emisji gazów cieplarnianych w cyklu życia bioetanolu paliwowego. *Naft. Gaz* **2012**, *68*, 1044–1049.
76. Smejtkova, A.; Vaculik, P. Comparison of power consumption of a two-roll mill and a disc mill. *Agron. Res.* **2018**, *16*, 1486–1492.
77. Boehm, A.; Meissner, P.; Plochberger, T. An energy based comparison of vertical roller mills and tumbling mills. *Int. J. Miner. Process.* **2015**, *136*, 37–41. [CrossRef]

© 2020 by the author. Licensee MDPI, Basel, Switzerland. This article is an open access article distributed under the terms and conditions of the Creative Commons Attribution (CC BY) license (http://creativecommons.org/licenses/by/4.0/).

Article

Assessment of Reducing Pollutant Emissions in Selected Heating and Ventilation Systems in Single-Family Houses [†]

Joanna Hałacz [1,*], Aldona Skotnicka-Siepsiak [2] and Maciej Neugebauer [1]

[1] Department of Electrical Engineering, Power Engineering, Electronics and Automation, Faculty of Technical Sciences, University of Warmia and Mazury in Olsztyn, Oczapowskiego str. 2, 10-719 Olsztyn, Poland; mak@uwm.edu.pl

[2] Department of Building Engineering, Faculty of Geodesy, Geospatial and Civil Engineering, University of Warmia and Mazury in Olsztyn, Oczapowskiego str. 2, 10-719 Olsztyn, Poland; aldona.skotnicka-siepsiak@uwm.edu.pl

[*] Correspondence: jhalacz@uwm.edu.pl; Tel.: +48-89-523-3-03

[†] This paper is an extended version of article presented at the 6th Scientific Conference Renewable Energy Systems, Engineering, Technology, Innovation ICoRES 2019, Krynica, Poland, 17–19 June 2019, to be published in E3S Web of Conferences.

Received: 23 January 2020; Accepted: 4 March 2020; Published: 6 March 2020

Abstract: The article presents the results of a study aiming to select the optimal source of heat for a newly designed single-family home. Commercial software was used to compare heating and ventilation systems involving a bituminous coal boiler, a condensing gas boiler, a biomass boiler, a heat pump with water and glycol as heat transfer media. The effectiveness of natural ventilation, mechanical ventilation with a ground-coupled heat exchanger, and solar heater panels for water heating were evaluated. The analysis was based on the annual demand for useful energy, final energy, and non-renewable primary energy in view of the pollution output of the evaluated heating systems. The analysis revealed that the heat pump with water and glycol was the optimal solution. However, the performance of the heat pump in real-life conditions was below its maximum theoretical efficiency. The biomass boiler contributed to the highest reduction in pollutant emissions (according to Intergovernmental Panel on Climate Change Change guidelines, carbon dioxide emissions have zero value), but it was characterized by the highest demand for final energy. Mechanical ventilation with heat recovery was required in all analyzed systems to achieve optimal results. The introduction of mechanical ventilation decreased the demand for final energy by 10% to around 40% relative to the corresponding heating systems with natural ventilation.

Keywords: exploitation of energy sources; heating and ventilation systems; pollutant emissions; reduction of pollutant emissions; single-family houses

1. Introduction

According to the definition provided by the World Health Organization (WHO), air pollution occurs where the chemical composition of the air may adversely affect the health of humans, animals, and plants as well as other elements of the environment, e.g. the soil, water, or climate [1,2]. The main air pollutants include gases and particulate matter. Pollution occurs when the pollutant content of the air is exceeded in relation to the background [3].

The gases considered to be most harmful include sulfur dioxide (SO_2), nitrogen oxide (NO_x), ammonia (NH_3), and non-methane volatile organic compounds (NMVOC) [4,5]. With regards to particulate matter, the most hazardous are those with a diameter of fewer than 10 micrometers (PM10) and 2.5 micrometers (PM2.5) [6].

The negative effects of air pollution on the human body have been repeatedly studied and confirmed. Research results have demonstrated increased mortality and a progressive reduction in human life expectancy [7,8]. The data indicate that approximately 3.7 million people die every year due to this reason [9]. It was demonstrated that air pollutants contribute inter alia to respiratory diseases including lung cancer and cardiovascular diseases [10–14]. An adverse effect of pollutants on the environment including water and soil was proven as well [15–18]. It is an important measure in this regard to conduct continuous research and to disseminate its findings, especially concerning the areas of consumer choices, whose consequences may affect the environment for decades to come [19,20].

Energy generation is one of the main sources of pollutant emissions. Households consume 12% of globally generated energy [21] and indoor heating accounts for a significant portion of energy consumption [22]. A review of the research conducted in recent decades and the proposed solutions reveals three main trends [23]. The first involves attempts to decrease the operating time of electrical devices by changing user habits. The second is the improvement in the thermal insulation of buildings, and the third involves the replacement of outdated equipment with new and energy-efficient devices [24]. Two trends predominated in recent years. First, attempts have been made to shift energy consumption to off-peak hours when the emissions from energy generation are the lowest [25,26]. Second, low-emission heating systems that rely on renewable energy sources have been popularized in newly developed or upgraded buildings [27,28].

Energy consumption patterns in households are influenced by climatic, demographic, economic, and lifestyle factors [29]. In European countries with a cold climate, including Poland, other than the Baltic States and Nordic countries, climatic factors play a key role because they prolong the heating season. Economic factors are also a very important consideration in Poland and other former Soviet block countries where income levels are lower, which explains the widespread use of conventional heating systems based on non-renewable fossil fuels [30]. These factors have been taken into consideration in this study. Efforts have also been made recently to lower household emissions by installing modern heating systems that fully meet the users' needs, decrease the heating carbon footprint, and are affordable.

Research studies addressing these topics are highly dispersed, and they have to account for regional factors, such as the availability of different energy sources and the relevant costs. In general, the energy market is highly diverse. In Finland, 48% of households have electric heating. Central heating systems are not popular, and most households are equipped with electric heaters [23]. In contrast, the heating systems in 80% of German households are fired with heating oil and gas [31]. The Polish heating market differs from other European countries in that it still relies heavily on coal. Therefore, research studies should account for local and regional factors, in particular in large countries such as Poland, to formulate the most effective recommendations for regions with similar climatic conditions and similar levels of economic growth. In Poland, the existing research is also dispersed, and most studies compare the emissions from modern heating systems that rely on renewable energy sources with conventional systems that are fired by fossil fuels. However, the number of the analyzed variants and/or pollutants are relatively small, which narrows down the scope of these analyses [30,32–34]. Some studies focus on a single energy source or a single pollutant [35,36]. Research into residential heating systems often fails to account for the impact of ventilation on pollutant emissions. For this reason, the study aimed to present and analyze the most popular combinations of heating and ventilation systems in view of the generated emissions and the associated costs.

2. Attempts to Solve the Pollutant Emission Problem

The European Environment Agency (EEA) data [37] indicate that the most polluted air in Europe is found in urban areas where 73% of the population lives. This value may increase to 82% in 2050 [38]. The problem is both important and complex, particularly in the context of the possibility of free spreading of the air and its pollutants over long distances from their emission sources.

The increasing awareness of hazards results in the development of environmental protection policy, and the establishment of relevant supervisory institutions [39,40]. The awareness of responsibility for future generations is increasing as well [41–44].

For many years, European Union institutions have made numerous efforts to improve air quality, with the aim of achieving a level that will have no adverse effects on human health or the environment [13,45]. One of the major initiatives is the "Clean Air" program which aims to raise the existing standards for emissions and air pollutant limits through legislative action. This strategy foresees comprehensive and consistent actions and measures to reduce pollutant emissions [46]. In 2015, EU Member States adopted Directives which set national emission limits for particulate matter (PM10 and PM2.5) and gaseous pollutants (sulfur and nitrogen oxides, ammonia, and non-methane volatile organic compounds).

The need for improvement in the air pollution situation is also responded to by the WHO which produced recommendations concerning the maximum daily values for the PM10 and PM2.5 particles that should not be exceeded more often than three times a year within a particular area [47].

Coordinated actions are slowly yielding results. According to the Eurostat studies [48], in the years 2006–2015 the average concentrations of most harmful substances in the air decreased. Despite the improvements, the situation is not resolved completely, and in many areas, the polluted air continues to harm human health and the environment. The European Union comprises 28 countries that are very diverse in terms of the economy, law, and climate. The needs and opportunities with regards to the policy concerning, e.g. energy production, which, according to study results, has a significant impact on pollutant emissions in particular countries, are also very different [49–51]. Therefore, the air pollution levels and the pollutant structure differ significantly [52,53]. As research shows, for example, countries with a small area and/or low population density may emit the same amounts of pollutants to the atmosphere as large densely populated ones [54]. In general terms, the greatest amounts of gaseous pollutants are emitted to the atmosphere by Germany, France, Italy, and Spain, while most fine particulate matter is emitted by France, Italy, and Poland [54]. In Poland, particulate matter concentration levels in the air can be eight (for PM10) and six (for PM2.5) times higher than those indicated by the WHO guidelines [55].

Air protection issues in Poland are governed by (mostly amended) regulations as well as guidelines of the Chief Inspectorate of Environmental Protection (GIOŚ). The most important of them include the Environmental Protection Law Act of 27 April 2001 [56], Regulation of the Minister of the Environment of 8 June 2018 on the assessment of substance levels in the air [57], Regulation of the Minister of Environment of 24 August 2012 on the levels of certain substances in the air [58], and "Guidelines for the development of the State Environmental Monitoring voivodeship programmes for the years 2016–2020" [59].

Based on the regulations and guidelines, Poland's report presenting the balance of air pollutant emissions reportable to the UN/ECE Convention for the years 2015–2017 was once again drawn up and published in 2019 [60].

The report indicates an upward trend for emissions of most of the main pollutants except sulfur dioxide. A list of emission levels and a comparison for the years 2015–2017 are presented in Table 1.

Table 1. A comparison of total pollutant emissions in Poland in the years 2015–2017 [60].

Pollutant	2015	2016	2017	2017/2015	2017/2016
		mg		%	
Sulfur dioxide (SO_2)	711,489	590,664	582,656	−18.1	−1.4
Nitrogen oxides (NO_x)	725,257	742,168	803,661	10.8	8.3
Non-methane volatile Organic compounds (NMVOC)	640,800	674,158	690,737	7.8	2.5
Ammonia (NH_3)	284,727	291,948	307,522	8.0	5.3
Carbon monoxide (CO)	2,342,630	2,456,468	2,543,251	8.6	3.5
Total suspended Particulate matter (TSP)	326,911	335,210	340,604	4.2	1.6
PM10 particulate matter	231,625	240,632	246,310	6.3	2.4
PM2.5 particulate matter	136,010	141,875	147,281	8.3	3.8
Black carbon (BC)	19,720	21,217	23,814	20.8	12.2
		kg		%	
Polychlorinated biphenyls (PCB)	563	578	578	2.8	0.1
Hexachlorobenzene (HCB)	4	4	4	1.2	−2.0
Polycyclic aromatic Hydrocarbons (PAH)	155,557	153,535	151,575	−2.6	−1.3

The table shows that compared to 2016, the emissions that increased the most in 2017 were nitrogen oxide (by 8.3%) and particulate matter (black carbon—BC by 12.2%). Sulfur dioxide emission decreased by 1.4%, while compared to 2015, by 18.1%. Emissions of most persistent organic pollutants also decreased, particularly those of hexachlorobenzene HCB (by approximately 2%); on the other hand, polychlorinated biphenyls (BCB) emissions increased slightly (by approximately 0.05%).

It should be stressed that for many pollutants (in particular nitrogen oxides, non-methane volatile organic compounds NMVOC, ammonia NH_3, carbon monoxide CO, black carbon BC, copper, and particulate matter), significant increases in emissions occurred over a two-year period [61].

Further analysis of the balance in 2017 indicates the problem associated with non-industrial combustion processes where the largest proportion of emissions results from the combustion in households. It was demonstrated that on a national scale, they emit 29.3% sulfur dioxide (SO_2), 10.7% NO_x and 22.3% black carbon (BC). The situation is even less advantageous regarding emissions of PM2.5 particulate matter (46.5%), carbon monoxide CO (59.2%), total suspended particulate matter TSP (44.6%), PM10 particulate matter (46.5%), and polychlorinated biphenyls PCB (68.4%) or polycyclic aromatic hydrocarbons PAH (83.7%) [58].

As research shows, electricity generation, particularly from fossil fuels, is most responsible for pollutant emissions [50,61]. Increasing research and efforts are focused on the development of a low-carbon economy. In addition to energy savings, one of the proposed solutions is to increase the use of energy from renewable sources. One study demonstrated that a net increase in energy production based on renewable sources by 1% results in a reduction in carbon dioxide, one of the most dangerous greenhouse gases, by 0.16% [62].

In Poland, one of the strategic objectives is a reduction in greenhouse gas emissions, supported by the use of renewable energy sources as well as pro-efficiency measures in the energy sector. In the years 2006–2015, while the energy sources based on primary energy gradually decreased, the use of renewable sources increased. In 2015, the proportion of energy from renewable sources amounted to 11.8% of total energy consumption, which ranked it 21st among EU countries.

However, hard coal and brown coal were still the major sources of energy in 2018, and their share of electricity generation reached 47.8% and 29.0%, respectively, marking a decrease of 9.8% from 2010. Renewable energy sources accounted for 12.7% of energy production, and their proportions in the power mix increased by 5.8% from 2010. The main sources of renewable energy were wind energy, biomass, and biogas. Solar energy had the smallest share of energy generation, but it was characterized by the highest growth dynamics [63]. However, a steady upward trend towards the use of renewable sources should be emphasized [64].

The article describes a study whose results will enable the selection of the optimal (also in terms of pollutant emissions) source of heat for a newly-designed single-family residential building. The analysis was developed based on research carried out in the city of Olsztyn, the capital of Warmińsko-Mazurskie Voivodeship.

Warmińsko-Mazurskie Voivodeship is the fourth largest voivodeship in Poland. Its area of 24,173 km^2 is inhabited by 1429 people (as of 31st December 2018) [65].

Compared to the rest of the country, a few industrial plants that are particularly harmful to the environment are located in this voivodeship. Emissions from point sources are mainly concentrated in localities which are seats of poviats (administrative units). Due to the poorly-developed transport network, a decrease in linear emissions is noticeable in localities and poviats in which several roads with high traffic volume intersect, and in the largest cities i.e., Elbląg and Olsztyn. The greatest pressure is exerted by surface emissions, i.e., those from individual heating systems. For example, almost 67% of PM10 particulate matter emitted within the voivodeship originate from this source, while approximately 8% each originate from farming, point sources, and road emissions. The relatively small proportion of surface emissions is within the city of Olsztyn and amounts to 18%. It is worth noting that the city has been implementing its own programs to combat air pollution. A recent initiative is the "Replace the Stove" project, which is a part of the "Clean Air Priority Programme." Its aim is to reduce or prevent emissions of particulate matter and other pollutants introduced into the atmosphere by single-family houses. Municipal authorities co-finance the replacement of heating sources in residential buildings with environmentally friendly sources. What is important is the program is addressed to both natural persons who own single-family houses and to persons holding a permit for the commencement of construction work. Financial support is also available to tenants' associations interested in obtaining grants for connecting properties to the district heating system [66].

The selection of the optimal heat/energy source in this particular region is especially justified due to the climatic conditions. Warmińsko-Mazurskie Voivodeship is located in a lowland area that is the coldest in the whole country. It is characterized by a high forestation rate and lake density. The climate is described as transitional, maritime/continental climate with average annual temperatures ranging from 6 to 8 °C [67,68].

This is the greatest area and one of the few in Poland where the growing season with the average daily temperature above 5 °C is the shortest. With the national average for the years 1971–2000 which amounts to 218 days of the growing season, in the Warmia and Mazury region it lasts less than 200 days, and in certain locations, it is shorter than 190 days [69].

Such climatic conditions result in a significant, compared to other regions, an extension of the heating system which, in turn, extends the period of increased pollutant emissions due to the increased demand for thermal energy. Given the relatively low levels of the region's industrialization and transport, the production of energy, including thermal energy in individual households is the major source of harmful substance emissions to the air.

The newly designed single-family building in areas similar, in climate terms, to the region of north-eastern Poland, must ensure unlimited access to domestic hot water (DHW) and fully satisfy the increased energy demand for thermal comfort.

3. Materials and Methods

The study aimed to identify the optimal combination of a heating system and a ventilation system characterized by low emissions for newly developed or upgraded single-family homes. Combinations involving heat sources that are most popular in the Polish market were analyzed. The energy efficiency of the analyzed systems was compared, and the results of theoretical analyses were confronted with real-world parameters to ensure that the selected solution is best adapted to local conditions.

A newly designed single-family building located in Olsztyn was analyzed. The building is a real-world object which is scheduled for construction in 2020. The floor space of the building is 152.76 m². The building is designed to be inhabited by four people. For the construction, materials were used that enable obtaining the maximum heat penetration coefficient values for space dividing elements in accordance with the Regulation of the Minister of Infrastructure and Construction of 2017 [70]; these amount to, respectively:

- for the ground slab: 0.30 (W/(m²K))
- for external walls: 0.20 (W/(m²K))
- for the flat roof: 0.15 (W/(m²K))
- for external windows: 0.9 (W/(m²K))
- for external doors: 1.3 (W/(m²K))

In the initial variant, gravity ventilation and a conventional hard coal boiler with 82% efficiency were used for the purposes of central heating and the preparation of domestic hot water (DHW).

ArCADia-TERMOCAD software by INTERsoft (ArCADiasoft, Łódź, Poland) [71] was used to draw up an energy performance certificate for the building, and the environmental effect for the heating and ventilation system variants presented in Table 2 was compared.

Table 2. Heating and ventilation system configuration variants in the analyzed building [19].

Variant	1	2	3	4	5	6	7	8
Hard coal boiler	X	X						
Natural gas condensation boiler			X	X				
Biomass boiler					X	X		
Heat pump							X	X
Gravity ventilation	X		X		X		X	
Mechanical ventilation with ground-coupled heat exchanger (GCHE)		X		X		X		X

X: option designation.

Additionally, for the purposes of domestic hot water generation, each of the variants considered the possibility of extending the installation system to include a system of flat-plate and vacuum collectors.

The analyzed heating and ventilation variants for the single-family home with an individual heating system were based on the most popular solutions in Poland, including a hard coal boiler, a condensing gas boiler, and a biomass boiler.

The heating and ventilation variant involving a heat pump was selected because the number of heat pump systems installed in Poland increased by 20% in 2018. At present, every seventh new building is heated by a heat pump [72]. Solar thermal collectors for water heating were included in the study due to the success of the "Support for dispersed renewable energy sources" program of the National Fund for Environmental Protection and Water Management [73]. Solar collectors with a combined area of 475,207 m² have been installed as part of the program [74]. In 2018, 1.29% of Polish households were equipped with solar collectors and 0.48% with heat pumps [75].

Single-family homes with natural ventilation are characterized by very high energy losses, and this fact was taken into consideration in the analyzed scenarios.

Based on the analysis of the determined value of the annual demand for usable energy (UE), final energy (FE), and non-renewable primary energy (PE), and the pollutant emission indices for particular systems (SO_2, NO_X, CO, CO_2, particulate matter, black carbon, and B-a-P), the optimal variant of the heating and ventilation system was indicated.

The values of usable energy (UE), final energy (FE), and primary energy (PE) were determined based on the methodology described by the Regulation of the Minister of Infrastructure and Development of

27 February 2015 on the methodology for calculating the thermal characteristics of a building or a part of a building and energy attribute certificates [76]. In heating systems, usable energy (UE) is defined as the energy that is evacuated from a building via the ventilation system, minus heat gain. In water heating systems, UE is defined as the energy evacuated from a building via sewage. Usable energy is calculated with the use of the below formulas:

$$UE = Q_u/A_f \tag{1}$$

$$Q_u = Q_{H,nd} + Q_{W,nd} + Q_{C,nd} \tag{2}$$

where:

Q_u—annual demand for usable energy, (kWh/year)
A_f—floor area of premises with controlled air temperature, (m^2)
$Q_{H,nd}$—annual demand for usable energy for heating and ventilation, (kWh/year)
$Q_{W,nd}$—annual demand for usable energy for water heating, (kWh/year)
$Q_{C,nd}$—annual demand for usable energy for cooling, (kWh/year)

Final energy (FE) is defined as the energy supplied to technical systems in a building, and it is calculated with the use of the below formulas:

$$FE = Q_f/A_f \tag{3}$$

$$Q_f = Q_{k,H} + Q_{k,W} + Q_{k,C} + Q_{k,L} + E_{el,pom} \tag{4}$$

where:

Q_f—annual demand for final energy supplied to technical systems in a building, (kWh/year)
$Q_{k,H}$—annual demand for final energy supplied to a building or a part of a building for heating purposes, (kWh/year)
$Q_{k,W}$—annual demand for final energy supplied to a building for water heating, (kWh/year)
$Q_{k,C}$—annual demand for final energy supplied to a building for cooling purposes, (kWh/year)
$Q_{k,L}$—annual demand for final energy supplied to a building for lighting common areas; this parameter is not calculated for single-family homes and apartments, (kWh/year)
$E_{el,pom}$—annual demand for auxiliary energy supplied to technical systems in a building, (kWh/year)

Non-renewable primary energy (EP) is defined as the energy which is stored in fossil fuels and has not been converted or transformed:

$$EP = Q_p/A_f \tag{5}$$

$$Q_p = Q_{p,H} + Q_{p,W} + Q_{p,C} + Q_{p,L} \tag{6}$$

where:

Q_p—annual demand for non-renewable primary energy for technical systems, (kWh/year)
$Q_{p,H}$—annual demand for non-renewable primary energy for heating systems, (kWh/year)
$Q_{p,W}$—annual demand for non-renewable primary energy for water heating, (kWh/year)
$Q_{p,C}$—annual demand for non-renewable primary energy for cooling systems, (kWh/year)
$Q_{p,L}$—annual demand for non-renewable primary energy for lighting common areas; this parameter is not calculated for single-family homes and apartments, (kWh/year)

Specific CO_2 emissions are calculated with the use of the below formulas:

$$E_{CO_2} = \left(E_{CO_2,H} + E_{CO_2,W} + E_{CO_2,C} + E_{CO_2,L} + E_{CO_2,pom}\right)/A_f \tag{7}$$

where:

$E_{CO_2,H}$—CO_2 emissions from fuel combustion in heating systems, (tCO_2/year)
$E_{CO_2,W}$—CO_2 emissions from fuel combustion in water heating systems, (tCO_2/year)
$E_{CO_2,C}$—CO_2 emissions from fuel combustion in cooling systems, (tCO_2/year)
$E_{CO_2,L}$—CO_2 emissions from fuel combustion for lighting installations, (tCO_2/year)

where each of the above components (n) is calculated with the use of the below Equation (8):

$$E_{CO_2,n} = 36 \times 10^{-7} \cdot Q_{k,n} \cdot W_{e,n} \tag{8}$$

$Q_{k,n}$—annual demand for final energy supplied to the analyzed system in a building, (kWh/year)
$W_{e,n}$—CO_2 emissions from the combustion of different types of fuel in the analyzed systems, (tCO_2/TJ)
$E_{CO_2,aux}$—CO_2 emissions from fuel combustion in heating systems (tCO_2/year) is calculated with the use of formula (10):
$E_{el,aux,n}$—annual demand for auxiliary energy supplied to the nth system in a building, where H is the heating system, W is the water heating system, and C is the cooling system, (kWh/year)
$W_{e,aux,C}$—CO_2 emissions from the combustion of different types of fuels in the nth system, where H is the heating system, W is the water heating system, and C is the cooling system, (tCO_2/TJ)

The annual consumption of energy carriers or electricity, district heating, solar energy, geothermal energy, wind energy, and gas is calculated with the use of formula (10):

$$C = Q_f / A_f \tag{9}$$

The consumption of different energy carriers and energy types than those indicated in formula (10) is calculated with the use of formula (11):

$$C = \frac{Q_f \cdot 3.6}{A_f \cdot W_o} \tag{10}$$

where:

W_o—heating value of fuel, (MJ/m^3 or MJ/kg)

Pollutant emissions (Table 3) are calculated based on the reference materials published by the Ministry of Environmental Protection, Natural Resources and Forestry [77] regarding emissions of air pollutants from fuel combustion, as well as the guidelines of the National Centre for Emissions Balancing and Management (KOBiZE) [78] with the use of the below formula:

$$E = B \cdot W \tag{11}$$

where:

B—fuel consumption, (m^3 or Mg)
W—pollutant emissions per unit of consumed fuel, (kg/Mg or kg/10-6 m^3 or kg/kWh)

Table 3. Pollutant emissions per unit of consumed fuel [78].

Type of Fuel	Pollutant Emissions per Unit of Consumed fuel							
	SO_2	NO_x	CO	CO_2	Particulate Matter	Black Carbon	BaP	SO_2
Hard coal (kg/Mg)	19.2	2.2	45	1850	7	3.5	0.014	19.2
Natural gas (kg/10^{-6} m^3)	1.88	1520	300	2,000,000	0.5	0	0	1.88
Biomass (kg/Mg)	0.11	1	26	1200	3	0	0	0.11
Electricity (kg/kWh)	0.009	0.0023	0.00069	0.812	0.0015	0.000003	0	0.0091

The efficiency and effectiveness of the devices installed at the University of Warmia and Mazury Research Laboratory and operating under actual weather conditions were taken into account. The feasibility of the assumptions of theoretical calculations was compared with the possibility for the use of a heat pump, mechanical ventilation with ground-coupled heat exchanger (GCHE), and flat-plate and vacuum collectors in the heating and ventilation systems.

4. Results and Discussion

The annual demand for usable energy for heating and ventilation in the analyzed building is 8075 kWh/year for the design variants with gravity ventilation (1, 3, and 5) and 1997 kWh/year for mechanical ventilation (variants 2, 4, and 6). Gravity ventilation enforces the extension of the period of heating system used in the building to 4675.88 h, compared to 3002.51h for mechanical ventilation systems. Heat losses for ventilation amount to slightly more than half the heat losses in the balance for the entire building for cases with natural ventilation, and approximately 6% where mechanical ventilation is used (Table 4). The comparison also shows the seasonal variability of the heating and ventilation system operation.

Table 4. A comparison of the demand for heat for heating purposes and DHW (domestic hot water) and the losses for ventilation for particular variants (own calculation).

Demand [kWh/Month] \ Month	1	2	3	4	5	6	7	8	9	10	11	12
Demand for heating purposes (case 1,3,5)	2040.0	1470.0	759.7	304.3	33.3	0.0	0.0	0.0	34.6	537.6	1233.5	1659.5
Losses for ventilation (case 1,3,5)	1463.0	1282.0	1085.0	870.0	564.0	276.0	143.0	217.0	432.0	849.0	1086.0	1271.0
Demand for DHW (case 1,3,5)	306.6	306.6	306.6	306.6	306.6	306.6	306.6	306.6	306.6	306.6	306.6	306.6
Demand for heating purposes (case 2,4,6)	701.0	358.1	75.1	10.9	0.2	0.0	0.0	0.0	0.3	44.9	295.3	510.8
Losses for ventilation (case 2,4,6)	107.0	94.0	79.0	64.0	41.0	20.0	10.0	16.0	32.0	62.0	79.0	93.0

Analysis of the EU values obtained in the study for particular cases (Figure 1) showed nearly twice its value for the variants with gravity ventilation compared to solutions with mechanical ventilation with GCHE. This results from low heat losses due to the advantageous heat penetration coefficient values for space dividing elements used in the building. Therefore, the proportion of losses related to the natural ventilation of the building is noticeable in the overall balance.

Systems with the lowest efficiency are characterized by the highest FE. In terms of this parameter, the most advantageous systems include heat pump systems (overall efficiency of 2.66), systems with a low-temperature natural gas condensation boiler (overall efficiency of 0.76), coal boiler systems (overall efficiency of 0.60), and biomass (wood) boiler systems (overall efficiency of 0.51). FE for the analyzed energy sources is lower by approximately 40% than for mechanical ventilation. Heat pump systems (case 7 and 8) for which the decrease is approximately 7% are an exception.

In terms of the demand for FE and the exploitation of non-renewable sources, biomass-using systems are the most advantageous. In this single case, the use of mechanical ventilation with the unchanged heat source in the building results in FE being higher than for the use of gravitational energy. In other systems, the introduction of mechanical ventilation results in a reduction in FE by approximately 10% for cases 7 and 8, by approximately 18% for cases 3 and 4, and by approximately

23% for cases 1 and 2. In terms of PE, systems with a coal boiler source are the most energy-intensive. In case 1, the demand for PE is higher by 76% than case 5 (the most advantageous one in this variant).

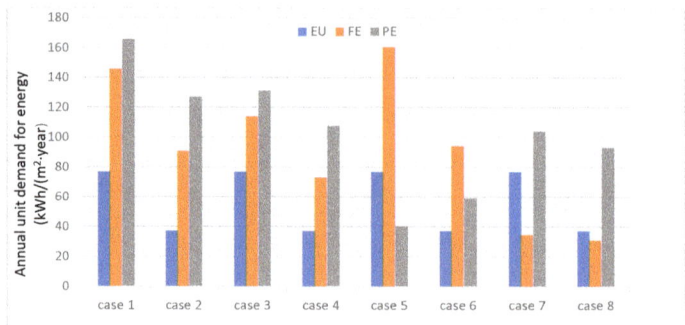

Figure 1. Useful (EU), final (FE), and primary (PE) energy in particular research variants (own elaboration).

High energy losses related to gravity ventilation encourage the use of mechanical ventilation with GCHE in the building, which enables the maintenance of energy-efficient construction standards and the heat recovery efficiency in the air handling unit at a level above 85%. The issue of the efficiency of heat extraction from the ground remains an open question. Testing on the ground-coupled pipe heat exchanger located at the Research Laboratory, conducted under the actual operational conditions, demonstrated that continuous operation under summer and transitional conditions is ineffective [79]. On the other hand, under winter conditions, the device enables efficient heat extraction from the ground [80]. Figure 2 shows the variability of temperatures at the input to and output from the GCHE in an exemplary month (October 2018). The device operated in a continuous mode without regeneration, and the average flowing airstream was approximately 155 m^3/h. During the analyzed period, 204.19 kWh of heat was extracted from the ground. During the period from September to March, average GCHE energy efficiency at a level of 35% was obtained, with experimentally-noted maximum hourly values of GCHE energy efficiency reaching approximately 90%.

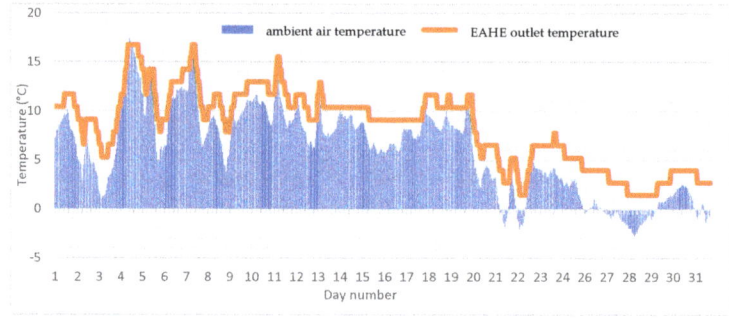

Figure 2. The distribution of external temperature and at the outlet from GCHE—November 2018 (own elaboration).

Compared to the results for actual operation, the use of the solar collector system in the DHW (domestic hot water) system should be considered (flat-plate and vacuum solar collectors connected to a 1000 L hot water storage tank were analyzed separately). The energy required for DHW purposes is constant throughout the year. In the analyzed variants with gravity ventilation, final energy for the DHW purposes amounts to approximately 38% (cases 1, 3, 5, and 7). On the other hand, for the variants with mechanical ventilation (cases 2, 4, 6, and 8) for domestic hot water preparation, it amounts to 40%–60% in the overall final energy balance for the building. Analysis of actual 2016 measurement data for flat-plate and vacuum collectors operating at the Research Laboratory at the University of Warmia and Mazury demonstrated that the average hourly efficiency of flat-plate collectors amounted to 22% and of vacuum collectors to 37% [81]. According to the Keymark test dependence for flat-plate collectors, an efficiency of 63% was obtained, while for vacuum collectors, the efficiency of 72% was obtained. The above results differ from the catalog efficiency of 74.3% for flat-plate collectors and of 78.9% for vacuum collectors, declared by the manufacturer. During the period of the most efficient operation of the above-mentioned devices (from May to September), unit heat from flat-plate collectors amounted to 77 kWh/m^2, while 188 kWh/m^2 was obtained from vacuum collectors. During the period from September to March, the collectors, irrespective of the type, operated sporadically, and they cannot be considered to be energy sources for the DHW [82].

The annual demand for usable energy (EU) was identical in water heating systems with and without solar collectors. The annual demand for final energy (FE) was higher in solar collectors which are less efficient than water heating systems based on a heat pump or a condensing gas boiler. In cases 7 and 8, where the use of solar collectors led to the greatest decrease in efficiency, relative to the highly efficient heat pump system, FE increased by 83% for flat-plate collectors and by 76% for vacuum collectors. In systems involving a condensing gas boiler (cases 3 and 4), FE increased by 4% when flat-plate collectors were added and by 2% when vacuum collectors were added. In the remaining cases, the introduction of flat-plate collectors decreased FE by 3%, and the installation of vacuum collectors decreased FE by 5%. If solar collectors were to supply 30% of the annual energy for water heating, the demand for non-renewable primary energy (PE) would decrease by 28% in cases 1 and 2, by 27% in cases 3 and 4, by 26% in cases 7 and 8, and by around 20% in cases 5 and 6.

During the winter period (from November to March), measurement data for the operation of a water–glycol compressor heat pump were also obtained, with the lower source in the form of vertical boreholes. The test results are presented in Figure 3. The result analysis demonstrated that the average coefficient of performance (COP) value was 3.36 and ranged from 2.24 to 4.46, while the COP catalog value declared by the manufacturer was 4.57.

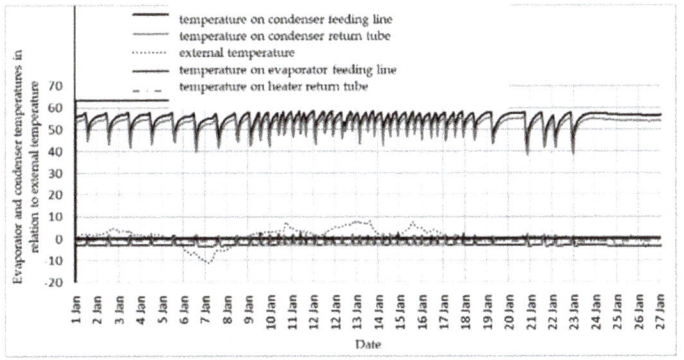

Figure 3. Heat pump operating parameters in January 2015 [83].

The emission analysis (Table 5) was conducted based on the indices of pollutant emission from fuel combustion in boilers with a nominal thermal power of up to 5 MW, according to the National Centre for Emissions Management (KOBiZE) [79]. The analysis results are presented in Table 4. The obtained results indicate that the use of a biomass (wood) boiler proved to be the least advantageous for the analyzed building due to carbon dioxide emissions. In this context, this solution appeared to be more emission-intensive than hard coal-fired boilers. It should be stressed, however, that in accordance with the IPCC guidelines [84], it is assumed in the emission trading scheme that the biomass combustion balance in this aspect has a zero value. It is assumed that the emission during combustion is compensated for CO_2 uptake in the photosynthesis process. In terms of CO_2 emissions, the most advantageous of the analyzed cases is the use of a low-temperature, natural gas condensation boiler, particularly in combination with mechanical ventilation with GCHE. A similar situation occurs for carbon monoxide emissions in the analyzed cases.

Table 5. Pollutant emissions for the heating and ventilation system and for domestic hot water [19].

Emissions Rate (kg/Year) \ Case	1	2	3	4	5	6	7	8
SO_2	58.36	48.94	4.06	19.89	4.68	20.21	48.07	43.13
NO_x	7.25	8.36	3.61	6.40	6.65	7.88	12.15	10.90
CO	127.57	69.60	0.82	1.78	146.60	75.75	3.64	3.27
Particulate matter	20.47	13.87	0.67	3.28	17.55	11.85	7.92	7.11
Black carbon	9.90	5.30	0.00	0.01	0.00	0.01	0.01	0.01
B-a-P	0.04	0.02	0.00	0.00	0.00	0.00	0.00	0.00
CO_2	5594.38	4574.19	3762.55	3576.96	7114.46	5201.41	4288.99	3848.32

Regarding the emissions of SO_2, black carbon, and benzo(a)pyrene, the systems based on natural gas or biomass combustion proved to be the most advantageous. A condition for achieving this effect is the least possible electricity consumption, even at the expense of excluding gravity ventilation. Regarding SO_2 emissions, the use of electricity feeding the heat pump (and auxiliary equipment), has consequences in emissions similar to those for a hard coal boiler. On the other hand, for the emissions of black carbon and benzo(a)pyrene, the use of hard coal is hundreds (B-a-P) or even thousands of times (black carbon) more disadvantageous.

Regarding NO_X emissions, the most disadvantageous energy source is electricity (heat pump), followed by the following sources, in order: hard coal boiler, biomass boiler, and natural gas boiler. It can also be noted that an increased proportion of NO_X emissions was found in the variants using mechanical ventilation requiring electricity supply (cases 2, 4 and 6), which confirms the significant effects of electricity use for the emissions of this pollutant.

Particulate matter emissions from hard coal combustion for the concerned building generates emissions higher by approximately 15% than a biomass boiler. This value is two (mechanical ventilation system) or three times (gravity ventilation source) higher than for the electricity using a heat pump. In terms of particulate matter emissions, the use of a natural gas boiler is the most advantageous.

In Table 6, the analyzed systems were classified based on pollutant emission levels, from the least to the most polluting cases.

Condensing gas boilers and biomass boilers ranked highest in the presented classification. The demand for grid electricity in the energy balance increased pollutant emissions, which was most apparent in mechanical ventilation systems with heat recovery in case 4 (condensing gas boiler and mechanical ventilation) and case 6 (biomass boiler with mechanical ventilation). Grid electricity (for example, for the operation of heat pumps in cases 7 and 8) does not generate pollution in the heated building, but it involves considerable loads and losses during energy production and transfer. However, individual users can reduce pollutant emissions by installing photovoltaic panels in the household.

Table 6. Classification of the analyzed systems based on pollutant emission levels, from the least to the most polluting [own elaboration].

SO$_2$ (*)		NOx		CO		CO$_2$		Particulate Matter		Black Carbon		B-a-P		Classification
kg/Year	Case no.	Kg/Year	Case no.	kg/Year	Case no.	kg/Year	Case no.	kg/Year	Case no.	Kg/Year	Case no.	kg/Year	Case no.	Case no.
4.06	3	3.61	3	0.82	3	3576.96	4	0.67	3	0.00	3	0.00	3	3
4.68	5	6.40	4	1.78	4	3762.55	3	3.28	4	0.00	5	0.00	5	5
19.89	4	6.65	5	3.27	8	3848.32	8	7.11	8	0.01	4	0.00	4	4
20.21	6	7.25	1	3.64	7	4288.99	7	7.92	7	0.01	6	0.00	6	6
43.13	8	7.88	6	69.60	2	4574.19	2	11.85	6	0.01	8	0.00	7	8

(*) the classification was based on SO$_2$ emissions in ambiguous cases.

5. Conclusions

Selected combinations of heating and ventilation systems were compared and analyzed in this study. Pollutant emissions from the presented solutions were determined. Unlike most research papers addressing the discussed topic, this study not only classified selected variants based on their emissions, but it also analyzed their energy efficiency. Special emphasis was placed on mechanical ventilation with heat recovery which can reduce emission levels. In the authors' opinion, mechanical ventilation with heat recovery is often unnecessarily disregarded or marginalized. This study proposes universal solutions that can be applied to heating and ventilation systems in countries with a similar climate.

The results of this study constitute highly valuable inputs for the Polish market because the analysed variants combine heating and ventilation systems that are most popular in single-family homes in Poland. The study also evaluated the effectiveness of solar heater panels for water heating which are increasingly popular among Polish homeowners. The theoretical efficiency of heating systems powered by renewable energy sources (solar heater panels, heat pump, ground heat exchanger) was compared with their performance in real-world conditions in north-eastern Poland.

The choice of an optimal heating and ventilation system for a single-family home is a complex problem. The real-world performance of system components is generally below their maximum theoretical efficiency. Even the most advanced systems based on renewable energy sources, such as heat pump circuits filled with a water–glycol mixture, have a theoretical COP of 4.57. In practice, the maximum COP was determined at 4.46, and the average COP during the heating season reached 3.36. Geothermal energy can be effectively used to preheat cold incoming air in the ground heat exchanger. This solution is far more effective for preheating cold air in winter than for cooling hot air in summer. However, even when hourly efficiency approximated 90% in winter, the average efficiency between September and March reached around 35%. In north-eastern Poland, the efficiency of solar thermal collectors in the optimal period of operation (May to September) was 11.3% below the theoretical value in flat-panel collectors and 6.9% below the theoretical value in vacuum collectors.

Energy consumption is an important consideration when selecting a heating and ventilation system. Usable energy (UE) values are influenced by structural partitions, the presence of thermal bridges and the airtightness of a building. The ventilation system also exerts a considerable influence on EU values. In the study, a highly efficient ventilation system with heat recovery (95%) decreased energy loss from the building by nearly 50%.

The consumption of usable energy in a building can be reduced not only through the conscious efforts of architects and developers, but also through legislative measures which introduce thermal insulation standards for structural partitions as well as airtightness standards. However, the existing legal regulations do not account for the impact and consequences of heat recovery in ventilation systems. In Poland, the minimum efficiency of heat recovery devices is set at 50% for mechanical ventilation systems and air-conditioning systems with an air exchange rate of 500 m^3/h and higher [85]. These solutions are characterized by low thermal efficiency, and they are mandatory in large buildings. Consequently, according to Statistics Poland data, only 0.26% of Polish households were equipped with mechanical ventilation systems with heat recovery in 2018 GUS (Statistics Poland) [75]. The measures aiming to improve the energy efficiency of single-family homes should not only increase the thermal performance of structural partitions (for example, through additional thermal insulation) but should also involve upgrades of heating systems. The improvement in the efficiency of central heating and water heating systems should be accompanied by the installation of mechanical ventilation systems with heat recovery.

Final energy (FE) is closely linked with the applied technical solutions and the efficiency of heating and ventilation systems. In the analyzed scenarios, systems based on heat pumps were characterized by the highest efficiency and the lowest FE values. In buildings equipped with heat pumps, mechanical ventilation with heat recovery can additionally decrease the demand for FE. This effect is most visible in buildings with low-efficiency heating and ventilation systems. Total efficiency was lowest in the system involving a biomass boiler, where the introduction of a ventilation system with heat recovery decreased the demand for FE by 41%. Final energy consumption is the key determinant of energy costs; therefore, this parameter should play the most important role for developers. However, primary energy (PE) is a vital parameter from the point of view of environmental protection and sustainable development because it denotes the amount of energy generated from fossil fuels, including mining operations and fuel transport. In this respect, case 5—where heat was supplied by a biomass boiler and additional energy was not required for mechanical ventilation—emerged as the optimal variant.

The study revealed that a condensing gas boiler combined with natural ventilation was characterized by the lowest pollutant emissions, excluding carbon dioxide. Carbon dioxide emissions were lower only in the system featuring a gas boiler and mechanical ventilation. A biomass boiler with mechanical ventilation was also a relatively low-emission solution.

The life cycle balance of the proposed solutions will be analyzed in future research.

Author Contributions: Conceptualization, J.H. and A.S.-S.; methodology, J.H. and A.S.-S.; software, A.S.-S.; validation, J.H., A.S.-S. and M.N.; formal analysis, J.H., A.S.-S. and M.N.; investigation, J.H. and A.S.-S.; resources, A.S.-S.; data curation, J.H. and A.S.-S.; writing—original draft preparation, J.H. and A.S.-S.; writing—review and editing, J.H.; visualization, J.H.; supervision, J.H.; A.S.-S. and M.N.; project administration, A.S.-S. All authors have read and agreed to the published version of the manuscript.

Funding: This research received no external funding.

Conflicts of Interest: The authors declare no conflict of interest.

References

1. *Systematic Review of Health Aspects of Air Pollution in Europe*; WHO Regional Office for Europe: Copenhagen, Denmark, 2004; Available online: http://apps.who.int/iris/bitstream/handle/10665/107571/E83080.pdf?sequence=1&isAllowed=y (accessed on 6 May 2019).
2. A Global Assessment of Exposure and Burden of Disease. 2016. Available online: http://apps.who.int/iris/bitstream/handle/10665/250141/9789241511353-eng.pdf?sequence=1 (accessed on 6 May 2019).
3. Azmi, S.Z.; Latif, M.T.; Ismail, A.S.; Juneng, L.; Jemain, A.A. Trend and status of air quality at three different monitoring stations in the Klang Valley, Malaysia. *Air Qual. Atmosphere Health* **2009**, *3*, 53–64. [CrossRef] [PubMed]
4. Anderson, N.; Strader, R.; Davidson, C. Airborne reduced nitrogen: Ammonia emissions from agriculture and other sources. *Environ. Int.* **2003**, *29*, 277–286. [CrossRef]
5. Chen, R.; Yin, P.; Meng, X.; Liu, C.; Wang, L.; Xu, X.; Ross, J.A.; Tse, L.A.; Zhao, Z.; Kan, H.; et al. Fine Particulate Air Pollution and Daily Mortality. A Nationwide Analysis in 272 Chinese Cities. *Am. J. Respir. Crit. Care Med.* **2017**, *196*, 73–81. [CrossRef] [PubMed]
6. Valavanidis, A.; Fiotakis, K.; Vlachogianni, T. Airborne Particulate Matter and Human Health: Toxicological Assessment and Importance of Size and Composition of Particles for Oxidative Damage and Carcinogenic Mechanisms. *J. Environ. Sci. Health Part C* **2008**, *26*, 339–362. [CrossRef] [PubMed]
7. Amann, I.; Bertok, J.; Cofala, F.; Gyarfas, C.; Heyes, C.; Klimont, Z.; Schopp, W.; Winiwarter, W. *Baseline Scenarios for the Clean Air for Europe (CAFE)*; Programme Final Report; CAFE Scenario Analysis Report No. 1; International Institute for Applied Systems Analysis: Laxenburg, Austria, 2005.
8. *The Benefits and Costs of the Clean Air Act from 1990 to 2020*; Technical Report for U.S.; Environmental Protection Agency Office of Air and Radiation: Washington, DC, USA, 2011. Available online: https://www.epa.gov/sites/production/files/2015-07/documents/summaryreport.pdf (accessed on 6 May 2019).
9. World Health Organisation. *Ambient (Outdoor) Air Quality and Health*; WHO: Geneva, Switzerland, 2014.
10. Künzli, N.; Kaiser, R.; Medina, S.; Studnicka, M.; Chanel, O.; Filliger, P.; Herry, M.; Horak, F., Jr.; Puybonnieux-Texier, V.; Quénel, P.; et al. Public-health impact of outdoor and traffic-related air pollution: A European assessment. *Lancet* **2000**, *356*, 795–801. [CrossRef]
11. Pope, C.A. Review: Epidemiological Basis for Particulate Air Pollution Health Standards. *Aerosol Sci. Technol.* **2000**, *32*, 4–14. [CrossRef]
12. Schwela, D. Air pollution and health in urban areas. *Rev. Environ. Health* **2000**, *15*, 13–42. [CrossRef]
13. Huang, J.; Pan, X.; Guo, X.; Li, G. Impacts of air pollution wave on years of life lost: A crucial way to communicate the health risks of air pollution to the public. *Environ. Int.* **2018**, *113*, 42–49. [CrossRef]

14. Ortega-García, J.A.; Sánchez-Solís, M.; Ferrís-Tortajada, J. Air pollution and children's health. *An. Pediatr.* **2018**, *89*, 77–79. [CrossRef]
15. Dellano-Paz, F.; Calvo-Silvosa, A.; Antelo, S.I.; Soares, I. The European low-carbon mix for 2030: The role of renewable energy sources in an environmentally and socially efficient approach. *Renew. Sustain. Energy Rev.* **2015**, *48*, 49–61. [CrossRef]
16. Ghaith, A.F.; Epplin, F.M. Consequences of a carbon tax on household electricity use and cost, carbon emissions, and economics of household solar and wind. *Energy Econ.* **2017**, *67*, 159–168. [CrossRef]
17. Cucchiella, F.; Gastaldi, M.; Trosini, M. Investments and cleaner energy production: A portfolio analysis in the Italian electricity market. *J. Clean. Prod.* **2017**, *142*, 121–132. [CrossRef]
18. Freedman, B. *Environmental Ecology: The Ecological Effects of Pollution, Disturbance, and Other Stresses*; Academic Press: Cambridge, MA, USA, 1995.
19. Hałacz, J.; Skotnicka-Siepsiak, A.; Neugebauer, M.; Nalepa, K.; Sołowiej, P. Assessment of options to reduce pollutant emissions in single-family houses in north-eastern Poland. In Proceedings of the Renewable Energy Sources-Engineering, Technology, Innovation, Krynica, Poland, 12–14 June 2019.
20. Hałacz, J.; Neugebauer, M.; Sołowiej, P.; Nalepa, K.; Wesołowski, M. Recycling Expired Photovoltaic Panels in Poland. In *Renewable Energy Sources: Engineering, Technology, Innovation*; Springer Proceedings in Energy; Wróbel, M., Jewiarz, M., Szlęk, A., Eds.; Springer: Cham, Switzerland, 2020; pp. 459–470.
21. EIA. *International Energy Outlook 2010*; US Energy Information Administration: Washington, DC, USA, 2010. Available online: https://www.stb.gov/STB/docs/RETAC/2010/March/EIA%20AEO%202010.pdf (accessed on 25 February 2020).
22. Huppes, H.; de Koning, A.; Suh, S.; Heijungs, R.; van Oers, L.; Nielsen, P.; Guinée, J.B. Environmental impacts of consumption in the European Union. *J. Ind. Ecol.* **2006**, *10*, 129–146. [CrossRef]
23. Mattinen, M.; Tainio, P.; Salo, M.; Jalas, M.; Nissinen, A.; Heiskanen, E. How building users can contribute to greenhouse-gas emission reductions in Finland: Comparative study of standard technical measures, user modifications and behavioural measures. *Energy Effic.* **2016**, *9*, 301–320. [CrossRef]
24. Abrahamse, W.; Steg, L.; Vlek, C.; Rothengatter, T. A review of intervention studies aimed at household energy conservation. *J. Environ. Psychol.* **2005**, *25*, 273–291. [CrossRef]
25. Moura, P.; López, G.; Moreno, J.I.; De Almeida, A.T. The role of Smart Grids to foster energy efficiency. *Energy Effic.* **2013**, *6*, 621–639. [CrossRef]
26. Weiß, T.; Fulterer, A.M.; Knotzer, A. Energy flexibility of domestic thermal loads—A building typology approach of the residential building stock in Austria. *Adv. Build. Energy Res.* **2019**, *13*, 122–137. [CrossRef]
27. Yang, Y.; Ren, J.; Solgaard, H.S.; Xu, D.; Nguyen, T.T. Using multi-criteria analysis to prioritize renewable energy home heating technologies. *Sustain. Energy Technol. Assessments* **2018**, *29*, 36–43. [CrossRef]
28. Bhati, N.; Kakran, S. Smart Home Energy Management with Integration of Renewable Energy. In Proceedings of the Second International Conference on Intelligent Computing and Control Systems (ICICCS), Madurai, India, 14–15 June 2018; pp. 1785–1789.
29. Olonscheck, M.; Holsten, A.; Kropp, J. Heating and cooling energy demand and related emissions of the German residential building stock under climate change. *Energy Policy* **2011**, *39*, 4795–4806. [CrossRef]
30. Wojdyga, K.; Chorzelski, M.; Różycka-Wrońska, E. Emission of pollutants in flue gases from Polish district heating sources. *J. Clean. Prod.* **2014**, *75*, 157–165. [CrossRef]
31. Michelsen, C.C.; Madlener, R. Switching from fossil fuel to renewables in residential heating systems: An empirical study of homeowners' decisions in Germany. *Energy Policy* **2016**, *89*, 95–105. [CrossRef]
32. Kaczmarczyk, M.; Sowiżdżał, A.; Tomaszewska, B. Energetic and Environmental Aspects of Individual Heat Generation for Sustainable Development at a Local Scale—A Case Study from Poland. *Energie* **2020**, *13*, 454. [CrossRef]
33. Miąsik, D.P.-; Rybak-Wilusz, E.; Rabczak, S. Ecological and Financial Effects of Coal-Fired Boiler Replacement with Alternative Fuels. *J. Ecol. Eng.* **2020**, *21*, 1–7. [CrossRef]
34. Krzysztof, N.; Maria, B.; Danuta, P.-M.; Sławomir, R. Emission of Air Pollutants in the Hot Water Production. *IOP Conf. Series Mater. Sci. Eng.* **2017**, *245*, 52032. [CrossRef]
35. Nowak, K.; Wojdyga, K.; Rabczak, S. Effect of coal and biomass co-combustion on the concentrations of selected gaseous pollutants. *IOP Conf. Series Earth Environ. Sci.* **2019**, *214*, 012130. [CrossRef]
36. Rabczak, S.; Miąsik, D.P. Effect of the type of heat sources on carbon dioxide emissions. *J. Ecol. Eng.* **2016**, *17*, 186–191. [CrossRef]

37. European Environmental Agency (EEA). *Urban Sprawl in Europe, Joint EEA-FOEN Report No. 11/2016*; EEA: Copenhagen, Denmark, 2016; Volume 20.
38. Kijewska, A.; Bluszcz, A. Analysis of greenhouse gas emissions in the European Union member states with the use of an agglomeration algorithm. *J. Sustain. Min.* **2016**, *15*, 133–142. [CrossRef]
39. Botta, E.; Kozluk, T. *Measuring Environmental Policy Stringency in OECD Countries: A Compsite Index Approach*; OECD Publishing: Paris, France, 2014.
40. Andersson, F.N.G. International trade and carbon emissions: The role of Chinese institutional and policy reforms. *J. Environ. Manag.* **2018**, *205*, 29–39. [CrossRef]
41. Zeng, S.; Nan, X.; Liu, C.; Chen, J. The response of the Beijing carbon emissions allowance price (BJC) to macroeconomic and energy price indices. *Energy Policy* **2017**, *106*, 111–121. [CrossRef]
42. Damsø, T.; Kjær, T.; Christensen, T.B. Implementation of local climate action plans: Copenhagen—Towards a carbon-neutral capital. *J. Clean. Prod.* **2017**, *167*, 406–415. [CrossRef]
43. Griggs, D.; Stafford-Smith, M.; Gaffney, O.; Rockström, J.; Öhman, M.C.; Shyamsundar, P.; Steffen, W.; Glaser, G.; Kanie, N.; Noble, I. Policy: Sustainable development goals for people and planet. *Nature* **2013**, *495*, 305–307. [CrossRef] [PubMed]
44. Wang, Q.; Dai, H.-N.; Wang, H. A Smart MCDM Framework to Evaluate the Impact of Air Pollution on City Sustainability: A Case Study from China. *Sustainability* **2017**, *9*, 911. [CrossRef]
45. GBD. 2016 Risk Factors Collaborators Global, regional, and national comparative risk assessment of behavioural, environmental and occupational, and metabolic risks or clusters of risks, 1990–2016: A systematic analysis for the Global Burden of Disease Study 2016. *Lancet* **2017**, *390*, 1345–1422.
46. Directives Regulating Emissions of Air Pollutants (EU) 2015/2193. 2015. Available online: https://eur-lex.europa.eu/legal-content/EN/TXT/PDF/?uri=CELEX:32015L2193&rid=4 (accessed on 8 May 2019).
47. Krzyzanowski, M.; Cohen, A. Update of WHO air quality guidelines. *Air Qual. Atmosphere Health* **2008**, *1*, 7–13. [CrossRef]
48. Eurostat European Environmental Agency (EEA). Emissions of Air Pollutants, EU-28 (1990–2016). Available online: http://ec.europa.eu/eurostat/statistics-explained/index.php/Air_pollution_statistics_-_emission_inventories (accessed on 8 May 2019).
49. Eurostat Homepage. Available online: https://ec.europa.eu/eurostat/en/web/products-manuals-and-guidelines/-/KS-GQ-15-009 (accessed on 9 May 2019).
50. Dellano-Paz, F.; Calvo-Silvosa, A.; Antelo, S.I.; Soares, I. Power generation and pollutant emissions in the European Union: A mean-variance model. *J. Clean. Prod.* **2018**, *181*, 123–135. [CrossRef]
51. Dellano-Paz, F.; Calvo-Silvosa, A.; Antelo, S.I.; Soares, I. Energy planning and modern portfolio theory: A review. *Renew. Sustain. Energy Rev.* **2017**, *77*, 636–651. [CrossRef]
52. Rodríguez, M.C.; Dupont-Courtade, L.; Oueslati, W. Air pollution and urban structure linkages: Evidence from European cities. *Renew. Sustain. Energy Rev.* **2016**, *53*, 1–9. [CrossRef]
53. Zimmer, A.; Koch, N. Fuel consumption dynamics in Europe: Tax reform implications for air pollution and carbon emissions. *Transp. Res. Part A Policy Pr.* **2017**, *106*, 22–50. [CrossRef]
54. Brodny, J.; Tutak, M. Analysis of the diversity in emissions of selected gaseous and particulate pollutants in the European Union countries. *J. Environ. Manag.* **2019**, *231*, 582–595. [CrossRef]
55. Jędrak, J.; Konduracka, E.; Badyda, A.J.; Dąbrowiecki, P. *The Impact of Air Pollution on Health Krakow Smog Alarm*; Kraków, Poland, 2017; Available online: https://krakowskialarmsmogowy.pl/files/images/ck/14882713101616070935.pdf (accessed on 5 March 2020).
56. ISAP Homepage. Available online: http://prawo.sejm.gov.pl/isap.nsf/DocDetails.xsp?id=WDU20170000519 (accessed on 9 May 2019).
57. ISAP Homepage. Available online: http://prawo.sejm.gov.pl/isap.nsf/DocDetails.xsp?id=WDU20180001119 (accessed on 9 May 2019).
58. ISAP Homepage. Available online: http://prawo.sejm.gov.pl/isap.nsf/DocDetails.xsp?id=WDU20120001031 (accessed on 9 May 2019).
59. GIOS Homepage. Available online: http://www.gios.gov.pl/images/dokumenty/pms/pms/PPMS_2016-2020.pdf (accessed on 10 May 2019).
60. Eur-Lex.Europa.Eu. Available online: https://eur-lex.europa.eu/legal-content/PL/TXT/PDF/?uri=CELEX:32016L2284&from=EN (accessed on 10 May 2019).

61. Kopidou, D.; Diakoulaki, D. Decomposing industrial CO_2 emissions of Southern European countries into production- and consumption-based driving factors. *J. Clean. Prod.* **2017**, *167*, 1325–1334. [CrossRef]
62. Zeng, S.; Liu, Y.; Liu, C.; Nan, X. A review of renewable energy investment in the BRICS countries: History, models, problems and solutions. *Renew. Sustain. Energy Rev.* **2017**, *74*, 860–872. [CrossRef]
63. GUS Homepage. Available online: https://stat.gov.pl/obszary-tematyczne/srodowisko-energia/energia/gospodarka-paliwowo-energetyczna-w-latach-2017-i-2018,4,14.html (accessed on 4 February 2020).
64. GUS Homepage. Available online: https://stat.gov.pl/obszary-tematyczne/srodowisko-energia/srodowisko/ochrona-srodowiska-2017,1,18.html (accessed on 10 May 2019).
65. GUS Olsztyn Homepage. Available online: https://olsztyn.stat.gov.pl/zakladki/zakladka1/ (accessed on 10 May 2019).
66. Olsztyn.eu Homepage. Available online: https://www.olsztyn.eu/gospodarka/program-ograniczania-niskiej-emisji.html (accessed on 10 May 2019).
67. Encyklopedia Warmii i Mazur Homepage. Available online: http://encyklopedia.warmia.mazury.pl/index.php/Wojew%C3%B3dztwo_warmi%C5%84sko-mazurskie (accessed on 10 May 2019).
68. Wikitravel Homepage. Available online: https://wikitravel.org/pl/Warmia-Mazury (accessed on 10 May 2019).
69. Nieróbca, A.; Kozyra, J.; Mizak, K.; Wróblewska, E. Changing length of the growing season in Poland. *Water Environ. Rural Areas* **2013**, *13*, 81–94.
70. ISAP Homepage. Available online: http://prawo.sejm.gov.pl/isap.nsf/DocDetails.xsp?id=WDU20170002285 (accessed on 10 May 2019).
71. Intersoft Homepage. Available online: https://www.intersoft.pl/cad/index.php?kat=certyfikat-energetyczny-audyt (accessed on 10 May 2019).
72. Raport Rynkowy Port PC: Pompy Ciepła 2019. Available online: http://portpc.pl/pdf/raporty/Raport_PORTPC_wersja_final_2019.pdf (accessed on 4 February 2020).
73. NFOSiGW Homepage. Available online: http://nfosigw.gov.pl/oferta-finansowania/srodki-krajowe/programy/doplaty-do-kredytow-na-kolektory-sloneczne/informacje-o-programie/ (accessed on 4 February 2020).
74. NFOSiGW Homepage. Available online: http://nfosigw.gov.pl/oferta-finansowania/srodki-krajowe/programy/doplaty-do-kredytow-na-kolektory-sloneczne/biezace-efekty-wdrazania/ (accessed on 4 February 2020).
75. GUS Homepage. Available online: https://stat.gov.pl/obszary-tematyczne/srodowisko-energia/energia/zuzycie-energii-w-gospodarstwach-domowych-w-2018-roku,2,4.html (accessed on 6 February 2020).
76. ISAP Homepage. Available online: http://prawo.sejm.gov.pl/isap.nsf/DocDetails.xsp?id=WDU20150000376 (accessed on 4 February 2020).
77. Wskaźniki Emisji Substancji Zanieczyszczających Wprowadzanych do Powietrza, z Procesów. Energetycznego Spalania Paliw. Available online: https://www.google.com/url?sa=t&rct=j&q=&esrc=s&source=web&cd=1&cad=rja&uact=8&ved=2ahUKEwjyx-H07L3nAhXBrIsKHa54AnoQFjAAegQIBhAB&url=http%3A%2F%2Fbip.umwd.dolnyslask.pl%2Fplik%2Cid%2C49873&usg=AOvVaw3eAHo51t1wcu2myZ5IQwPR (accessed on 6 February 2020).
78. KOBIZE Homepage. Available online: https://krajowabaza.kobize.pl/docs/male_kotly.pdf (accessed on 10 May 2019).
79. Skotnicka-Siepsiak, A.; Wesołowski, M.; Piechocki, J. Validating Models for Calculating the Efficiency of Earth-to-Air Heat Exchangers Based on Laboratory Data for Fall and Winter 2016 in Northeastern Poland. *Pol. J. Environ. Stud.* **2019**, *28*, 3431–3438. [CrossRef]
80. Skotnicka-Siepsiak, A.; Wesołowski, M.; Piechocki, J. Experimental and Numerical Study of an Earth-to-Air Heat Exchanger in Northeastern Poland. *Pol. J. Environ. Stud.* **2018**, *27*, 1255–1260. [CrossRef]
81. Skotnicka-Siepsiak, A.; Wesołowski, M.; Neugebauer, M.; Piechocki, J.; Sołowiej, P. The Influence of Weather Conditions and Operating Parameters on the Efficiency of Solar Power Collectors Based on Empirical Evidence. In *Renewable Energy Sources: Engineering, Technology, Innovation*; Springer: Cham, Switzerland, 2018; pp. 95–106.
82. Skotnicka-Siepsiak, A.; Wesołowski, M.; Konopacki, P. Raport z badań porównawczych instalacji płaskich i próżniowych cieczowych kolektorów słonecznych funkcjonujących w takich samych warunkach roboczych. *Ciepłownictwo Ogrzewnictwo Wentylacja* **2015**, *46*, 478–481.

83. Chojnowka, J. Badanie Efektywności Funkcjonowania Pompy Ciepła Grunt-Woda. Master's Thesis, University of Warmia and Mazury, Olsztyn, Poland, 22 June 2015.
84. IPCC Homepage. Available online: https://www.ipcc-nggip.iges.or.jp/public/gl/invs6.html (accessed on 10 May 2019).
85. ISAP Homepage. Available online: http://prawo.sejm.gov.pl/isap.nsf/download.xsp/WDU20190001065/O/D20191065.pdf (accessed on 7 February 2020).

© 2020 by the authors. Licensee MDPI, Basel, Switzerland. This article is an open access article distributed under the terms and conditions of the Creative Commons Attribution (CC BY) license (http://creativecommons.org/licenses/by/4.0/).

Article

Decision Support System for the Production of Miscanthus and Willow Briquettes

Sławomir Francik [1,*], Adrian Knapczyk [1], Artur Knapczyk [2] and Renata Francik [3]

[1] Department of Mechanical Engineering and Agrophysics, University of Agriculture in Krakow, Balicka 120, 31-120 Kraków, Poland; adrian.knapczyk@urk.edu.pl
[2] Private Researcher, Kurow 26, 34-233 Kurow, Poland; artur.knapczyk@gmail.com
[3] Institute of Health, State Higher Vocational School, Staszica 1, 33-300 Nowy Sącz, Poland; rfrancik@pwsz-ns.edu.pl
* Correspondence: slawomir.francik@urk.edu.pl; Tel.: +48-12-662-4641

Received: 31 January 2020; Accepted: 10 March 2020; Published: 15 March 2020

Abstract: The biomass is regarded as a part of renewable energy sources (RES), which can satisfy energy demands. Biomass obtained from plantations is characterized by low bulk density, which increases transport and storage costs. Briquetting is a technology that relies on pressing biomass with the aim of obtaining a denser product (briquettes). In the production of solid biofuels, the technological as well as material variables significantly influence the densification process, and as a result influence the end quality of briquette. This process progresses differently for different materials. Therefore, the optimal selection of process' parameters is very difficult. It is necessary to use a decision support tool—decision support system (DSS). The purpose of the work was to develop a decision support system that would indicate the optimal parameters for conducting the process of producing *Miscanthus* and willow briquettes (pre-comminution, milling and briquetting), briquette parameters (durability and specific density) and total energy consumption based on process simulation. Artificial neural networks (ANNs) were used to describe the relationship between individual parameters of the briquette production process. DSS has the form of a web application and is opened from a web browser (it is possible to open it on various types of devices). The modular design allows the modification and expansion the application in the future.

Keywords: decision support system; briquettes production; willow; *Miscanthus*; artificial neural network; multilayer perceptron

---

1. Introduction

The usage of energy around the world is constantly rising. This is due to the increase in population and increasing demands of society [1,2]. The major part of energy today energy is obtained from burning fossil fuels. This is detrimental to the environment mainly because of greenhouse gas emissions [3]. The need for reducing the emission of greenhouse gases, and restricting the usage of fossil fuels with simultaneous increase in energy demand caused an increase in the amount of research regarding renewable energy sources (RES) [4,5].

The biomass is regarded as a RES of high potential, which can satisfy energy demands of contemporary society both in developed and developing countries throughout the world [6–9]. The interest in biofuels created from different biomass types including agricultural energy waste and energy cultivations is growing [10,11]. The use of biomass to produce energy is around 50% of all RES in the world, and in Europe it is 70% [3]. Biomass can also be a raw material for the production of liquid and gaseous solid biofuels [11–13].

The growing global bioenergy market requires new biomass sources. Logging wood and wood biomass waste is insufficient. It is necessary to gradually replace it with agricultural biomass (energy plantations, aquatic plants and algae, agricultural waste, food processing waste and municipal solid waste [8,14,15]). The use of lignocellulosic biomass, e.g., *Miscanthus*, willow, poplar, acacia or paulownia as a source is particularly interesting [5,11,16].

Biomass obtained from plantations is characterized by low bulk density, which increases transport and storage costs. For this reason, the shredded material is subjected to compaction using high pressure [8,17]. Briquetting is one of the basic technologies for thickening and converting biomass into solid biofuel—bio-briquettes intended for direct combustion [8,10,18]. Briquetting is a process that comprises several stages, such as initial shredding, milling and densifying. Sometimes drying the raw material or cooling the briquettes is required [5,19,20].

Main qualitative parameters of briquettes are related to density, porosity and durability. Moisture content and compressive strength also influence the quality of briquette. These parameters are very significant from the viewpoint of logistic processes, combustion processes and so on [21,22]. Briquettes of bulk density above 1000 kg·m^{-3} are biofuels of high quality [23]. In the production of solid biofuels, the technological as well as material variables significantly influence the densification process, and as a result influence end quality of briquettes [24]. These factors can be divided into preproductive (material properties: particle size and moisture content), productive (impact of drying, storage or shredding on the properties of the material and the densification process) and post productive (storing, transport and trans-shipment conditions) [19,23,24]. The size of the particles and their distribution are included in the main factors determining the physical and mechanical properties of the briquette. Finer milling means higher density, hardness and durability, but also causes higher production costs [10]. Many researches were dedicated to describing and explaining the processes of biomass densification. The efficacy of the densification process stems from the quality of bonds between particles [10]. The process of densifying the fragmented biomass occurs under high pressure, which causes the contact area to increase, so adhesive forces lead to permanent connections between those particles [17]. The work delivered by pressure is partially converted into heat. During densification, the friction and shear in between the particles and between the particles and the briquetting machine, as well as the molecular adhesive forces cause the temperature of the briquette to increase [9].

Various types of biomass are used for the production of briquettes. The most widely and commonly used type for solid biofuel production is wood biomass (when considering commercial purposes, as it has high content of lignin—a natural binder). Others other than wood types may also be valid for such purposes [8]. Various species of *Miscanthus* and bamboo have great potential for bioenergy production due to high yield, high cellulose and hemicellulose content, high calorific value and low ash content [6,25]. Additionally, waste biomass from harvesting lines (poppy, oat, wheat and rice hulls) or corn waste can be used as raw materials for briquette production [8,26,27].

As shown above, both the characteristics of the raw material and operational parameters are responsible for the quality of briquette. Therefore, in the production of solid biofuel (briquette), it is very important to understand the impact of these main factors on the properties of briquette. With this knowledge, product quality can be improved [9,21].

Unfortunately, biomass as a raw material has vastly different properties. Therefore, there has not been a single, consistent mathematical description of the briquetting process so far. This process progresses differently for different materials. Therefore, the optimal selection of process' parameters (machine settings included in the production line) is very difficult. It is thus necessary to use decision support tools.

Operational research methods are classic decision support tools. However, due to the increasing complexity of decision-making problems and the growing possibilities of IT systems, in recent decades, a new type of tools supporting decision-makers has been developed: the so-called decision support systems (DSS). The definition formulated by Sprague and Carlson was [28,29]: "DSS comprises a class of information system that draws on transaction processing systems and interacts with the other parts of the overall information system to support the decision-making activities of managers and other knowledge workers in organizations". In recent years, there has been a rapid development of research on DSS. This is confirmed by the rapidly growing number of publications [29,30].

DSS is composed of three basic components, which are: the database management subsystem, the model base management subsystem and the dialog generation and management subsystem (user interface), which handles the interaction of the user with the system [28,30]. DSSs are used in many areas, e.g., IT, production engineering, transport, biosystem engineering and many more. In agricultural engineering, DSSs are used in pest management, crop production, biomass production, operational planning machine activities and pig products chains [29]. Decision support systems cover various parts of the supply chain. They are mostly used to deal with suppliers, optimize shipments, transport and production. The industries that most often use DSSs in the supply chain are manufacturing (e.g., medicine and fertilizer production, etc.) and agriculture. DSSs in the supply chain are also used in automotive processes, computers, construction, e-commerce, fisheries, food, forestry, logistics, medicine and petroleum [31–33]. DSSs are used to supplement energy-related decision processes [34]. Decision support systems were also a research subject in the field of RES. For example, DSSs have been developed for:

- Evaluation of energy and economic benefits of power to gas/heat technologies [35].
- The renewable energy management (a small scale photovoltaic energy production) based on the existing geographic information systems [36].
- Achieving energy balance in a low-voltage microgrid with RES (photovoltaic panels and wind turbines) [37].
- Supplementing the selection of optimal sites for grid-connected photovoltaic power plants using an environmental DSS [38].
- Setting priorities regarding the selection of bioenergy home heating sources in Southern Europe (DSS uses MCDA—multicriteria decision analysis—methods) [39].

One of the components of DSSs is a model. It is important that they describe as accurately as possible the processes on which a DSS is expected to provide the decision-maker with information on the best solutions. The desire to increase the possibility of developing accurate decision variants has resulted in the emergence of intelligent DSSs, in which methods of artificial intelligence are used. In order to find dependencies between individual parameters describing the briquette production process artificial neural networks (ANNs), which are one of the methods of artificial intelligence, can be used.

The term artificial neural network refers to a computational and machine learning technique [40–43]. One type of neural networks are multilayer perceptron neural networks (MLPs). An MLP consists of an input layer, a hidden layer (one or two) and an output layer. Each layer consists of neurons with a non-linear activation function, which are connected to each other [44–46]. ANNs are universal nonlinear approximators. Implementation of artificial neural networks for different research proved to be very efficient and accurate [20,40,43,44,47–52].

ANNs are widely employed to exploit the empirical knowledge. Particularly, ANNs are applied in modeling, classification and clustering tasks, prediction processes, decision-making processes and management of industrial production systems [53]. ANNs are used in various scientific fields including, for example, bioinformatics, biochemistry, medicine, meteorology, economic sciences, robotics, aquaculture, food security and climatology. ANNs are also used in agriculture, agrophysics or agricultural engineering [20,40,44,47–49,51,54]. For example, an ANN was used to model paper production. An ANN was implemented to predict the fibers' length [53]. ANNs are also employed to control and predict variables in drying processes (a tomato drying model, tobacco and willow woodchips drying processes) [20,55]. Many applications of ANNs involve forecasting and prediction in agriculture. The ANNs with MLP topology to forecast winter rapeseed yields and winter wheat yield were developed [46,56]. The neural model forecasting temperature changes inside the heated foil tunnel enables the optimization of decisions regarding the control of heating system. Consequently, this allows one to lower the energy usage needed to ensure optimal internal climate in heated foil tunnel [44]. The ANN model can be also a part of a more complex DSS designated for internal climate management.

Some authors propose artificial neural networks (ANNs) for energy demand prediction and optimization (energy demand in smart grid; heating, ventilation and air conditioning energy savings in office buildings) [34]. Many ANN applications are related to renewable energy sources (different uses of ANN models for better energy production predictions). Research addresses for example the creation and use of ANNs to forecast solar radiation (the main problem for the best use of photovoltaic systems) and wind power forecasting [41,45,57–59]. ANNs are applied for forecasting building energy usage and demand [42]. The diversification of ANNs applications is vast, demonstrating the importance of this tool. ANNs are a powerful tool for making predictions based on a large number of interrelated experimental data [20,44,47–50,52,55,60–63].

The aim of the work was to create a DSS, which could approximate optimal parameters for producing briquettes from *Miscanthus* and willow. Such parameters are durability and specific density (briquette parameters) and these associated with production processes (precomminution, milling and briquetting parameters).

2. Materials and Methods

As part of the work, a decision support system was designed, made and tested on a selected production line of plant biomass briquettes. The proposed decision support system is a continuation of a larger research cycle. The whole work consisted of 4 stages (Figure 1): I—conducting experimental tests, II—creating simulation models using ANN, III—performing simulation experiments and creating a database and IV—developing the inference module for the proposed DSS.

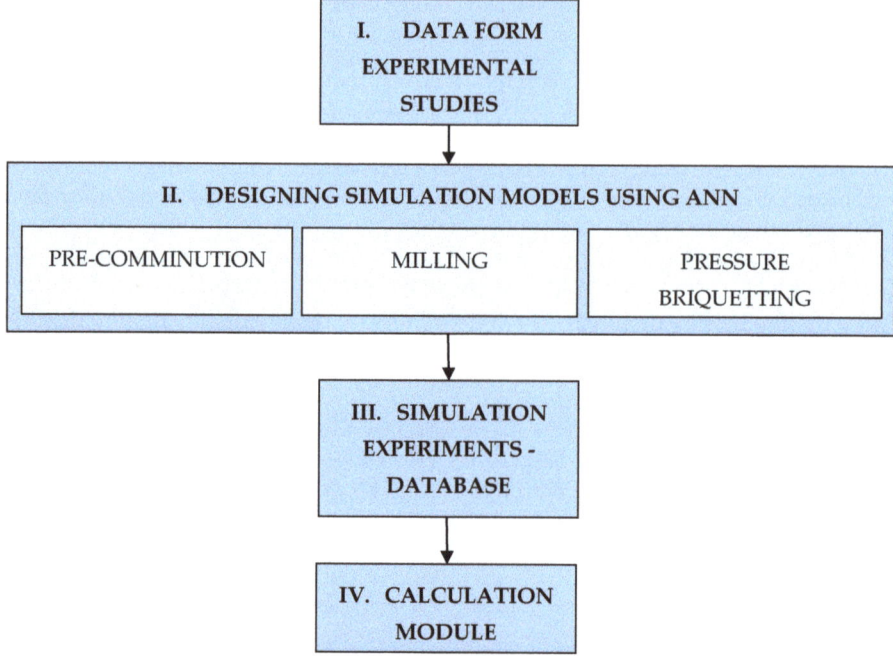

Figure 1. Stages of work.

2.1. Testing the Briquetting Process

Stage I involved testing the briquetting process of the line located in the laboratory of the University of Agriculture in Krakow. The research material was obtained from the experimental plots of the Energy Plants Collection located at the Faculty of Production and Energy Engineering. The line diagram and parameters of the briquette production line are shown in the Figure 2. Briquettes obtained in the production process had a nominal diameter of 50 mm and a length of about 50 mm.

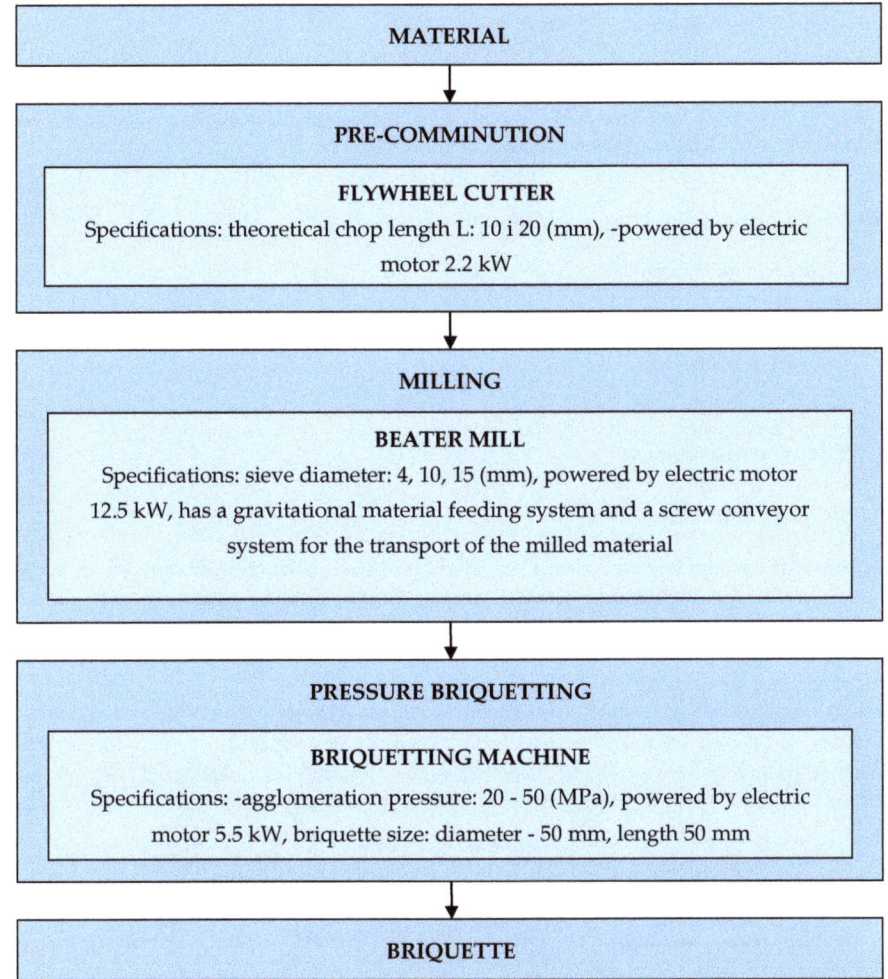

Figure 2. Diagram and parameters of the production line.

The resulting briquette quality was assessed according to the guidelines set in standards EN 15234-3:2012 (Solid biofuels—Fuel quality assurance—Part 3: Wood briquettes for non-industrial use). In accordance with the requirements in norm, a specific density of briquettes was set using a kit for determining the specific density—RADWAG - WPS 510/C/1. Next, the mechanical durability (DU) of the obtained briquettes was specified in accordance with the standard EN 15210-2:2011 (Solid biofuels—determination of mechanical durability of pellets and briquettes—Part 2: Briquettes). DU was calculated from the formula [64]:

$$DU = \frac{m_A}{m_B} \cdot 100\% \quad (1)$$

where:
DU—mechanical durability of briquette, %;
m_A—mass of the sample after the test, g;
m_B—mass of the sample before the test, g.

2.2. Developing ANN Models

Stage II consisted of developing ANN models for each stage of briquette production. Output parameters from individual stages were treated as input to the next.

It has been assumed that individual models will enable the following parameters to be determined:

1. For the precomminution process and the milling process:
   - Energy consumption,
   - Bulk density,
   - Granulometric composition of the comminuted and the milled material (share of individual fractions).
2. For the briquetting process:
   - Energy consumption,
   - Specific density,
   - Briquette durability.

Analysis of the production process and factors affecting the quality of the briquette allowed one to determine the inputs for individual models. For creating the models and the learning process of ANNs, the function "Automatic Designer" of Statistica was used. The back-propagation learning algorithm and then the conjugate gradient algorithm were used for each ANN. An ANN of the MLP type was chosen to create neural models.

The basic element of MLP is the so-called artificial neuron (Figure 3). The values of input variables and the threshold signal are fed to each neuron. These values are multiplied by the weights of individual inputs and then the products calculated in this way are added together. The resulting sum is transformed by a non-linear activation function resulting in an output signal at the neuron's output.

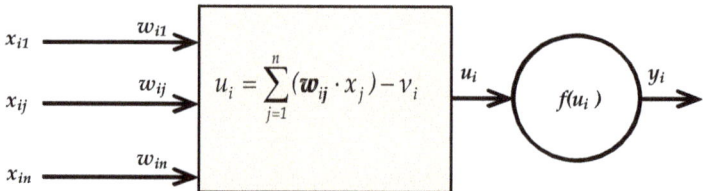

Figure 3. Diagram of an artificial neuron.

The structure of MLP networks is shown in Figure 4. These networks have a layered structure: input layer, hidden layers and output layer. Each layer has a different number of artificial neurons. Neurons of adjacent layers are connected to each other (no connections between neurons of the same layer). The value of the output from each neuron is passed to the inputs of all neurons of the next layer. In the case of output layer neurons, the values obtained are the result of calculations. There is a summary ANN response to the input values.

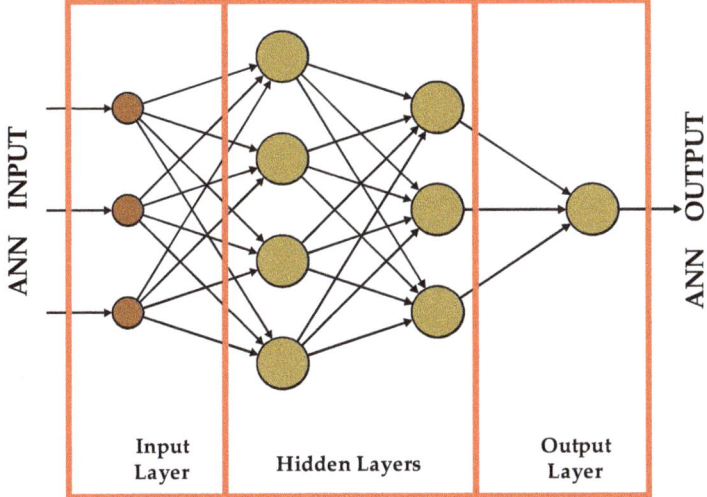

Figure 4. Diagram of the multilayer perceptron (MLP) type neural network.

ANNs are adapted to the modeled phenomenon during the learning process (weight changes for individual neurons occur). Gradient methods are used to teach MLP networks. Two such methods were used in the work: the back-propagation learning algorithm and then the conjugate gradient algorithm.

The values of the root mean-square error (RMSE) and mean absolute percentage error (MAPE) calculated for testing data set were the criterion of choice. RSME and MAPE are commonly used statistical errors to evaluate the model's performance [20,44–46,65].

The selection criteria were the root mean-square error (RMSE) and mean absolute percentage error (MAPE) calculated for the test data set. RSME and MAPE are commonly used statistical errors to assess model performance.

The RMSE and the MAPE were calculated from the following formulas:

$$RMSE = \sqrt{\frac{1}{n} \cdot \sum_{i=1}^{n} (Y_i - Y_{ANN,i})^2} \qquad (2)$$

$$MAPE = \frac{1}{n} \cdot \left| \frac{Y_i - Y_{ANN,i}}{Y_i} \right| \cdot 100\% \qquad (3)$$

where:
n—number of observations,
Y_i—values obtained during research,
$Y_{ANN,i}$—calculated by the ANN value.

Sensitivity analyses were performed for the created models in order to reject input variables that did not improve the accuracy of individual ANNs. The learning process has been repeated many times to obtain the best ANN. For each model, 100 neural networks of different architectures were tested, of which the best was selected.

2.3. Performing Simulation Experiments and Creating A Database

The following assumptions were made in the computer simulation process:

1. The production of briquettes from *Miscanthus* and willow proceeds in three stages continuously—the material passes through subsequent devices without interoperational storage.
2. The moisture content decreases by 2% at every stage of production.
3. During the simulation, the humidity of the chipped material varied from 13% to 21%.
4. Precomminution allows one to obtain chopped straw with a theoretical length of 10 and 20 mm.
5. Milling is carried out in one step—only one sieve (sieve diameters: 15, 10 and 4 mm).
6. Briquetting takes place at an adjustable pressure in the range of 20–56 MPa, every 2 MPa.

Based on computer simulations a database was created (Figure 5) for selected energy crops (*Miscanthus*, willow).

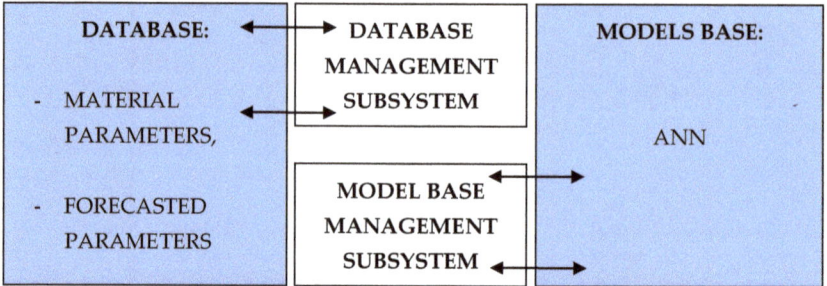

Figure 5. Organization chart of decision support system (DSS) elements (databases, database management subsystem, models databases and models database management subsystem).

2.4. Developing the Inference Module for the Proposed DSS

The final stage was the design and implement the inference module. This module was designed to calculate the optimal product and process parameters based on the total energy consumption. This module was an integral part of the computer program, which has also been designed and made as part of the work.

Assumptions for the designed application:

- Application in web technology,
- Responsive work mode,
- The ability to generate, modify and save reports,
- The ability to add more energy crops,
- Modular nature of the application,
- Expandable.

3. Results

In Figure 6 the essence of the DSS is shown. The ANN models used in the developed DSS are shown on Figures 7–11. Inputs and outputs of individual ANNs and connections between them are marked.

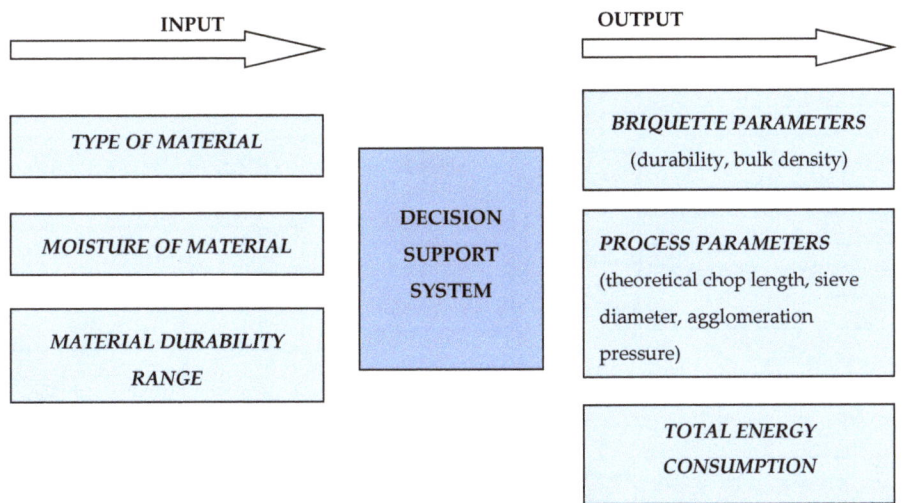

Figure 6. The essence of the DSS.

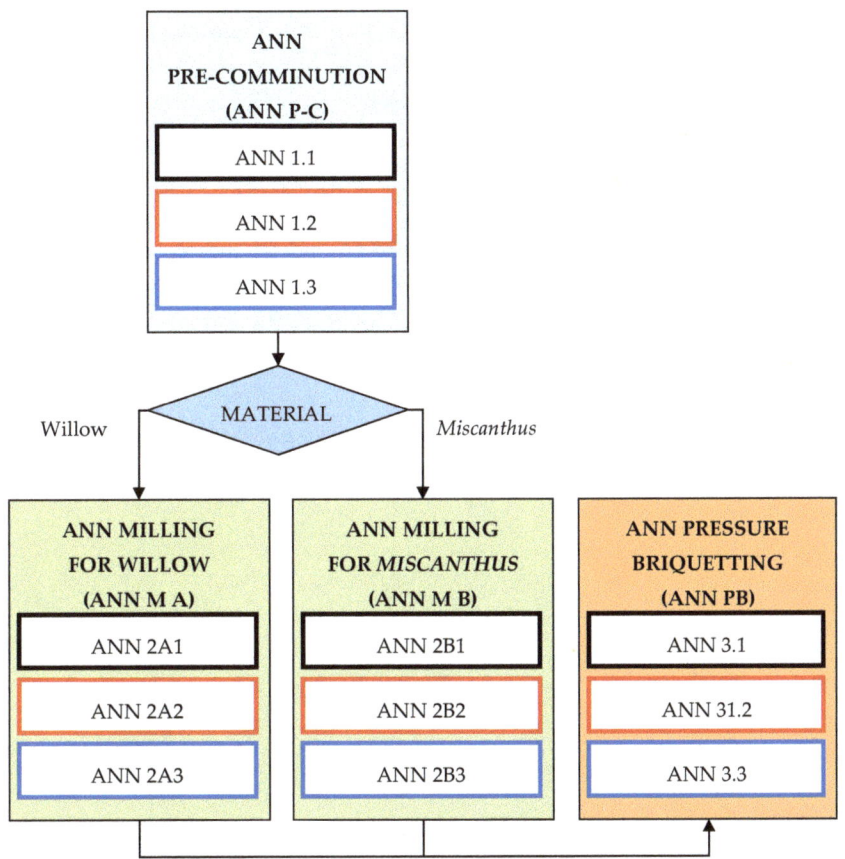

Figure 7. Models obtained for individual stages of the briquettes production process.

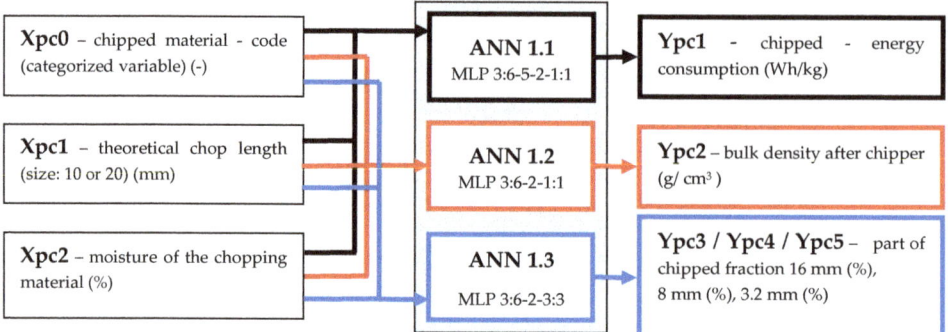

Figure 8. Inputs and outputs of individual artificial neural networks (ANNs) of the precomminution process.

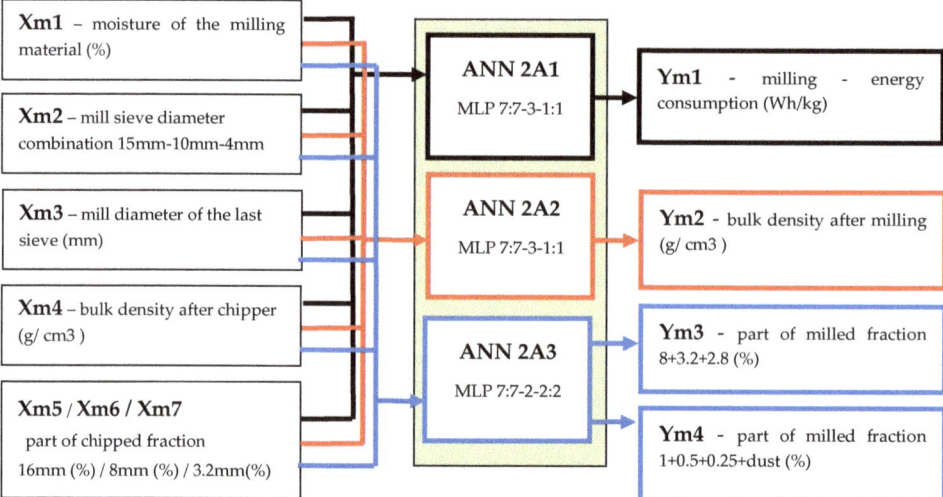

Figure 9. Inputs and outputs of individual ANNs of the milling process—willow.

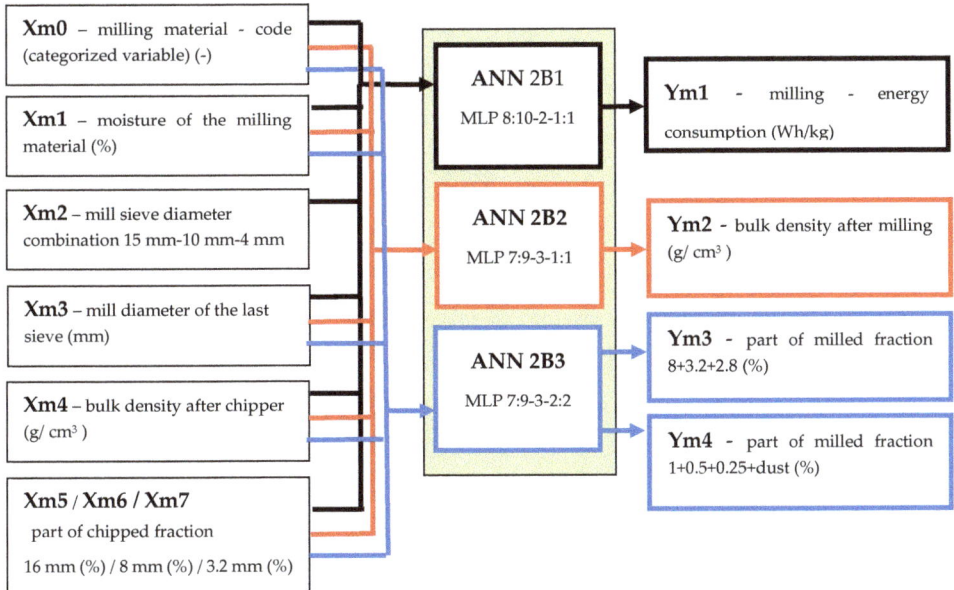

Figure 10. Inputs and outputs of individual ANNs of the milling process—*Miscanthus*.

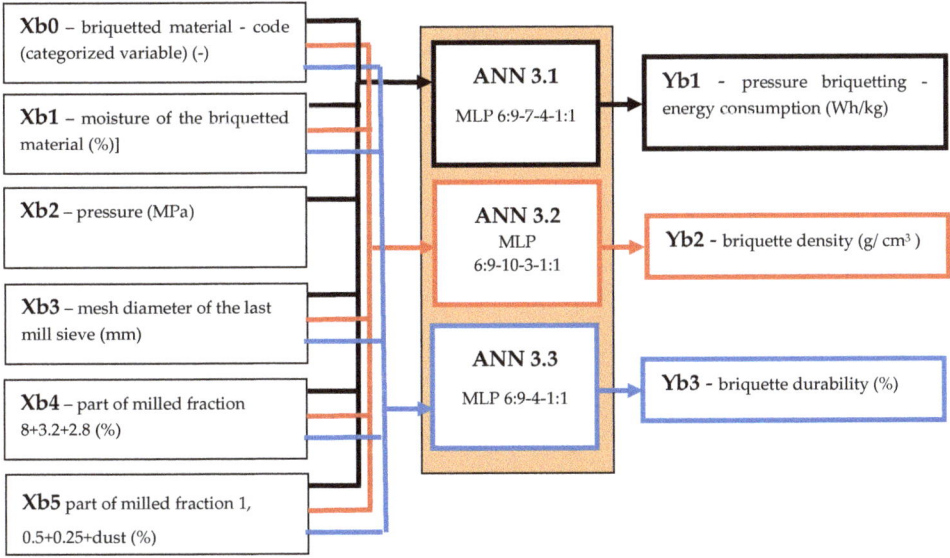

Figure 11. Inputs and outputs of individual ANNs of the briquetting process.

Tables 1–3 present the RMSE and MAPE error values calculated for ANN models describing the precomminution, the milling and the briquetting processes. Figure 12 shows a comparison of parameter values describing the briquetting process obtained using ANNs with average values measured during experimental tests for a selected example for the diameter of the last mill sieve (diameter = 15 mm).

Table 1. Root mean-square error (RMSE) and mean absolute percentage error (MAPE) error values calculated for ANNs describing the precomminution process.

ANN	Energy Consumption		Bulk Density		Parts of Chipped Fractions	
	RMSE (Wh/kg)	MAPE (%)	RMSE (g/cm^3)	MAPE (%)	RMSE (%)	MAPE (%)
Willow	0.17	10.27	0.000642	0.21	2.50	10.08
Miscanthus	0.11	3.73	0.001187	1.28	1.72	8.40

Table 2. RMSE and MAPE error values calculated for ANNs describing the milling process.

ANN	Energy Consumption		Bulk Density		Parts of Milled Fractions	
	RMSE (Wh/kg)	MAPE (%)	RMSE (g/cm^3)	MAPE (%)	RMSE (%)	MAPE (%)
Willow	3.38	5.48	0.00212	1.31	2.79	5.90
Miscanthus	2.44	6.56	0.00027	0.24	1.30	3.79

Table 3. RMSE and MAPE error values calculated for ANNs describing the briquetting process.

ANN	Energy Consumption		Briquette Density		Briquette Durability	
	RMSE (Wh/kg)	MAPE (%)	RMSE (g/cm^3)	MAPE (%)	RMSE (%)	MAPE (%)
Willow	1.22	3.43	0.0174	1.76	1.42	1.31
Miscanthus	1.63	3.26	0.0156	1.43	1.02	0.91

The detailed algorithm of the DSS inference module operation is presented in Figure 13.

The application has a web form and is opened from a web browser. The technologies used were the project in the Model-View-Controller (MVC) design pattern, Java programming language and elements of the SQL programming language. The following libraries were used: Spring MVC, Spring Web, Spring Security and Hibernate. The application has been scaled to one database server. Postgre SQL server version 9.4 was chosen. Coding was done using the NetBeans programming environment.

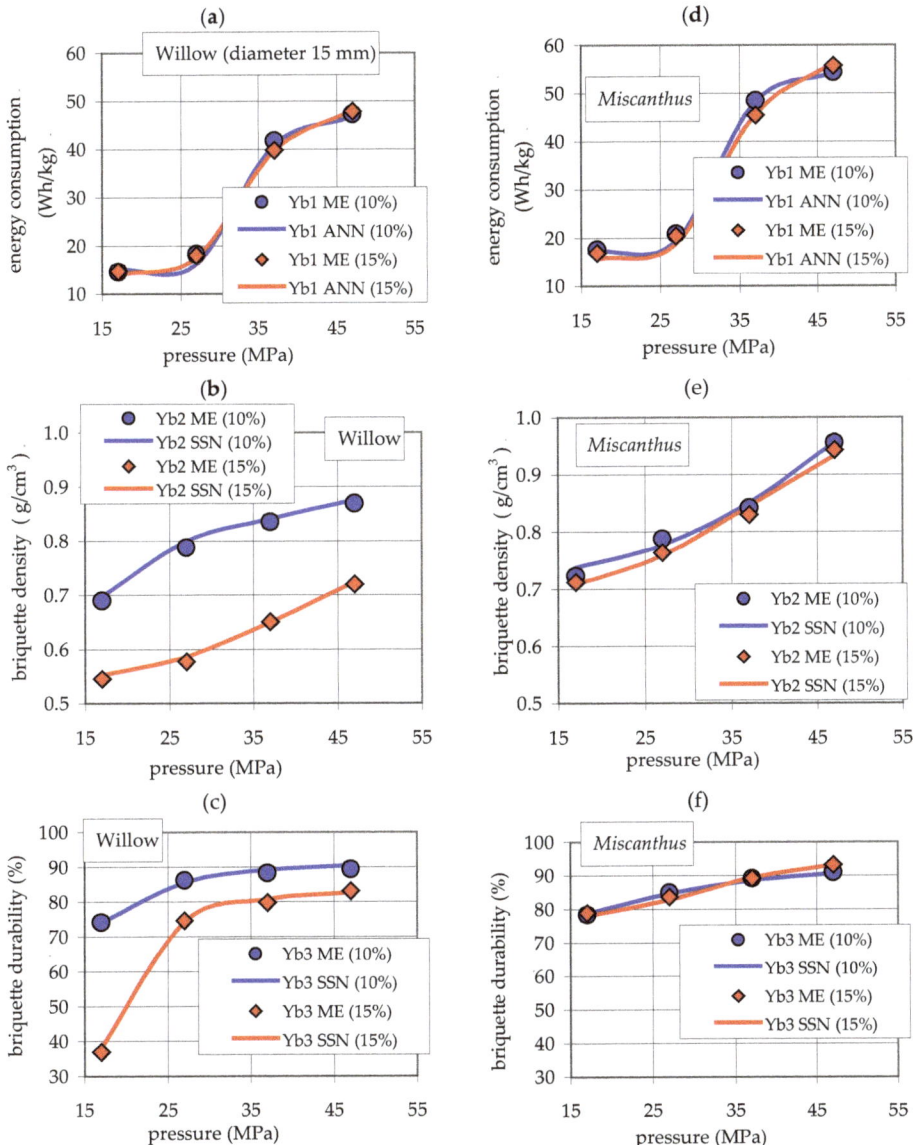

Figure 12. Comparison of parameter values describing the briquetting process obtained using ANNs with average values measured during experimental tests (for the diameter of the last mill sieve of 15 mm): (**a**) energy consumption for Willow; (**b**) briquette density for Willow; (**c**) briquette durability for Willow; (**d**) energy consumption for *Miscanthus*; (**e**) briquette density for *Miscanthus*; (**f**) briquette durability for *Miscanthus*.

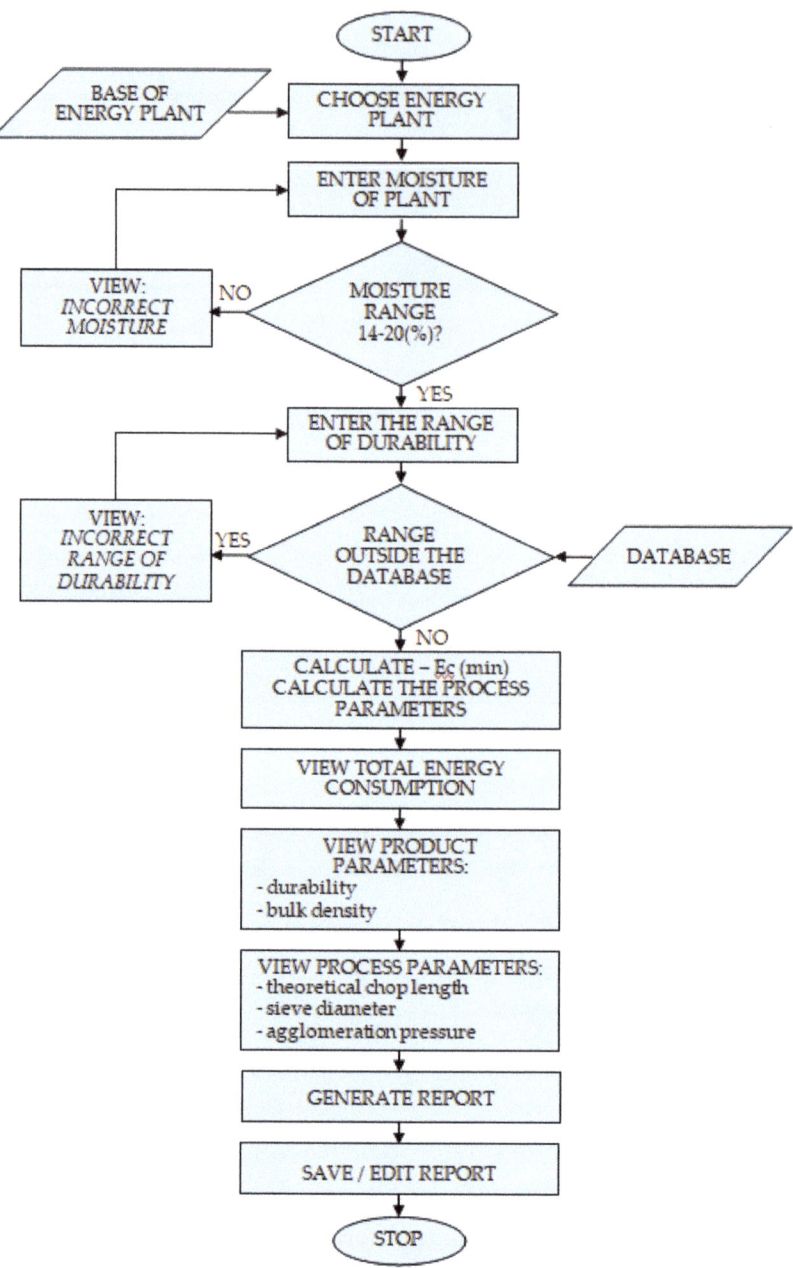

Figure 13. The algorithm of the inference module.

The system was developed using elements of graphical notation of the UML (unified modeling language). The program also meets the principles used in software engineering. Two sample schemes were selected, namely a use case diagram (Figure 14) and an implementation diagram (Figure 15).

Energies 2020, 13, 1364

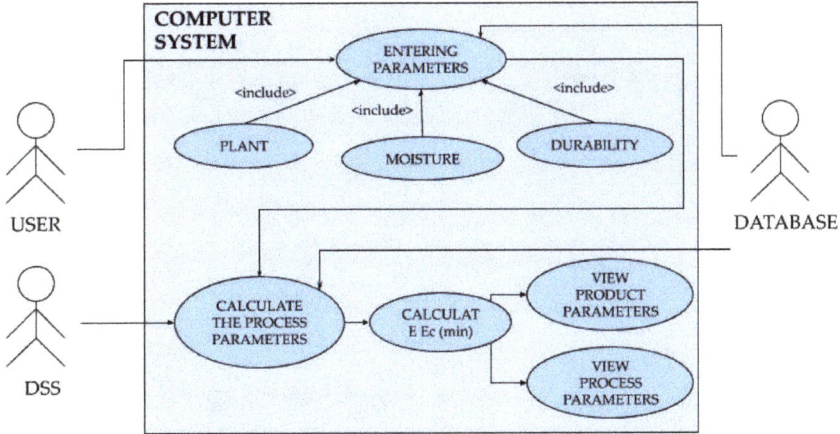

Description:

Designations used in accordance with UML notation:

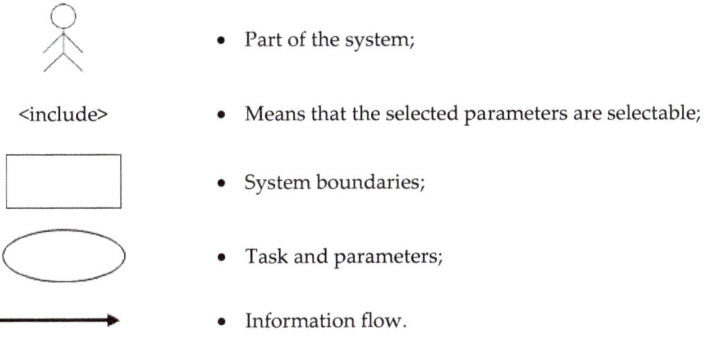

- Part of the system;

<include>

- Means that the selected parameters are selectable;

- System boundaries;

- Task and parameters;

- Information flow.

Figure 14. Use case diagram.

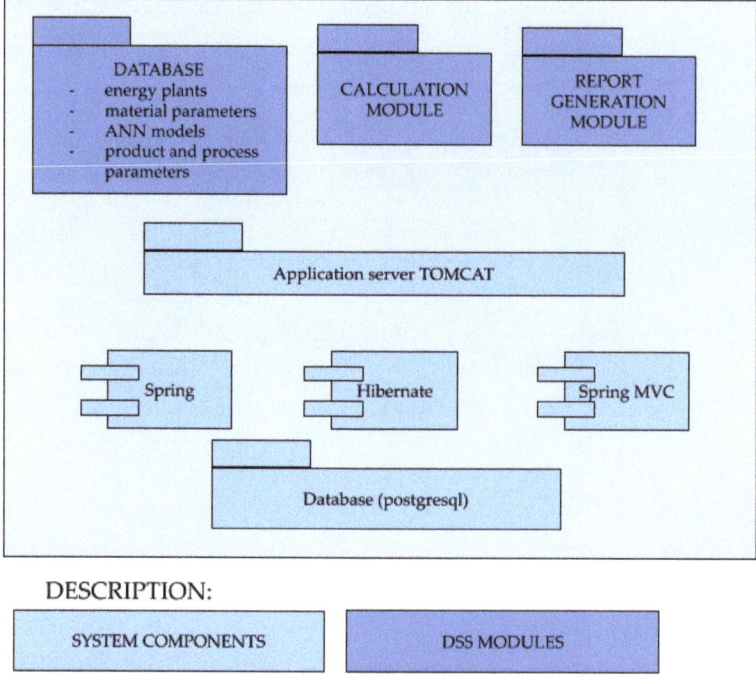

Figure 15. Implementation diagram.

4. Discussion

As part of this work, a DSS (Figure 6), which is designed to assist the user in the selection of process parameters for a given line producing briquettes from energy plants based on input parameters based on data from simulations has been designed and made.

The requirement for the proper operation of DSSs is the accuracy of models describing individual processes. ANNs models used in our DSS show high accuracy of calculations (independent models were developed to calculate particular output variables).

For the precomminution process (Table 1) the MAPE error values varied from 0.21% to 10.27%. The lowest errors occurred for bulk density (MAPE = 0.21% and 1.28%). The ANN model calculating the values of the parts of chipped fractions variable was the least accurate (MAPE = 10.08% for willow and MAPE = 8.40% for *Miscanthus*). For variables "energy consumption" and "parts of chipped fractions" ANNs were more accurate for *Miscanthus*. The ANN accuracy for bulk density is very high for both plants.

For the milling process (Table 2) the MAPE error values varied from 0.24% to 6.56%. Additionally, in this case, the lowest errors occurred for bulk density (MAPE = 1.31% for willow and MAPE = 0.24% for *Miscanthus*). The error values for energy consumption and parts of milled fractions did not exceed a few percent and show the high accuracy of the models (MAPE = 5.48% and 5.90% for willow and MAPE = 6.45% and 3.79% for *Miscanthus*).

For the briquetting process (Table 3), the MAPE error values varied from 0.91% to 3.43%. The highest error values occurred for energy consumption (MAPE = 3.43% for willow and MAPE = 3.26% for *Miscanthus*). For the other two output variables, these errors were much lower: MAPE = 1.76% and 1.43% for briquette density and MAPE = 1.31% and 0.91% for briquette durability. No major differences in the accuracy of neural models were observed between willow and *Miscanthus* briquetting.

Figure 12 shows examples of using ANNs to simulate the briquetting process (this is the most important of the briquette production processes that determines the quality of the final product and energy consumption). The results for one type of sieve (diameter = 15 mm) and for two selected humidity levels (10% and 15%) are summarized. The comparison of average real values (obtained from measurements) with the values of three output variables calculated by ANNs confirmed the high accuracy of the developed models for the entire range of applied briquetting pressures. It can be observed that in the case of *Miscanthus*, changes in humidity had a small effect on the values of the output parameters of briquetting (Figure 12d–f). However, for willow, the change in humidity caused significant differences in the case of briquette density and briquette durability (Figure 12b,c). Higher humidity reduced the density and durability of the briquette at the same briquetting pressures. It is worth noting that at 17 MPa pressure, the durability of willow briquettes made from raw material with humidity of 15% was much lower (by 37%) than 10% biomass briquettes. Despite the large non-linearity of changes, the ANN describing the durability of the briquette simulates the process very well.

A detailed operation algorithm of the DSS inference module is shown in the Figure 13. This diagram can be divided into two parts: the parameters specified by the user and the parameters calculated by the DSS inference module.

The user specifies the type of energy plant (*Miscanthus* or willow), moisture content of the energy plant and the expected durability of the briquette. If the plant has too much humidity (>20%), additional drying of the plant material is necessary.

Based on the given parameters, the DSS inference module calculates the process parameters (settings for the ax chopper, beater mill and piston briquetting press) and determines the forecast parameters of the briquette (durability and specific density). The objective function is to minimize total energy consumption. From the selected durability range and set humidity for the selected plant material, the data set with the lowest total energy consumption was selected. This solution gave the opportunity to perform simulations for many variants of interest to potential users.

The parameters characterizing the quality of produced briquettes are mechanical durability and specific density. These parameters are commonly used as indicators of briquette quality [8,17,21–23,64].

Durability and density are related to each other. For briquettes made of a specific type of biomass and under certain conditions, the higher the density, the greater the durability. The values of these parameters are very important for briquette producers—they affect the costs of transport and storage of briquettes. The higher durability and density of the briquettes means that they have higher transport and load compliance—transport and storage costs are lower. For transport, the limit is the maximum transport volume of the transport means (e.g., trucks), not its load capacity. The higher density of briquettes means that a larger mass can be transported with one transport means. Additionally, higher density briquettes have lower costs of the storage process—a certain mass takes up less storage space, it is better to use storage devices, etc. The greater durability of the briquettes means that they have a higher loadability (stacking resistance), which allows better use of the surface of the transport means and storage space. Briquettes with higher durability are less damaged (crushed) during transport, handling and storage processes.

Examples of user interface screens of the designed DSS are shown in the Figures 16 and 17.

As the DSS is intended for briquette producers in Poland, the user interface is in the Polish language. By asking the DSS question via the user interface (Figure 16), the user enters via the keyboard information: regarding the selected energy plant ("Select Plant"), minimum ("Minimum durability (%)") and maximum ("Maximum durability (%)") durability of the briquette and humidity of the material ("Humidity (%) Range: 14-20"). After entering the information, the decision-maker can click the calculate button and will be taken to the screen with the DSS response (Figure 17). In case of entering incorrect data, the "Clear" button, which will delete the entered data, may also be clicked.

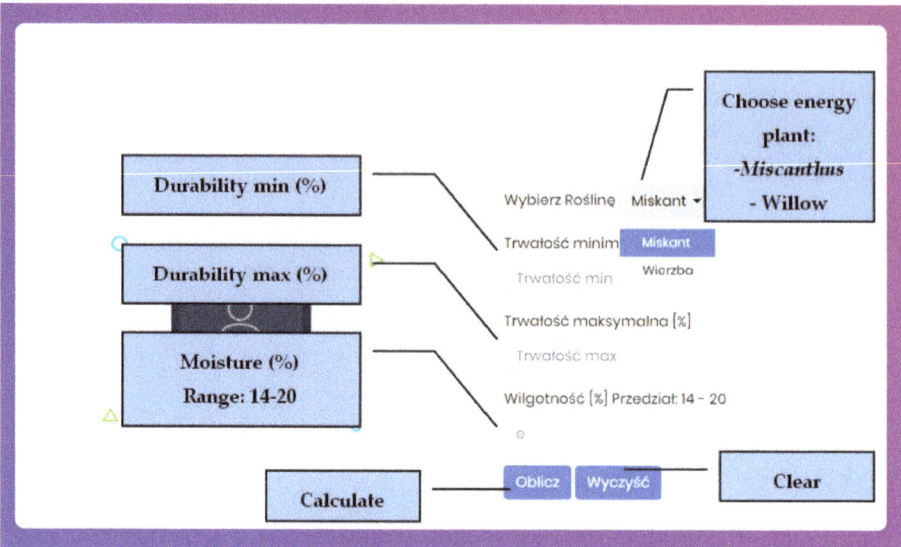

Figure 16. User interface screen of the designed DSS for data entry.

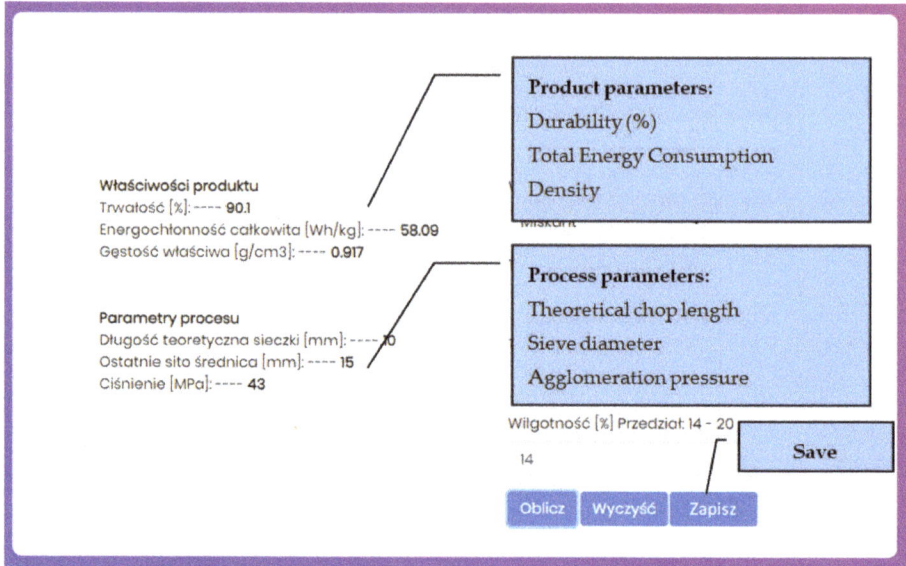

Figure 17. DSS user interface screen with system response.

The screen with the DSS response presents the results of the calculations. The results were divided into two groups: product properties and process parameters. As a part of the process properties, the following are displayed: durability (%), total energy consumption (Wh/kg) and specific density (g/cm^3). As a part of the process, the following parameters are displayed: theoretical chop length (mm), diameter of the last sieve (mm) and agglomeration pressure (MPa). The decision-maker can save several subsequent unit tests in the form report to compare and choose the best solution. The generated report can be saved and modified.

The use case diagram (Figure 14) illustrates the main functionality of the system, which consists of the calculating product, input and process parameters, total energy consumption and displaying them. This diagram also shows the division of permissions/tasks between parts of the systems. The user has the authority to enter data and generate, modify and save reports. DSS is designed to interpret input data and perform calculations according to the designed algorithm. The database stores data on input parameters (material properties), ANN models database and forecasted output parameters (product and process).

The implementation diagram (Figure 15) reflects the physical structure of the system. The modules of the web application itself are also distinguished. Due to the use of ready-made libraries (Spring, Hibernate and Spring MVC), the process of analyzing data was significantly shortened.

System operation was confirmed by a series of unit tests. Selected tests are presented in the Tables 4 and 5. The selected unit tests parameters reflected the upper briquette durability ranges.

Table 4. Simulation results for selected input data—*Miscanthus*.

Durability Range (%)	Moisture (%)	Theoretical Chop Length (mm)	Diameter of the Last Sieve (mm)	Briquetting Pressure (MPa)	Durability (%)	Density (g/cm^3)	Energy Consumption (Wh/kg)
90-100	14	10	15	43	90.1	0.917	58.09
80-90	14	10	15	20	81.0	0.748	24.51
90-100	15	10	15	36	90.1	0.827	53.71
80-90	15	10	15	20	80.9	0.748	24.28
90-100	16	10	15	36	90.4	0.825	53.02
80-90	16	10	15	20	80.6	0.751	24.19
90-100	17	10	15	39	90.2	0.857	54.41
80-90	17	10	15	22	80.1	0.757	24.31
90-100	18	10	15	38	90.4	0.857	53.81
80-90	18	10	15	22	80.2	0.747	24.3
90-100	19	10	15	39	90.0	0.864	57.05
80-90	19	10	15	23	80.4	0.736	24.75
90-100	20	10	10	33	90.0	0.796	62.18
80-90	20	10	15	28	80.4	0.743	36.98

Table 5. Simulation results for selected input data—willow.

Durability Range (%)	Moisture (%)	Theoretical Chop Length (mm)	Diameter of the Last Sieve (mm)	Briquetting Pressure (MPa)	Durability (%)	Density (g/cm3)	Energy Consumption (Wh/kg)
90-100	14	10	15	42	90.0	0.859	53.83
80-90	14	10	15	22	81.8	0.760	26.73
90-100	15	10	15	42	90.3	0.854	53.06
80-90	15	10	15	20	80.3	0.709	26.06
90-100	16	10	15	38	90.3	0.840	50.22
80-90	16	10	15	20	91.1	0.661	26.02
90-100	17	10	15	44	90.2	0.832	54.12
80-90	17	10	15	22	81.5	0.640	26.02
90-100	18	10	10	32	90.4	0.795	73.38
80-90	18	10	15	26	80.9	0.631	27.08
90-100	19	10	10	38	90.4	0.785	78.27
80-90	19	10	15	34	80.4	0.636	51.07
90-100	20	20	4	36	90.6	0.853	115.24
80-90	20	10	10	24	81.1	0.724	70.4

The decision maker, having to determine the parameters of the briquette production process, had information about the type of plant (Willow or *Miscanthus*) and biomass moisture level. He also knew what durability the briquette he wanted to achieve (introduces the minimum and maximum durability value to DSS). On this basis the DSS, using ANN models, calculated the parameters of individual stages of the briquette production process (theoretical chop length, diameter of the last sieve and briquetting pressure) and briquette quality parameters (durability and density) with minimal energy consumption.

For example, for *Miscanthus* with 14% humidity and the 90%-100% durability range, the DSS process parameters: theoretical chop length= 10 mm, diameter of the last sieve = 15 mm and briquetting pressure = 43 MPa will ensure briquette durability of 90.1% with minimal energy consumption of 58.09 Wh/kg (Table 4, row 1). If the decision maker states that 80% durability is sufficient, DSS will calculate new process parameter values (only briquetting pressure = 20 MPa changes) that will ensure briquette durability of 81.0% with a minimum energy consumption of 24.51 Wh/kg (Table 4, row 2).

It can be seen that reducing the briquette quality requirements from 90% to 80% results in a significant reduction in energy consumption (by about 50%) for both *Miscanthus* and willow (Tables 4 and 5).

5. Conclusions

The DSS designed and made in this work supported the decision-maker/user in the selection of process parameters for the selected line producing briquettes from energy plants (*Miscanthus* and willow). The developed DSS is unique—in the literature review no similar decision support tools were found in the processes of producing briquettes from energy plants.

The developed DSS belongs to the class of Intelligent DSSs, as it uses artificial intelligence methods—ANNs. The completed application is fully functional and is ready for implementation in a real production system. Due to the use of web technology, it can be run on a desktop computer, tablet or smartphone. DSS can be used by small enterprises, in which decision-makers usually do not have expertise in the operation of complex computer systems. In Polish conditions, the production of briquettes for energy purposes is most often carried out in small and medium-sized enterprises, therefore there is a demand for this type of DSSs.

The modularity of the system allows future development of the designed system. It is possible to improve the DSS through, e.g., supplementing or improving the model database. The developed DSS is only applicable to the briquetting process. However, due to the open modular structure of the program, after minor program modifications and supplementing the model database with neural networks describing the pellet production process, the DSS can also be used for pellet production. Further research will seek to increase this DSS's functionality. It is planned to supplement it with other types of energy crops and to introduce other types of production lines. In the future, it is planned to add a module for creating production schedules.

Author Contributions: Conceptualization, S.F. and A.K. (Adrian Knapczyk); Methodology, S.F. and A.K. (Adrian Knapczyk); Software, A.K. (Adrian Knapczyk) and A.K. (Artur Knapczyk); Validation, S.F.; Formal analysis, S.F.; Investigation, S.F.; Writing-Original Draft Preparation, S.F., A.K. (Adrian Knapczyk) and R.F.; Writing-Review & Editing, S.F., A.K. (Adrian Knapczyk) and R.F.; Visualization, A.K. (Adrian Knapczyk). All authors have read and agreed to the published version of the manuscript.

Funding: Article processing charges were financed from the subsidy of the Ministry of Science and Higher Education for the Agricultural University of Hugo Kołłątaj in Krakow for the year 2020.

Conflicts of Interest: The authors declare no conflict of interest.

References

1. Ioannou, K.; Tsantopoulos, G.; Arabatzis, G.; Andreopoulou, Z.; Zafeiriou, E. A Spatial Decision Support System Framework for the Evaluation of Biomass Energy Production Locations: Case Study in the Regional Unit of Drama, Greece. *Sustainability* **2018**, *10*, 531. [CrossRef]
2. Brunerová, A.; Roubík, H.; Brožek, M. Bamboo fiber and sugarcane skin as a bio-briquette fuel. *Energies* **2018**, *11*, 2186. [CrossRef]

3. Kozina, T.; Ovcharuk, O.; Trach, I.; Levytska, V.; Ovcharuk, O.; Hutsol, T.; Mudryk, K.; Jewiarz, M.; Wrobel, M.; Dziedzic, K. Spread Mustard and Prospects for Biofuels. In *Renewable Energy Sources: Engineering, Technology, Innovation*; Mudryk, K., Werle, S., Eds.; Springer Proceedings in Energy; Springer: Berlin/Heidelberg, Germany, 2018; pp. 791–799. ISBN 978-3-319-72371-6.
4. Francik, S.; Knapczyk, A.; Wójcik, A.; Ślipek, Z. Optimisation Methods in Renewable Energy Sources Systems-Current Research Trends. In *Renewable Energy Sources: Engineering, Technology, Innovation*; Wróbel, M., Jewiarz, M., Szlęk, A., Eds.; Springer: Berlin/Heidelberg, Germany, 2020; pp. 841–852. ISBN 978-3-030-13887-5.
5. Gentil, L.V.; Vale, A.T. Energy balance and efficiency in wood sawdust briquettes production. *Floresta* **2015**, *45*, 281–288.
6. Knapczyk, A.; Francik, S.; Wojcik, A.; Bednarz, G. Influence of Storing Miscanthus x gigantheus on Its Mechanical and Energetic Properties. In *Renewable Energy Sources: Engineering, Technology, Innovation*; Mudryk, K., Werle, S., Eds.; Springer Proceedings in Energy; Springer: Berlin/Heidelberg, Germany, 2018; pp. 651–660.
7. Ivanova, T.; Mendoza Hernández, A.H.; Bradna, J.; Cusimamani, E.F.; Montoya, J.C.G.; Espinel, D.A.A. Assessment of Guava (*Psidium guajava* L.) wood biomass for briquettes' production. *Forests* **2018**, *9*, 613. [CrossRef]
8. Brunerová, A.; Roubík, H.; Brožek, M.; Herák, D.; Šleger, V.; Mazancová, J. Potential of tropical fruit waste biomass for production of bio-briquette fuel: Using Indonesia as an example. *Energies* **2017**, *10*, 2119. [CrossRef]
9. Adzic, M.M.; Savic, R.A. Cooling of wood briquettes. *Therm. Sci.* **2013**, *17*, 833–838. [CrossRef]
10. Chaloupková, V.; Ivanova, T.; Ekrt, O.; Kabutey, A.; Herák, D. Determination of particle size and distribution through image-based macroscopic analysis of the structure of biomass briquettes. *Energies* **2018**, *11*, 331. [CrossRef]
11. Moiceanu, G.; Paraschiv, G.; Voicu, G.; Dinca, M.; Negoita, O.; Chitoiu, M.; Tudor, P. Energy consumption at size reduction of lignocellulose biomass for bioenergy. *Sustainability* **2019**, *11*, 2477. [CrossRef]
12. Knapczyk, A.; Francik, S.; Fraczek, J.; Slipek, Z. Analysis of research trends in production of solid biofuels. In Proceedings of the Engineering for Rural Development, Jelgava, Latvia, 22–24 May 2019; Volume 18, pp. 1503–1509.
13. Hebda, T.; Brzychczyk, B.; Francik, S.; Pedryc, N. Evaluation of suitability of hazelnut shell energy for production of biofuels. *Eng. Rural Dev.* **2018**, *17*, 1860–1865.
14. Mudryk, K.; Wrobel, M.; Jewiarz, M.; Pelczar, G.; Dyjakon, A. Innovative Production Technology of High Quality Pellets for Power Plants. In *Renewable Energy Sources: Engineering, Technology, Innovation*; Mudryk, K., Werle, S., Eds.; Springer Proceedings in Energy; Springer: Berlin/Heidelberg, Germany, 2018; pp. 701–712. ISBN 978-3-319-72371-6; 978-3-319-72370-9.
15. Francik, S.; Knapczyk, A.; Francik, R.; Slipek, Z. Analysis of Possible Application of Olive Pomace as Biomass Source. In *Renewable Energy Sources: Engineering, Technology, Innovation*; Mudryk, K., Werle, S., Eds.; Springer Proceedings in Energy; Springer: Berlin/Heidelberg, Germany, 2018; pp. 583–592.
16. Ivanyshyn, V.; Nedilska, U.; Khomina, V.; Klymyshena, R.; Hryhoriev, V.; Ovcharuk, O.; Hutsol, T.; Mudryk, K.; Jewiarz, M.; Wrobel, M.; et al. Prospects of Growing Miscanthus as Alternative Source of Biofuel. In *Renewable Energy Sources: Engineering, Technology, Innovation*; Mudryk, K., Werle, S., Eds.; Springer Proceedings in Energy; Springer: Berlin/Heidelberg, Germany, 2018; pp. 801–812.
17. Wrobel, M.; Mudryk, K.; Jewiarz, M.; Glowacki, S.; Tulej, W. Characterization of Selected Plant Species in Terms of Energetic Use. In *Renewable Energy Sources: Engineering, Technology, Innovation*; Mudryk, K., Werle, S., Eds.; Springer Proceedings in Energy; Springer: Berlin/Heidelberg, Germany, 2018; pp. 671–681.
18. Xu, J.; Chang, S.; Yuan, Z.; Jiang, Y.; Liu, S.; Li, W.; Ma, L. Regionalized techno-economic assessment and policy analysis for biomass molded fuel in China. *Energies* **2015**, *8*, 13846–13863. [CrossRef]
19. Wróbel, M. Assessment of Agglomeration Properties of Biomass-Preliminary Study. In *Renewable Energy Sources: Engineering, Technology, Innovation*; Wróbel, M., Jewiarz, M., Szlęk, A., Eds.; Springer: Berlin/Heidelberg, Germany, 2020; pp. 411–418. ISBN 978-3-030-13887-5.
20. Francik, S.; Łapczyńska-Kordon, B.; Francik, R.; Wójcik, A. Modeling and Simulation of Biomass Drying Using Artificial Neural Networks. In *Renewable Energy Sources: Engineering, Technology, Innovation.*; Mudryk, K., Werle, S., Eds.; Springer International Publishing AG; Springer: Berlin/Heidelberg, Germany, 2018; pp. 571–581. ISBN 9783319723716.

21. Wrobel, M.; Mudryk, K.; Jewiarz, M.; Knapczyk, A. Impact of raw material properties and agglomeration pressure on selected parmèters of granulates obtained from willow and black locust biomass. *Eng. Rural Dev.* **2018**, *17*, 1933–1938.
22. Safana, A.A.; Abdullah, N.; Sulaiman, F. Bio-char and bio-oil mixture derived from the pyrolysis of mesocarp fibre for briquettes production. *J. Oil Palm Res.* **2018**, *30*, 130–140.
23. Brunerová, A.; Brožek, M.; Šleger, V.; Nováková, A. Energy Balance of Briquette Production from Various Waste Biomass. *Sci. Agric. Bohem.* **2018**, *49*, 236–243. [CrossRef]
24. Krizan, P.; Matus, M.; Soos, L.; Beniak, J. Behavior of Beech Sawdust during Densification into a Solid Biofuel. *Energies* **2015**, *8*, 6382–6398. [CrossRef]
25. Brand, M.A.; Balduino Junior, A.L.; Nones, D.L.; Gaa, A.Z.N. Potential of bamboo species for the production of briquettes [Potencial de espécies de bambu para a produção de briquetes]. *Pesqui. Agropecu. Trop.* **2019**, *49*, 236–243. [CrossRef]
26. Islam, S.; Ahiduzzaman, M. Assessment of Rice Husk Briquette Fuel Use as an Alternative Source of Woodfuel. *Int. J. Renew. Energy Res.* **2016**, *6*, 1601–1611.
27. Mazurkiewicz, J.; Marczuk, A.; Pochwatka, P.; Kujawa, S. Maize Straw as a Valuable Energetic Material for Biogas Plant Feeding. *Materials* **2019**, *12*, 3848. [CrossRef]
28. Felsberger, A.; Oberegger, B.; Reiner, G. A review of decision support systems for manufacturing systems. In Proceedings of the CEUR Workshop Proceedings, Graz, Austria, 18–19 October 2016; Volume 1793.
29. Knapczyk, A.; Francik, S.; Wróbel, M.; Jewiarz, M.; Mudryk, K. Decision support systems for scheduling tasks in biosystems engineering. In Proceedings of the E3S Web of Conferences, Czajowice, Poland, 19–20 September 2019; Volume 132.
30. Hasan, M.S.; Ebrahim, Z.; Wan Mahmood, W.H.; Ab Rahman, M.N. Decision support system classification and its application in manufacturing sector: A review. *J. Teknol.* **2017**, *79*, 153–163. [CrossRef]
31. Behmel, S.; Damour, M.; Ludwig, R.; Rodriguez, M. Optimization of river and lake monitoring programs using a participative approach and an intelligent decision-support system. *Appl. Sci.* **2019**, *9*, 4157. [CrossRef]
32. Han, H.; Huang, M.; Zhang, Y.; Liu, J. Decision support system for medical diagnosis utilizing imbalanced clinical data. *Appl. Sci.* **2018**, *8*, 1597. [CrossRef]
33. Teniwut, W.A.; Hasyim, C.L. Decision support system in supply chain: A systematic literature review. *Uncertain Supply Chain Manag.* **2020**, *8*, 131–148. [CrossRef]
34. Kaklauskas, A.; Dzemyda, G.; Tupenaite, L.; Voitau, I.; Kurasova, O.; Naimaviciene, J.; Rassokha, Y.; Kanapeckiene, L. Artificial Neural Network-Based Decision Support System for Development of an Energy-Efficient Built Environment. *Energies* **2018**, *11*, 1994. [CrossRef]
35. Badami, M.; Fambri, G.; Manco, S.; Martino, M.; Damousis, I.G.; Agtzidis, D.; Tzovaras, D. A Decision Support System Tool to Manage the Flexibility in Renewable Energy-Based Power Systems. *Energies* **2020**, *13*, 153. [CrossRef]
36. Besser, A.; Kazak, J.K.; Świąder, M.; Szewrański, S. A Customized Decision Support System for Renewable Energy Application by Housing Association. *Sustainability* **2019**, *11*, 4377. [CrossRef]
37. Stamatescu, I.; Arghira, N.; Fagarasan, I.; Stamatescu, G.; Iliescu, S.S.; Calofir, V. Decision Support System for a Low Voltage. *Energies* **2017**, *10*, 118. [CrossRef]
38. Aran Carrion, J.; Espin Estrella, A.; Aznar Dols, F.; Zamorano Toro, M.; Rodriguez, M.; Ramos Ridao, A. Environmental decision-support systems for evaluating the carrying capacity of land areas: Optimal site selection for grid-connected photovoltaic power plants. *Renew. Sustain. Energy Rev.* **2008**, *12*, 2358–2380. [CrossRef]
39. Martín-Gamboa, M.; Dias, L.C.; Quinteiro, P.; Freire, F.; Arroja, L.; Dias, A.C. Multi-Criteria and Life Cycle Assessment of Wood-Based Bioenergy Alternatives for Residential Heating: A Sustainability Analysis. *Energies* **2019**, *12*, 4391. [CrossRef]
40. Tamouridou, A.A.; Pantazi, E.X.; Alexandridis, T.; Lagopodi, A.; Kontouris, G.; Moshou, D. Spectral Identification of Disease in Weeds Using Multilayer Perceptron with Automatic Relevance Determination. *Sensors* **2018**, *18*, 2770. [CrossRef]
41. Sampaio, G.S.; de Aguiar Vallim Filho, A.R.; da Silva, L.S.; da Silva, L.A. Prediction of Motor Failure Time Using An Artificial Neural Network. *Sensors* **2019**, *19*, 4342. [CrossRef]
42. Runge, J.; Zmeureanu, R. Forecasting Energy Use in Buildings Using Artificial Neural Networks: A Review. *Energies* **2019**, *12*, 3254. [CrossRef]

43. Kasantikul, K.; Yang, D.; Wang, Q.; Lwin, A. A Novel Wind Speed Estimation Based on the Integration of an Artificial Neural Network and a Particle Filter Using BeiDou GEO Reflectometry. *Sensors* **2018**, *18*, 3350. [CrossRef] [PubMed]
44. Francik, S.; Kurpaska, S. The Use of Artificial Neural Networks for Forecasting of Air Temperature inside a Heated Foil Tunnel. *Sensors* **2020**, *20*, 652. [CrossRef] [PubMed]
45. Liu, Y.; Zhang, S.; Chen, X.; Wang, J. Artificial Combined Model Based on Hybrid Nonlinear Neural Network Models and Statistics Linear Models—Research and Application for Wind Speed Forecasting. *Sustainability* **2018**, *10*, 4601. [CrossRef]
46. Niedbała, G.; Nowakowski, K.; Rudowicz-Nawrocka, J.; Magdalena, P.; Tomczak, R.J.; Tyksiński, T.; Pinto, A.A. Multicriteria Prediction and Simulation of Winter Wheat Yield Using Extended Qualitative and Quantitative Data Based on Artificial Neural Networks. *Appl. Sci.* **2019**, *9*, 2773. [CrossRef]
47. Mudryk, K.; Francik, S.; Fraczek, J.; Slipek, Z.; Wrobel, M. Model of actual contact area of rye and wheat grains with flat surface. In Proceedings of the Engineering for Rural Development, Jelgava, Latvia, 23–24 May 2013; pp. 292–296.
48. Wrobel, M.; Fraczek, J.; Francik, S.; Slipek, Z.; Mudryk, K. Modelling of unit contact surface of bean seeds using Artificial Neural Networks. In Proceedings of the Engineering for Rural Development, Jelgava, Latvia, 23–24 May 2013; pp. 287–291.
49. Francik, S.; Ślipek, Z.; Frączek, J.; Knapczyk, A. Present Trends in Research on Application of Artificial Neural Networks in Agricultural Engineering. *Agric. Eng.* **2016**, *20*, 15–25. [CrossRef]
50. Vlontzos, G.; Pardalos, P.M. Assess and prognosticate green house gas emissions from agricultural production of EU countries, by implementing, DEA Window analysis and artificial neural networks. *Renew. Sustain. Energy Rev.* **2017**, *76*, 155–162. [CrossRef]
51. Liakos, K.G.; Busato, P.; Moshou, D.; Pearson, S.; Bochtis, D.D. Machine Learning in Agriculture: A Review. *Sensors* **2018**, *18*, 2674. [CrossRef]
52. Iddio, E.; Wang, L.; Thomas, Y.; McMorrow, G.; Denzer, A. Energy efficient operation and modeling for greenhouses: A literature review. *Renew. Sustain. Energy Rev.* **2020**, *117*, 109480. [CrossRef]
53. Almonti, D.; Baiocco, G.; Tagliaferri, V.; Ucciardello, N. Artificial Neural Network in Fibres Length Prediction for High Precision Control of Cellulose Refining. *Materials* **2019**, *12*, 3730. [CrossRef]
54. Tamouridou, A.A.; Alexandridis, T.K.; Pantazi, X.E.; Lagopodi, A.L.; Kashefi, J.; Kasampalis, D.; Kontouris, G.; Moshou, D. Application of Multilayer Perceptron with Automatic Relevance Determination on Weed Mapping Using. *Sensors* **2017**, *17*, 2307. [CrossRef]
55. Martinez-Martinez, V.; Baladron, C.; Gomez-Gil, J.; Ruiz-Ruiz, G.; Navas-Garcia, L.M.; Aguiar, J.M.; Carro, B. Temperature and Relative Humidity Estimation and Prediction in the Tobacco Drying Process Using Artificial Neural Networks. *Sensors* **2012**, *12*, 14004–14021. [CrossRef] [PubMed]
56. Niedbała, G. Application of Artificial Neural Networks for Multi-Criteria Yield Prediction of Winter Rapeseed. *Sustainability* **2019**, *11*, 533. [CrossRef]
57. Bermejo, J.F.; Fernandez, J.F.G.; Polo, F.O.; Marquez, A.C. A Review of the Use of Artificial Neural Network Models for Energy and Reliability Prediction. A Study of the Solar PV, Hydraulic and Wind Energy Sources. *Appl. Sci.* **2019**, *9*, 1844. [CrossRef]
58. Tina, G.M. Special Issue on Applications of Artificial Neural Networks for Energy Systems. *Appl. Sci.* **2019**, *9*, 3734. [CrossRef]
59. Zhou, J.; Xu, X.; Huo, X.; Li, Y. Forecasting Models for Wind Power Using Extreme-Point Symmetric Mode Decomposition and Artificial Neural Networks. *Sustainability* **2019**, *11*, 650. [CrossRef]
60. Yu, H.; Chen, Y.; Hassan, S.G.; Li, D. Prediction of the temperature in a Chinese solar greenhouse based on LSSVM optimized by improved PSO. *Comput. Electron. Agric.* **2016**, *122*, 94–102. [CrossRef]
61. Rodrigues, E.; Gomes, Á.; Gaspar, A.R.; Henggeler Antunes, C. Estimation of renewable energy and built environment-related variables using neural networks—A review. *Renew. Sustain. Energy Rev.* **2018**, *94*, 959–988. [CrossRef]
62. Reynolds, J.; Ahmad, M.W.; Rezgui, Y.; Hippolyte, J.L. Operational supply and demand optimisation of a multi-vector district energy system using artificial neural networks and a genetic algorithm. *Appl. Energy* **2019**, *235*, 699–713. [CrossRef]
63. Zheng, M.; Leib, B.; Wright, W.; Ayers, P. Neural models to predict temperature and natural ventilation in a high tunnel. *Trans. ASABE* **2019**, *62*, 761–769. [CrossRef]

64. Wrobel, M.; Fraczek, J.; Francik, S.; Slipek, Z.; Mudryk, K. Influence of degree of fragmentation on chosen quality parameters of briquette made from biomass of cup plant *Silphium perfoliatum* L. In Proceedings of the Engineering for Rural Development, Jelgava, Latvia, 23–24 May 2013; pp. 653–657.
65. Byliński, H.; Sobecki, A.; Gębicki, J. The Use of Artificial Neural Networks and Decision Trees to Predict the Degree of Odor Nuisance of Post-Digestion Sludge in the Sewage Treatment Plant Process. *Sustainability* **2019**, *11*, 4407. [CrossRef]

© 2020 by the authors. Licensee MDPI, Basel, Switzerland. This article is an open access article distributed under the terms and conditions of the Creative Commons Attribution (CC BY) license (http://creativecommons.org/licenses/by/4.0/).

Article

Effect of Compaction Pressure and Moisture Content on Quality Parameters of Perennial Biomass Pellets

Jakub Styks *, Marek Wróbel, Jarosław Frączek and Adrian Knapczyk

Department of Mechanical Engineering and Agrophisics, University of Agriculture in Kraków, Balicka 120, 30-149 Kraków, Poland; marek.wrobel@urk.edu.pl (M.W.); jaroslaw.fraczek@urk.edu.pl (J.F.); adrian.knapczyk@urk.edu.pl (A.K.)
* Correspondence: jakub.styks@urk.edu.pl

Received: 7 February 2020; Accepted: 4 April 2020; Published: 11 April 2020

Abstract: In Poland the use of solid biomass obtained from intentional plantations of energy plants is increasing. This biomass is most often processed into solid fuels. There are growing indications that renewable energy sources, in particular biomass production, will continue to develop, so the better we know the raw material, the more effectively we will be able to use it. The results of tests that determine the impact of compaction pressure on selected quality parameters of pellets made from selected biomass types are presented. Material from plants such as Giant miscanthus (*Miscanthus* × *giganteus* Greef et Deu), Cup plant (*Silphium perfoliatum* L.), Virginia mallow (*Sida hermaphrodita* (L.) Rusby) was studied. The compaction process was carried out using the SIRIO P400 hydraulic press with a closed chamber with a diameter of 12 mm. Samples were made in four pressures: 131; 196; 262; 327 MPa and three moisture levels: 8%, 11%, 14%. It was found that with increasing compaction pressure and moisture content up to a certain point, the density and durability of the pellets also increased. Each of the materials is characterized by a specific course of changes in the parameters tested.

Keywords: pressure compaction; moisture content; solid density; mechanical durability; Cup plant; Virginia mallow; Giant miscanthus; perennial biomass

1. Introduction

Biomass is a raw material that can come from woody crops, it can be biomass of whole plants, e.g., poplars [1,2] or a part of it, so far treated as waste [3,4]. The second source is agricultural crops in the form of targeted crops e.g., switchgrass [5] or virginia mallow and miscanthus [6] and crop residue e.g., bean straw [7]. With the currently existing restrictions on the use of firewood from forests and waste wood from industry, wider use is required for the production of solid biofuels from biomass from agriculture, mainly straw of different plants [8,9]. The solution to this problem may be deliberate plantations of energy plants, i.e., those which by definition obtain large mass gains in a relatively short time, suitable mainly for burning and obtaining a lot of thermal energy [10]. Among the species of plants, perennials especially *Helianthus tuberosus* [11], *Phalaris arundinacea* [12], *Rudbeckia laciniata* [13] and *Miscanthus* [14], shrubs, trees like *Pinus sylvestris* [15], *Quercus* [16] or *Fagus sylvatica* [17] are gaining a special attention for energy use. Perennial plantations are becoming increasingly popular in Poland. Perennial plants can be divided into monocotyledonous (Reed canary, Giant miscanthus) and dicotyledonous (Cup plant, Virginia mallow) [14,18]. These are species with rapid growth, high yields, and high resistance to diseases and pests [19,20]. These plants have low soil requirements and, therefore, can be grown in lower-class soils without competing with food crops. Research on energy crops is being conducted in many research centers where the process of densification is studied [21,22], as well as the impact of other biomass additives [23,24] but also the combustion process [25]. Observations indicate an increase in popularity of using renewable energy sources (RES) in the form of biomass converted into solid biofuels [1,20,26].

The definition of biomass is solid or liquid substances of plant and animal origin, obtained from products, waste and residues from agricultural and forestry production, as well as the industry processing their products and parts of other waste that are biodegradable [27]. Biomass resources for energy purposes, estimated in various scenarios and strategic documents, are the highest among all available renewable energy sources in Poland. Their use, compared to other renewable energy RES, is dominant in all energy sectors of the country [28]. The most popular and significant energy plants in Poland are willow [29], poplar [30], miscanthus [18], alder [31], black locust [29], pine [32], and beech [32].

The rational use of plant biomass is closely related with the processing of raw material in the form of loose mass into pressure-compacted granules [22,29]. The problem of raw material is its considerable volume, i.e., low bulk density. Compaction may be the solution to this problem, among others. The compaction process is a process whose presence can be observed in many industrial branches. It is used, among others, by food, pharmaceutical, feed, zoological and biofuel concerns. The reason for the popularity of the material compaction process are mainly: volume reduction, maximization of density, unified dose composition, unified shapes, reduction of storage costs, easier storage, dust reduction or reduction of transport costs.

Both in industrial and laboratory processes compaction as a process that consists of three stages characteristic of this process. The first is the movement of particles, during which particles of material move relative to each other, reducing free space, and maximizing the contact surface with each other. The second stage is elasto-plastic deformation. In this phase, an increase in pressure to the maximum value can be observed. The particles during this stage are already tightly packed, and establish intermolecular connections with each other. The last, third stage of the thickening process is stabilization. At this stage, the pressure decreases, and the material that is compacted expands, i.e., a slight increase in its volume occurs [33,34].

Factors directly affecting the compaction process, i.e., technological factors, are factors that we are able to control during the compaction process or before the start of the process. Each of them has a direct impact on the quality of the resulting product, in this case pellets. Sources [33,35] say that the best moisture that is used in the compaction process of biomass oscillates around the level of 7.8%–15%. This is the extent to which the material thickens, it is neither too dry and does not crumble or crumbs, nor is it too wet, so it does not "stick" or stratify.

For the production of pellets, the moisture range 10%–15% is the most common [36–38]. Quality standards for solid biofuels require moisture below 10% (pellet class A1) and 12% (briquette class A1). The upper moisture limit of lower quality classes never exceeds 15%.

The above recommendations apply to the pelletising process, where apart from moisture and pressure, the quality obtained is also influenced by temperature, which is considered together with moisture to be the most important factor determining the pelletising process and which extends the technological moisture range to the mentioned values.

In the basic research, the influence of moisture without temperature which does not include the thickening process is determined. The process which often takes place below the lignite plasticizing temperature (about 85 °C [39]) is briquetting in which only pressure and moisture determine the quality of the agglomerate obtained. Studies of this process indicate that the value of optimal moisture content depends on the compacted material. Križan [40], when studying the wood dust compaction process, showed that in the moisture range 5%–23%, the highest density is achieved by a material with 15% moisture. According to Brożek [41], for the biomass of plane, the optimal moisture level is about 8%.

The degree of material refinement and the unification of its geometry allows us to get rid of too much space between the material particles, which increases the contact area of the material [42]. Too low a compaction pressure of the material can result in too high expansion and thus delamination and general decomposition. Pressure and compaction time directly affect the durability of the pellet, because it allows the appropriate "packing" of the material in a specific volume, which is determined by the matrices [34,35]. The temperature allows melting lignins, natural binders, vitrification of the material, which also translates into product durability [33].

The literature on the subject lacks clear, unequivocal evidence of the susceptibility of the aforementioned energy plant species to the pressure compaction process [2,24,25] especially species relatively recently recognized as energetic. In the process of producing compact biofuels, the value of the compacting pressure used is important [43]. The purpose of the work is to determine the impact of biomass compaction pressure and moisture content of raw material of selected plants on main quality parameters of pellets.

2. Materials and Methods

The material that was used for the research was obtained from a plantation of energy plants located in University of Agriculture in Krakow. The chosen material was shoots of Giant miscanthus *Miscanthus × giganteus* Greef et Deu, Cup plant *Silphium perfoliatum* L. and Virginia mallow *Sida hermaphrodita* (L.) Rusby. Shoots were harvested after the growing season, in November 2018.

The compaction tests were carried out on material (which firstly was ground to grain size below 1 mm) with three moisture levels: 8%, 11% and 14%. A compaction process was carried out for each moisture level at four different pressures (131; 196; 262; 327 MPa), three replicates per trial with three samples were prepared for each material at each of the four pressures and each of the three moistures, which resulted in 108 samples. The test of mechanical durability (DU) and solid density (DE) was carried out 24 h after the production of pellets. Details of the research process are presented in Figure 1.

2.1. Seasoning

First of all, the whole shoots of collected material were cut into 5 cm pieces, then material was subjected to natural drying to reach a moisture content of about 10%. It is necessary to carry out this process before further preparation of the material for testing.

The moisture content was determined in accordance with EN ISO 18134-1 [44]. A sample of the material was placed in a weighing vessel and dried in the laboratory dryer (SLW 115, Pol-Eko, Wodzisław Śląski, Poland). Drying took place at 105 ± 2 °C until the material was stabilized. Total moisture M_{ad} was determined according to the formula:

$$M_{ar} = \frac{(m_2 - m_3) - (m_4 - m_5)}{(m_2 - m_1)} * 100 \qquad (1)$$

M_{ar}—total moisture (%),
m_1—mass of the empty tray used for the portion (g),
m_2—mass of the tray and test portion before drying (weight in room temp) (g),
m_3—mass of the tray and test portion after drying (weight when still hot) (g),
m_4—mass of the reference tray before drying (weight at room temp) (g),
m_5—mass of the tray after drying (weight when still hot) (g).

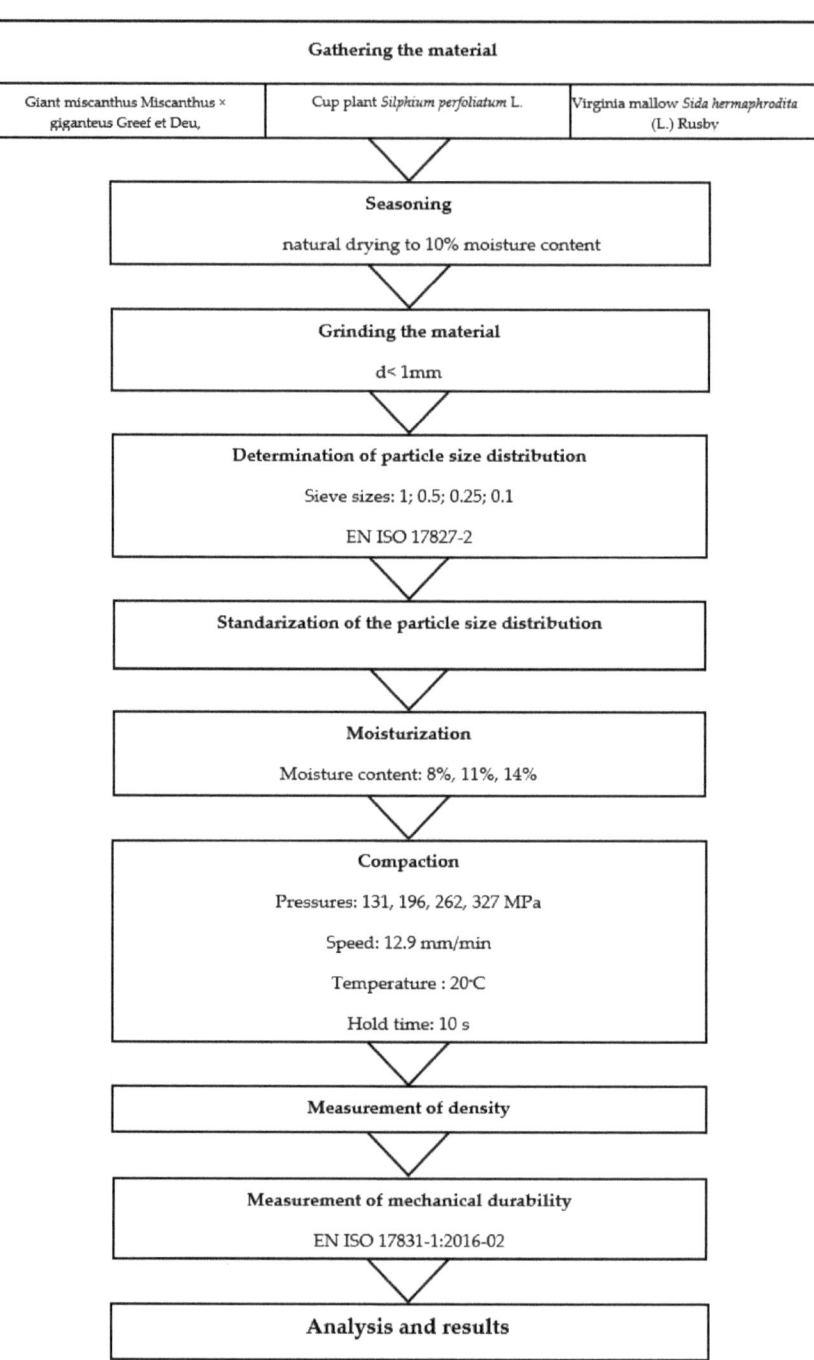

Figure 1. Research schedule.

2.2. Grinding

The dried material to be tested was ground to obtain the raw material in grain size below 1 mm, which will be subjected to the compaction process in subsequent stages of the test. Grinding was carried out in two stages. The first stage involved grinding the material into a fraction below 6 mm using a knife mill (Testchem LMN-100, Pszów, Poland). In the second stage, the material was ground using a flail mill (PX-MFC 90D, Polymix, Kinematika, Luzern, Switzerland). The grinder was equipped with a sieve with 1 mm hole diameter.

2.3. Determination of Particle Size Distribution

The determination of the particle size distribution was carried out in accordance with EN ISO 17827-2 [45]. This standard is intended for biomass in the form of dust and sawdust; 3.15 mm diameter sieves and # 2,8 sieves are required; 2; 1.4; 1; 0.5; 0.25; 0.1 mm. The test material was ground to the fraction d < 1 mm, so the samples were sieved using a shaker (LPzE-4e, Morek Multiserw, Marcyporęba, Poland), with a set consisting of woven sieves with mesh size 1; 0.5; 0.25; 0.1 mm. In this way, the test material samples were divided into 4 dimensional fractions, called sieve classes:

- C_1: 0.1—grain diameter d ≤ 0.1 mm,
- C_2: 0.25—grain diameter in terms 0.1 < d ≤ 0.25 mm,
- C_3: 0.5—grain diameter in terms 0.25 < d ≤ 0.5 mm,
- C_4: 1—grain diameter in terms 0.5 < d ≤ 1 mm.

The next step was to calculate d_{50} (3) (middle value of grain size), which showes a significant difference between each material particle size distribution. The difference between the fractional composition of the samples was demonstrated to eliminate the impact, and it was decided to standardize the composition. Influence of material composition has already been demonstrated, inter alia, in studies of Jewiarz [46], Wróbel [42] or Manouchehrinejad [47].

A mix of material with a particle size distribution proposed by Wróbel was used for the tests [42].

$$d_{50} = C_{<50} + (50 - S_{<50}) \cdot \frac{C_{>50} - C_{<50}}{S_{>50} - S_{<50}} \quad (2)$$

where:

$S_{<50}$—the highest cumulative share of the S fraction but not exceeding the value 50%,
$S_{>50}$—the lowest cumulative share of fraction S, however, exceeding the values 50%,
$C_{<50}$—sieve class corresponding $S_{<50}$
$C_{>50}$—sieve class corresponding $S_{>50}$

2.4. Reaching Specific Moisture

To obtain the desired level of moisture content of raw material the climate chamber (KBF-S 115, Binder, Tuttlingen, Germany) was used. The material was moisturized successively to three levels of moisture: 8%, 11% and 14%. This kind of appliance gives the possibility of precise moistening of the material to the desired moisture, which is important in the accuracy of the results. Before pressure compaction, it was decided to determine the analytical moisture (2):

$$M_{ad} = \frac{(m_2 - m_3)}{(m_2 - m_1)} * 100 \quad (3)$$

M_{ad}—total moisture (%),
m_1—dish weight with lid (g),
m_2—dish weight with lid and sample before drying (g),
m_3—dish weight with lid and sample after drying (g).

2.5. Pressure Compaction

The pressure compaction process was carried out according to methodology created by Wróbel [42]. The pressure compaction process of base mixtures was carried out with the use of the hydraulic press (P400, Sirio, Meldola, Italy, Figure 2), having a compaction set placed between the upper and lower planes of the press head. This set includes: a dia, a piston and a counter piston. After preparing the station, the sleeve, counter-piston and piston formed a closed chamber (Ø 12 mm) in which a sample (1 g) of previously prepared material was compacted. For the test were used four levels of pressure: 131; 196; 262; 327 MPa, constant temperature 20 °C and constant speed of compaction 12.9 mm/min. The produced pellets were placed in tightly closed containers for 24 h.

Figure 2. Compaction station: (**a**) hydraulic press SIRIO P400, (**b**) compaction chamber and piston, (**c**) samples.

2.6. Pellets Solid Density

After 24 h solid density DE was determined using a quassic liquid pycnometer (GeoPyc 1360, Micromeritics Instrument Corp., Norcross, GA, USA, Figure 3). The resulting samples were placed in a measuring cylinder, which is filled with powder with a particle size of 250 µm. This powder enables eliminating the phenomenon of wetting and soaking the liquid into the pores of the tested material, as would be the case if the liquid was used. The detailed measurement method is described by Wróbel [42].

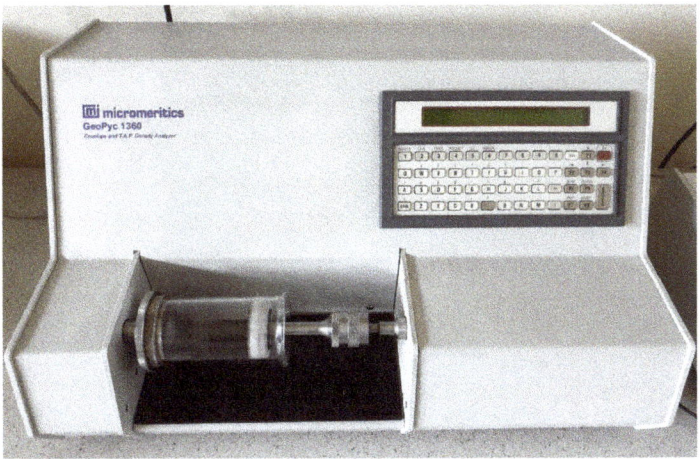

Figure 3. GeoPyc 1360 density determination station.

2.7. Mechanical Durability

To measure the mechanical durability (DU) of the pellets, a mechanical tester (Figure 4.) was used, whose construction and operating principle complies with the modified European standard EN ISO 17831-1 [48]. The modification introduced to the methodology involved the use of ballast material (made of polystyrene) with a density similar to the samples (approx. 1070 kg/m^3). The weight of the ballast material used was 500 g and the weight of the samples subjected to the durability test did not exceed 10 g, so the total weight of the sample was consistent with standard guidelines. Another modification of the research method was the reduction of the rotation of the chamber in which the material is located. Studies have shown that at standard revolutions, ballast material can cause destruction of samples, so based on preliminary tests carried out by Wróbel [42], revolutions of 250 rpm/min were assumed. The granule durability test lasted 10 min. After stopping the tester, samples were taken from polystyrene ballast material. Each of the samples was weighed and their dimensions checked. Then DU was counted for each granule. Calculations of the mechanical durability of the pellet were made according to the formula (2) defined by standard EN ISO 17831-1:

$$DU = \frac{m_E}{m_A} \times 100 \qquad (4)$$

DU—mechanical durability of the pellets (%),
m_E—mass of the sample of the tested pellet before the test (g),
m_A—mass of the sample of the tested pellet after the test (g).

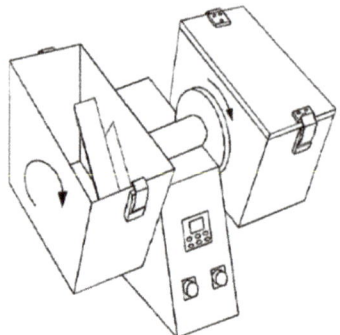

Figure 4. Mechanical durability determination station [48].

The EN ISO 17225-6 [49] the standard indicates two quality groups of pellets made from straw A (DU ≥ 97.5) and B (DU ≥ 96.5).

3. Results

Studies have shown a relationship between increasing compaction pressure, moisture content in the material, and mechanical durability and density of the pellets produced. It was shown that with the parameters used, the pellet density required by the standard was achieved, while the mechanical durability was lower than required by the standard.

3.1. Particle Size Distribution

To reject the impact of material fragmentation, material particle size distribution was determined.

The analysis showed differences in percentage shares. Research on the particle size distribution showed differences between the grain size of individual fractions of the tested material (Figure 5). Differences was observed between the mass distribution of the material on individual screens among all tested materials. The most numerous group turned out to be a group with grain size C_3 (0.1< d ≤ 0.25) mm and C_4 (0.5 < d ≤ 1 mm).

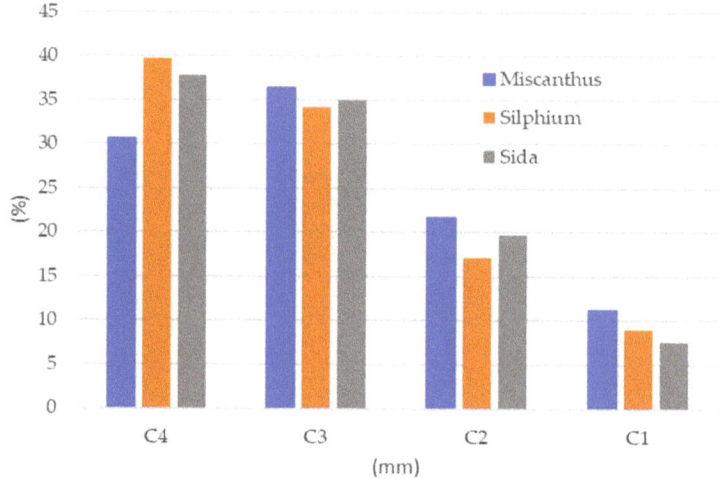

Figure 5. Particle size distribution.

To better present the differences, a cumulative graph was made that shows that Miscanthus differs from other materials (Figure 6).

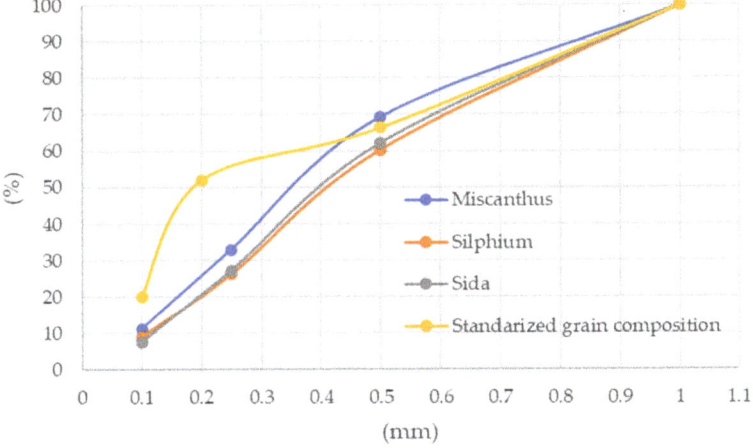

Figure 6. Cumulative particle size distribution.

Based on the above chart, it can be seen that particle size distribution of Miscanthus is significantly different from the other materials, which determined the d_{50}.

In order to show significant differences between individual fraction groups, the d_{50} was calculated, which was determined for each of the tested materials and amounted to 0.37 mm for Miscanthus, 0.42 mm for Cup plant and 0.41 mm for Virginia mallow. In order to avoid the influence on the results, the tested material was unified (Table 1).

Table 1. Standarized particle size distribution.

Fraction (mm)	C_1	C_2	C_3	C_4
Value (%)	20	32	14.4	33.6

3.2. Density and Durability

3.2.1. Miscanthus

The diagram on Figure 7a presents a graph of the solid density variation of the finished pellet sample depending on the compaction pressure and moisture content. As the moisture content of the raw material increases, the density of the resulting pellets decreases. The increase in agglomeration pressure causes the expected increase in density. However, only for material with a moisture content of 8% after exceeding 262 was a density above 1000 kg/m³ achieved—this value is the threshold value characterized by high-quality pellets [34]. The material with 11% moisture obtained max DE at the level of 940 kg/m³. In the case of 14% moisture material after exceeding the pressure of 196 MPa, changes in density are slight and the density obtained does not exceed 880 kg/m³.

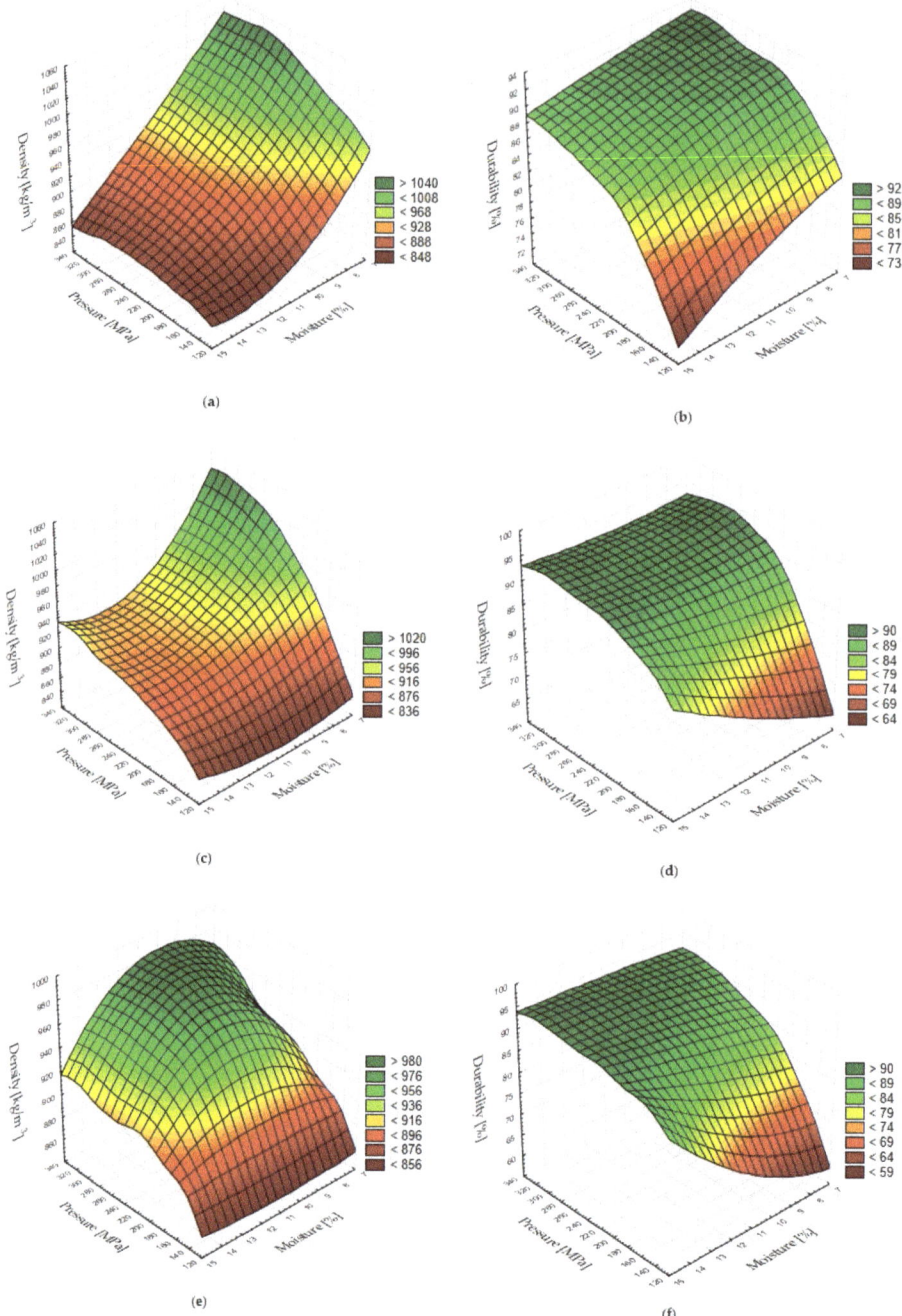

Figure 7. Influence of moisture and pressure on: (**a**) Miscanthus (solid density, DE), (**b**) Miscanthus (mechanical durability, DU), (**c**) Silphium DE, (**d**) Silphium DU, (**e**) Sida DE, (**f**) Sida DU.

In the case of durability similar dependencies were observed. An increase in moisture causes a decrease in durability. The increase in pressure causes an increase in durability, while relatively more significant increases were observed during an increase in pressure from 131 to 196. A further increase in pressure resulted in a relatively lower increase in the value of durability.

3.2.2. Silphium

In Figure 7c one can observe the results of the process of compaction of material originating from Cup plant similar to the Miscanthus case. The most favorable moisture level for the compacting process of this material is 8% of moisture content. The highest density (1000 kg/m^3) was recorded in the granules, which were compacted using a pressure of 327 MPa. Here, also, the increase in density is clear until the pressure reaches 262 MPa.

The course of density changes for material with 11% and 15% moisture is similar. This density from about 850 kg/m^3 at a pressure of 131 increases to 900–920 kg/m^3 at a pressure of 196. A further increase in pressure results only in an average value of 930 kg/m^3. At 131, the material's moisture content does not affect the density achieved.

Figure 7d shows the results of the test of mechanical durability of granules made of crushed Silphium. Compared to Miscanthus, moisture causes differences in density only at the lowest compaction pressure. In this case, unlike the miscanthus, an increase in moisture causes an increase in durability from about 68% to about 83%. An increase in pressure to a value of 196 causes an increase in durability to an average level of about 89% regardless of material moisture. A further increase results in a slight increase in durability (up to approx. 93%). without differences due to moisture. The most resistant to the test of mechanical durability turned out to be granules produced at a pressure of 327 MPa from a material with a moisture content of 14%. The least durable granules were those made of a material with a water content of 8% using a pressure of 131 MPa. The highest durability was 93.3% and this was higher than the lowest durability by 35.33%.

3.2.3. Sida

Figure 7e presents the results of the process of compaction of a sample of material from Virginia mallow.

In this case, the course of density changes is still different (compared to Miscanthus and Sylphia). In no case was the threshold value DE 1000 kg/m^3 achieved—the maximum density achieved was 980 kg/m^3. In this case, the highest density values were obtained for materials with a moisture content of 11%. In this case, as for Silphium, at 131 pressure the effect of moisture on density is not noticeable. The highest density of the resulting granules was observed at moisture of 11%. The pressure that turned out to be the best for compacting this material was 262 MPa. In turn, the lowest density was observed for granules made of a material with a moisture content of 11%, compacted with a pressure of 131 MPa. The highest density was 980.1 kg/m^3 and this was higher than the lowest density by 13.19%.

Figure 7f contains the results of the test of mechanical durability of granules made of crumbled Sida. In this case, the course of changes is similar to sylph pellets. Differences in durability caused by moisture can only be seen at a pressure of 131, an increase in pressure reduces the effect of moisture on durability.

The most resistant to the test of mechanical durability turned out to be granules produced at a pressure of 327 MPa from a material with a moisture content of 14%. The least durable granules were those made of a material with a water content of 8% using a pressure of 131 MPa. The highest durability was 94.27% and this was higher than the lowest durability by 46.9%. Photographs of the samples after the endurance test are shown below (Figure 8).

Figure 8. Samples after mechanical durability test-Sida.

3.3. Analysis

Based on the above results, statistical analysis was conducted. A detailed statistical analysis was made for two hypotheses:

- differences in the moisture content of the material have a significant impact on the quality parameters of the pellets at compaction pressures levels.
- differences in the compaction pressure significantly affect the quality parameters of the pellets in investigated values of raw material moisture content.

The selected quality parameters were solid density (DE) and mechanical durability (DU). The analyses were carried out for all three examined perennial biomass.

In this paper two-way analysis of variance (ANOVA) was carried out. Normality of decomposition was checked (Shapiro–Wilk test). For all cases, the distribution was normal. Then the assumption of the equality of variance was also checked (Brown–Forsythe test). For all cases, this equality has been met. The next step was to carry out one-way ANOVA and post-hoc analysis (Scheffé's test), which allowed to indicate between which groups there are statistically significant differences.

The tables (Table 2,Table 3,Table 4,Table 5,Table 6,Table 7) show two-way ANOVA results for the significance level $\alpha = 0.05$. The individual values mean: SS-sum of squares, df—degrees of freedom, MS—mean square, F Value—F test value, p—Value-probability value. The p—Value indicator shows the probability of obtaining results as extreme as the observed results of a statistical hypothesis test. The F Value indicator is the value of a statistical hypothesis test value. The results are presented for single factors (Pressure, Moisture) and for combined factors (Pressure × Moisture).

Table 2. Two-way analysis of variance (ANOVA)—DE.

	SS	df	MS	F Value	p-Value
Intercept	30.76175	1	30.76175	439,073.2	0.000000
Pressure	0.01765	3	0.00588	84.0	0.000000
Moisture	0.08335	2	0.04168	594.9	0.000000
Pressure × Moisture	0.00245	6	0.00041	5.8	0.000737
Error	0.00168	24	0.00007		

Table 3. Two-way ANOVA—DU.

	SS	df	MS	F Value	p-Value
Intercept	273,118.3	1	273,118.3	456,654.2	0.000000
Pressure	704.2	3	234.7	392.5	0.000000
Moisture	119.1	2	59.5	99.6	0.000000
Pressure × Moisture	18.1	6	3.0	5.0	0.001785
Error	14.4	24	0.6		

Table 4. Two-way ANOVA—DE.

	SS	df	MS	F Value	p-Value
Intercept	30.41612	1	30.41612	376,098.5	0.000000
Pressure	0.05424	3	0.01808	223.6	0.000000
Moisture	0.01930	2	0.00965	119.3	0.000000
Pressure × Moisture	0.00656	6	0.00109	13.5	0.000001
Error	0.00194	24	0.00008		

Table 5. Two-way ANOVA—DU.

	SS	df	MS	F Value	p-Value
Intercept	275,507.9	1	275,507.9	458,302.1	0.000000
Pressure	1817.4	3	605.8	1007.8	0.000000
Moisture	92.8	2	46.4	77.2	0.000000
Pressure × Moisture	190.3	6	31.7	52.8	0.000000
Error	14.4	24	0.6		

Table 6. Two-way ANOVA—DE.

	SS	df	MS	F Value	p-Value
Intercept	31.32602	1	31.3202	808,303.5	0.000000
Pressure	0.05172	3	0.01724	444.9	0.000000
Moisture	0.00420	2	0.00210	54.2	0.000000
Pressure × Moisture	0.00289	6	0.00048	12.4	0.000002
Error	0.00093	24	0.00004		

Table 7. Two-way ANOVA—DU.

	SS	df	MS	F Value	p-Value
Intercept	267,292.0	1	267,292.0	266,914.1	0.000000
Pressure	2347.1	3	782.4	781.3	0.000000
Moisture	381.2	2	190.6	190.3	0.000000
Pressure × Moisture	278.9	6	46.5	46.4	0.000000
Error	24.0	24	1.0		

3.3.1. Miscanthus

Tables 2 and 3 shows the two-way ANOVA results for the significance level α = 0.05.

The two-way ANOVA results for Miscanthus showed that pressure and moisture as well as their joint effect significantly affect both tested qualitative features (DE and DU).

After one-way ANOVA and post-hoc analysis (Scheffé's test), it was found that:

- Statistical analysis showed that the best density and durability were obtained for a material with a moisture content of 8%.
- For each pressure, except for 131 MPa, the influence of moisture on the density obtained was significant (Figure 9a). The pressure of 131 MPa significantly affected the density of the material with a moisture of 8%.
- As the moisture increases, the significance of the differences between pressures on the density obtained decreases, while the significance of the differences between the pressures on the achieved durability decreases as the moisture increases (Figure 9b).

- An increase in pressure results in a decrease in the significance of differences between moisture for the durability obtained.

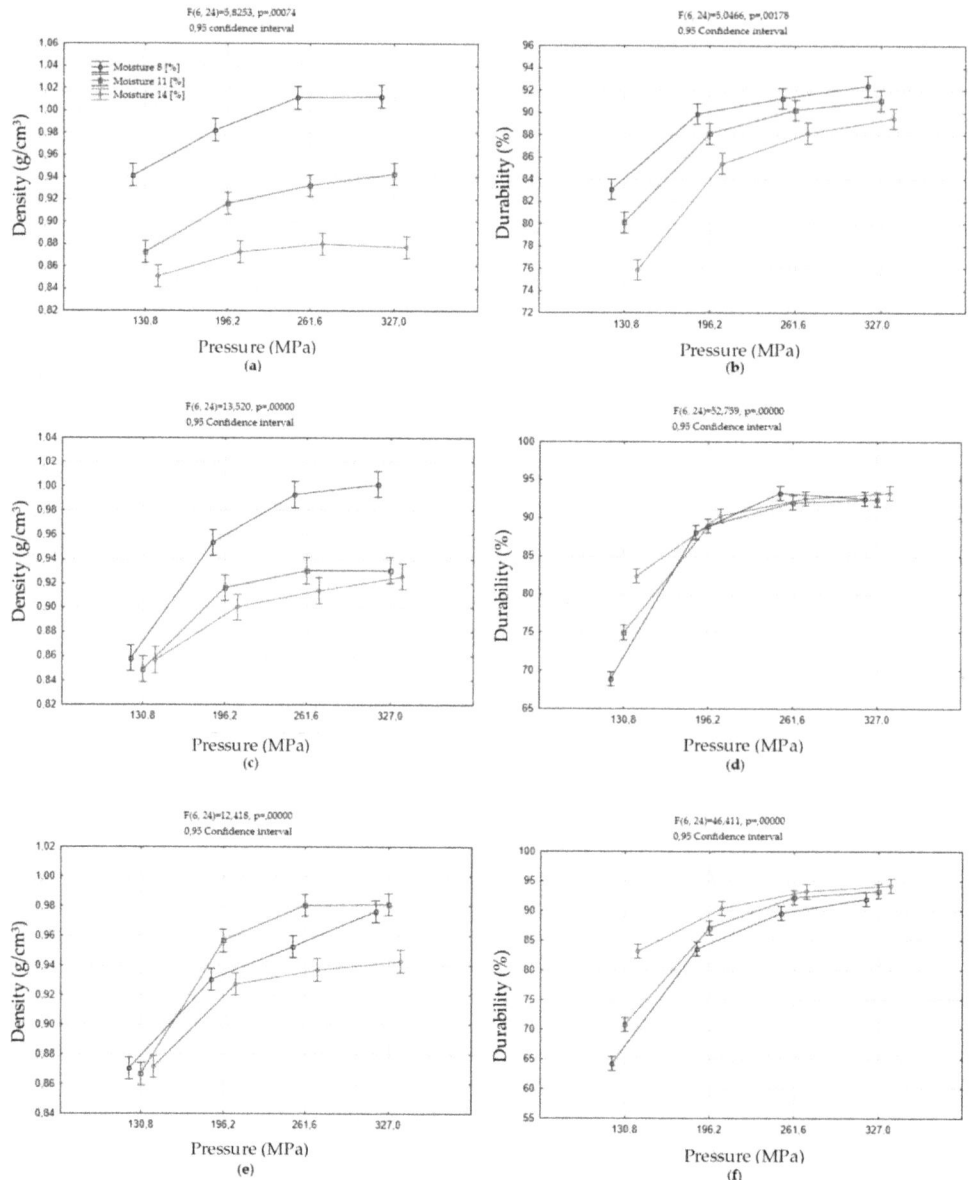

Figure 9. Influence of moisture and pressure on: (**a**) Miscanthus DE, (**b**) Miscanthus DU, (**c**) Silphium DE, (**d**) Silphium DU, (**e**) Sida DE, (**f**) Sida DU.

3.3.2. Silphium

Tables 4 and 5 shows the two-way ANOVA results for the significance level $\alpha = 0.05$.

Two-way ANOVA results for the Silphium showed that pressure and moisture and their joint effect significantly affect both tested qualitative features (DE and DU).

After one-way ANOVA and post-hoc analysis (Scheffé's test), it was found that:

- Statistical analysis showed that for each pressure, except for 131 MPa, only for 8% moisture there were significant differences in the density obtained (Figure 9c). At 131 MPa, there were no significant differences between moistures.
- The use of 8% moisture significantly affects the density obtained in the entire pressure range, resulting in the best results. At 11% and 14% moisture, there are no significant differences between the pressures on the density obtained.
- For each pressure, except for 131 MPa, moisture does not affect the durability of the granules produced (Figure 9d).
- For each moisture value, significant differences in the granule durability occur only at a pressure of 131 MPa (lower values).

3.3.3. Sida

Tables 6 and 7 shows the two-way ANOVA results for the significance level $\alpha = 0.05$.

The ANOVA two-way results for mallow have shown that pressure and moisture as well as their joint effect significantly affect both tested qualitative features (DE and DU).

After one-way ANOVA and post-hoc analysis (Scheffé's test), it was found that:

- Statistical analysis of the results showed that for all given pressure values, except for 131 MPa, the best granule density was obtained for moisture 11% (Figure 9e).
- For 14% moisture, there are no significant differences between pressures, except for a pressure of 131 MPa in DE.
- Density significantly different from the others was obtained each time using a pressure of 131 MPa for each moisture.
- For each pressure, except for 131 MPa, there were significant differences between moisture in durability (Figure 9f).
- For any pressure, except 131 MPa, moisture does not affect the durability obtained. For moisture of 11% and 14%, significant differences in the durability obtained occur only at a pressure of 131 MPa (lower values).

4. Discussion

The results show how complex the process is and how many factors can affect the quality parameters of the pellets obtained. The paper shows that both pressure and moisture of the compacted material have a significant impact on the density and durability of the obtained pellets. The aim of the study was to show only the influence of the compaction pressure and moisture of the material on the indicated quality parameters. Therefore, other factors such as grain composition, process temperature, compaction time and speed were unified. It was shown that depending on the material, the influence of moisture and pressure varies. In the case of Miscanthus, in the whole range of pressure tested, the highest values of DE were obtained for a material with 8% moisture (Figure 7a). A similar relationship was observed for Sylphium, however, in this case, it occurred after the pressure of 262 MPa was exceeded (Figure 7c). At 132 MPa the influence of moisture was insignificant. A different relation occurs in the case of Sida, in this case with the moisture of 11%, at a pressure above 262 MPa, allows to obtain the highest values of DE (Figure 7e).

For DU for the lowest value of the test pressure (132 MPa), the course of changes is characterized by the greatest differences for the materials tested. For Miscanthus, with the increase in material moisture, the durability of pellets decreases from 82.8% to 76.84% level (Figure 7b). Different trends were observed for Sylphium and Sida. In these cases the durability increases from 70.49% to 83.98% for Sylphium (Figure 7d) and from 66.21% to 83.85% for Sida (Figure 7f). However, when a certain pressure value (characteristic for the material) is exceeded, the resulting DU value stabilizes and does not depend on moisture.

Literature data indicate that the technological moisture range of the raw material should be between 8%–18% [50,51]. Obernberger and Thek [52] report that pellet production is possible in the moisture range of 8%–12%, however, their research included evaluation of the quality of wood pellets whose moisture content ranged from 5.7%–7.7% and straw pellets with moisture content of 5.6%–7.2%. Pellets came from producers from the European Union (EU) countries and were produced on professional production lines. Consequently, their quality was influenced not only by pressure and moisture but also by the type of material, temperature and type of pelletiser.

The research conducted showed that the increase in moisture in the case of Miscanthus and Sylphium causes a decrease in the DE values of the obtained granules. In the case of the slime the highest density was obtained at 11% moisture. Similar trends are also found in other studies [39] that studied the change in density of briquettes made from pine and oak biomass (moisture ranging from 5% to 15%) depending on the densification pressure. For both pine and oak, the specific density of the briquettes decreases with increasing moisture.

A similar relationship was observed for briquettes made from beech biomass [53]. With an increase in moisture from 4.7% to 19.5%, the specific gravity of the briquettes obtained decreases. The relative decrease of specific density depends on the densification pressure. Studies on the process of pellet production from spruce biomass indicate similar relationships [54]. However, studies carried out for acacia biomass and tobacco stems showed an inverse relationship [55,56]. This means that the type of material has an influence on how moisture influences the course of the densification process.

In the case of DU, at a pressure of 132 MPa, the differences between Miscanthus and other materials were noted. As the moisture content increases, DU of miscanthus pellets decreases, in the case of Sylphium and slime the situation is the opposite. The influence of raw material on the course of DU changes can be found in literature.

Samuelson [57] showed that the durability of pellets made of pine biomass with 8.1% moisture content was characterized by mechanical durability at 91.8%, for 10.7% DU it was 96.9%, and for 13.1% DU it was noted that DU dropped to almost 70%.

Colley [58] showed that in the 6%–8% moisture range the durability of millet pellets was approximately 96.0% and with increasing DU it decreased to 78% at 17%.

Ishii [59] in his study presented that by increasing the moisture content of the material from 15% to 20% the mechanical durability of the pellets can be increased. By comparing this data with the results obtained in the study, it can be concluded that depending on the course of the study, the number of factors taken into account or the choice of material, the influence of the material moisture on the density and durability may vary. Therefore, it is not possible to compare materials compacted under different conditions, of different origin, because it has been clearly shown that the materials compacted in different, characteristic ways.

Analyzing the results obtained, the highest values of both tested quality parameters can be obtained at the moisture and pressure levels characteristic for the tested material. For Miscanthus and Silphium, it is the material moisture at 8% and minimum pressure of 262 MPa, for the glacier at 11% moisture and 262 MPa. Density is more dependent on material moisture and compaction pressure in comparison to DU, whose value, for a given material, after overturning the characteristic pressure level, does not correlate with moisture. In the cases studied only for Miscanthus and Silphium the assumed density thresholds of 1000 kg/m^3 were obtained. In the case of DU the assumed threshold was not obtained for any of the tested materials. The tested moisture range after exceeding a certain pressure value is insignificant and further pressure increase does not cause a DU increase. This means that the density is possible to obtain with the tested factors, whereas to obtain the threshold durability it is necessary to consider the influence of further factors such as the speed of the densification process and the process temperature. As the literature indicates, the mechanical durability of the pellets is also dependent on the temperature of the process. It is an important factor because it allows activation of binders naturally present in the material in the form of lignin, lignocellulose or resin. The temperature has been kept constant at 20 °C in the tests, resulting in a lack of activation of the binders. Kaliyan [60] describes in his research that both temperature and moisture of the material can activate the binders, which gives the possibility to reduce process costs. From the literature it can be concluded that higher moisture ranges (above 10% Colley [58]) do not always increase durability (on the contrary), so another factor must be considered to increase durability to a minimum 97.5%.

Usually the influence of material moisture on the durability of the pellet is tested in combination with the temperature of the process. As the study presents basic research, temperature as one of many influencing factors has been omitted. It should be noted that the tests were carried out for pellets, but the results can also be used in the briquetting process, where temperature is not always taken into account.

The results obtained and their juxtaposition with the literature data suggest that the number of factors influencing the quality parameters of pellets will increase, which will be continued in subsequent studies.

5. Conclusions

In this research, we followed the effect of material moisture content and compression pressure on the final specific density and mechanical durability of pellets with a standardized particle size distribution. This impact was followed by compaction of three materials: Giant miscanthus, Cup plant and Virginia mallow. The results indicate that the pressure but also the moisture of the material have a significant impact on the final density and mechanical durability of the pellets and it is different depending on the tested material.

It turns out that to achieve Giant miscanthus pellets with the best parameters, just use a material with a moisture content of 8% and compress it at a pressure of 196 MPa (for durability) and 262 for density. Above this level of pressure, the differences are not so significant and this allows energy saving in the processing processes.

For Silphium, the course of changes in DE and DU is slightly different. The highest DE values were obtained for material with a moisture content of 8%. At a pressure of 262 pellets achieve values close to the assumed threshold (1000 kg/m^3), a further increase in pressure does not cause a significant change in DE. In the case of DU, 14% moisture allows the best values of this parameter to be obtained at a pressure of 132 MPa, compared to 11% and 8% moisture. At higher pressures, the influence of moisture is negligible and the obtained DU value stabilizes at around 90%–93%. To obtain the maximum values of both quality parameters it is therefore necessary to compact the material with a moisture of 8% with a pressure of 262 MPa.

Similar properties were noted in pellets made of Cup plant. The material moisturize to 8% has the highest density. The compaction pressure that gave the best result in terms of mechanical durability is 262 MPa. No significant differences in mechanical durability were noticed using the pressure of 327 MPa, so the lower level is completely sufficient and also allows for energy savings and thus a reduction in production costs.

The results of the process of compression pellets made of Virginia mallow indicate that the best level of moisture is 11%. However, the difference between the highest density for 8% and 11% moisture at 372 MPa is not significant, but the same DE level can be obtained for a material with a moisture content of 11% already at a pressure of 262 MPa. The lowest density was found in pellets made of material with a moisture content of 14%. In turn, testing of mechanical durability showed that the highest durability parameters were achieved by pellets made at a pressure of 327 MPa, while this durability is not significantly different from the durability obtained at a pressure of 262 MPa. To obtain the maximum values of both quality parameters, it is therefore necessary to thicken the material with a moisture content of 11% to adjust 262 MPa.

Studies have shown that material of different origin shows differences in the results obtained and moisture and pressure does not eliminate it. The expressive differences are between the grass-coated

Miscanthus and the Cup plant and Virginia mallow which represents the dicotyledonous perennials. Characteristic is the different course of DU changes with increasing moisture at a pressure of 132 MPa. For Miscanthus, DU decreases as moisture increases, unlike for other materials.

In the future, it is planned to perform further tests enriched with factors such as: compaction speed and temperature. It is also planned to demonstrate the effect of material moisture and compaction pressure on the elastic spring-back of finished pellets.

Author Contributions: Conceptualization, J.S. and M.W., methodology, M.W., J.S.; formal analysis, J.F.; data curation, A.K.; writing—original draft preparation, J.S.; writing—review and editing, J.S., M.W.; visualization, A.K.; All authors have read and agreed to the published version of the manuscript.

Funding: This research was supported by the Ministry of Science and Higher Education of the Republic of Poland (of Production and Power Engineering, University of Agriculture in Kraków).

Conflicts of Interest: The authors declare no conflict of interest.

References

1. Karbowniczak, A.; Hamerska, J.; Wróbel, M.; Jewiarz, M.; Nęcka, K. Evaluation of selected species of woody plants in terms of suitability for energy production. In *Renewable Energy Sources: Engineering, Technology, Innovation*; Springer Proceedings in Energy: Cham, Switzerland, 2018.
2. Civitarese, V.; Acampora, A.; Sperandio, G.; Assirelli, A.; Picchio, R. Production of wood pellets from poplar trees managed as coppices with different harvesting cycles. *Energies* **2019**, *12*, 2973. [CrossRef]
3. Relova, I.; Vignote, S.; León, M.A.; Ambrosio, Y. Optimisation of the manufacturing variables of sawdust pellets from the bark of *Pinus caribaea* Morelet: Particle size, moisture and pressure. *Biomass Bioenergy* **2009**, *33*, 1351–1357.
4. Dołżyńska, M.; Obidziński, S.; Kowczyk-Sadowy, M.; Krasowska, M. Densification and combustion of cherry stones. *Energies* **2019**, *12*, 3042. [CrossRef]
5. Tumuluru, J. Pelleting of pine and switchgrass blends: Effect of process variables and blend ratio on the pellet quality and energy consumption. *Energies* **2019**, *12*, 1198. [CrossRef]
6. Kowalczyk-Juśko, A.; Kulig, R.; Laskowski, J. The influence of moisture content of selected Energy crops on the briqueting process parameters. *TEKA Comm. Mot. Energ. Agric.* **2011**, *11*, 189–196.
7. Ovcharuk, V.; Boyko, O.; Horodyska, O.; Vasulyeva, O.; Mudryk, K.; Jewiarz, M.; Wróbel, M.; Styks, J. Prospects for the production of biofuels from crop residues bean and its environmental and technological characteristics. In *Renewable Energy Sources: Engineering, Technology, Innovation*; Springer: Cham, Switzerland, 2020.
8. Kościk, B. *Rośliny Energetyczne*; Wydawnictwo Akademii Rolniczej: Lublin, Poland, 2003.
9. Kulig, R.; Laskowski, J. The effect of preliminary processing on compaction parameters of oilseed rape straw. *TEKA Comm. Mot. Energ. Agric.* **2011**, *11*, 209–217.
10. Artyszak, D. Rośliny energetyczne—Charakterystyka podstawowych gatunków i ich wykorzystanie w polskiej energetyce. In Proceedings of the Nowoczesna Energetyka Europy Środkowo-Wschodniej 2015, Warsaw, Poland, 19–28 October 2015.
11. Czeczko, R. Porównanie stopnia uwodnienia różnych części *Helianthus tuberosus* aspekcie ich przydatności jako biopaliwa. *Ochrona Środowiska Zasobów Naturalnych* **2011**, *49*, 521–524.
12. Księżak, J.; Faber, A. Ocena możliwości pozyskiwania biomasy z mozgi trzcinowatej na cele energetyczne. *Łąkarstwo W Polsce* **2007**, *11*, 141–148.
13. Mudryk, K.; Frączek, J.; Ślipek, Z.; Francik, S.; Wróbel, M. Chosen physico-mechanical properties of cutleaf coneflower (*Rudbeckia laciniata* L.) shoots. In Proceedings of the 12th International Scientific Conference Engineering for Rural Development, Jelgava, Latvia, 23–24 May 2013.
14. Ivanyshyn, V.; Nedilska, U.; Khomina, V.; Klymyshena, R.; Hryhoriev, V.; Ovcharuk, O.; Hutsol, T.; Mudryk, K.; Jewiarz, M.; Wróbel, M.; et al. Prospects of growing *Miscanthus* as alternative source of biofuel. In *Renewable Energy Sources: Engineering, Technology, Innovation*; Springer Proceedings in Energy: Cham, Switzerland, 2018.
15. Wąsik, R.; Michalec, K.; Mudryk, K. Research reports variability in static bending strength of the "Tabórz" scots pine wood (*Pinus sylvestris* L.). *Drewno Prace Naukowe Doniesienia Komunikaty* **2016**, *59*. [CrossRef]

16. Li, Y.; Liu, H. High pressure densification of wood residues to form an upgraded fuel. *Biomass Bioenergy* **2000**, *19*, 177–186. [CrossRef]
17. Križan, P.; Matúš, M.; Šooš, J.; Beniak, J. Behavior of beech sawdust during densification into a solid biofuel. *Energies* **2015**, *8*, 6382–6398. [CrossRef]
18. Knapczyk, A.; Francik, S.; Wójcik, A.; Bednarz, G. Influence of storing *Miscanthus* × *gigantheus* on its mechanical and energetic properties. In *Renewable Energy Sources: Engineering, Technology, Innovation*; Springer: Cham, Switzerland, 2018; pp. 651–660.
19. Czeczko, R. Uprawy wybranych roślin energetycznych. Energia odnawialna w nauce i praktyce. In Proceedings of the EKOENERGIA '2012: Energia odnawialna w nauce i praktyce, Lublin, Poland, 26–27 October 2012; pp. 170–172.
20. Kieć, J. Wady i zalety roślin energetycznych. In *Energetyka Alternatywna*; Wydawnictwo Dolnośląskiej Wyższej Szkoły Przedsiębiorczości i Techniki w Polkowicach: Polkowice, Poland, 2011.
21. Ovcharuk, O.; Hutsol, T.; Ovcharuk, O.; Rudskyi, V.; Mudryk, K.; Jewiarz, M.; Wróbel, M.; Styks, J. Prospects of use of nutrient remains of corn plants on biofuels and production technology of pellets. In *Renewable Energy Sources: Engineering, Technology, Innovation*; Springer: Cham, Switzerland, 2020; pp. 293–300.
22. Hejft, R. Ciśnieniowa aglomeracja materiałów roślinnych. In *Białystok Wydaw*; I Zakład Poligrafii Instytutu Technologii Eksploatacji: Radom, Poland, 2002; ISBN 83-7204-251-9.
23. Anukam, A.; Berghel, J.; Frodeson, S.; Bosede Famewo, E.; Nyamukamba, P. Characterization of pure and blended pellets made from Norway spruce and Pea starch: A comparative study of bonding mechanism relevant to quality. *Energies* **2019**, *12*, 4415. [CrossRef]
24. Moliner, C.; Lagazzo, A.; Bosio, A.; Botter, R.; Arato, E. Production, Characterization, and Evaluation of Pellets from Rice Harvest Residues. *Energies* **2020**, *13*, 4415. [CrossRef]
25. Greinert, A.; Mrówczyńska, A.; Grech, R.; Szefner, W. The use of plant biomass pellets for energy production by combustion in dedicated furnaces. *Energies* **2020**, *13*, 463. [CrossRef]
26. Francik, S.; Knapczyk, A.; Francik, R.; Ślipek, Z. Analysis of possible application of olive pomace as biomass source. In *Renewable Energy Sources: Engineering, Technology, Innovation*; Springer: Cham, Switzerland, 2018; pp. 583–592.
27. ISO, BSEN. *Solid Biofuels—Terminology, Definitions and Descriptions*; ISO: Geneva, Switzerland, 2014. (In Polish)
28. Janowicz, L. Biomasa w Polsce. *Energ. I Ekol.* **2006**, *8*, 601–604.
29. Wrobel, M.; Mudryk, K.; Jewiarz, M.; Knapczyk, A. Impact of raw material properties and agglomeration pressure on selected parmeters of granulates obtained from willow and black locust biomass. In Proceedings of the 17th International Scientific Conference Engineering for Rural Development, Jelgava, Latvia, 23–25 May 2018; pp. 1933–1938.
30. Niemczyk, M.; Kaliszewski, A.; Jewiarz, M.; Wróbel, M.; Mudryk, K. Productivity and biomass characteristics of selected poplar (*Populus* spp.) cultivars under the climatic conditions of northern Poland. *Biomass Bioenergy* **2018**, *111*, 46–51. [CrossRef]
31. Saletnik, B.; Puchalski, C.; Zaguła, G.; Bajcar, M. Porównanie użyteczności energetycznej wybranych brykietów z biomasy. *Inżynieria Rolnicza* **2013**, *3*, 341–347.
32. Stolarski, M.; Szczukowski, S.; Tworkowski, J. Characteristics of selected biofuels produced from solid biomass. *Problemy Inzynierii Rolniczej* **2007**, *15*, 21–26.
33. Pietsch, W. *Agglomeration Processes: Phenomena, Technologies, Equipment*; John Wiley & Sons: Hoboken, NJ, USA, 2008.
34. Stelte, W.; Holm, J.; Sanadi, A.; Barsberg, S.; Ahrenfeldt, J.; Henriksen, U. Fuel pellets from biomass: The importance of the pelletizing pressure and its dependency on the processing conditions. *Fuel* **2011**, *90*, 3285–3290. [CrossRef]
35. Križan, P.; Svátek, M.; Matúš, M.; Beniak, J.; Lisý, M. Determination of compacting pressure and pressing temperature impact on biomass briquettes density and their mutual interactions. In Proceedings of the 4th SGEM GeoConference on Energy and Clean Technologies, Albena, Bulgaria, 19–25 June 2014; pp. 133–140.
36. Abdoli, M.A.; Golzary, A.; Hosseini, A.; Sadeghi, P. *Wood Pellet as a Renewable Source of Energy from Production to Consumption*; Springer: Cham, Switzerland, 2018.
37. Hu, Q.; Shao, J.; Yang, H.; Yao, D.; Wang, X.; Chen, H. Effects of binders on the properties of bio-char pellets. *Appl. Energy* **2015**, *157*, 508–516. [CrossRef]

38. Tumuluru, J. Effect of moisture content and hammer mill screen size on the briquetting characteristics of woody and herbaceous biomass. *KONA Powder Part. J.* **2019**. [CrossRef]
39. Križan, P.; Šooš, L.; Matúš, M.; Beniak, J.; Svátek, M. Research of significant densification parameters influence on final briquettes quality. *Wood Res.* **2015**, *60*, 301–316.
40. Križan, P. *The Densification Process of Wood Waste*; Capello, E., Boyd, M., Eds.; Walter de Gruyter GmbH & Co. KG: Berlin, Germany, 2015.
41. Brožek, M. The effect of moisture of the raw material on the properties briquettes for energy use. *Acta Univ. Agric. Silv. Mendel. Brun.* **2016**, *64*, 1453–1458. [CrossRef]
42. Wróbel, M. *Zagęszczalność i Kompaktowalność Biomasy Celulozowej*; Polskie Towarzystwo Inżynierii Rolniczej: Kraków, Poland, 2019.
43. Mani, S.; Tabil, S.; Sokhansanj, S. Effects of compressive force, particle size and moisture content on mechanical properties of biomass pellets from grasses. *Biomass Bioenergy* **2006**, *30*, 648–654. [CrossRef]
44. ISO, BSEN. *Solid Biofuels—Determination in Moisture Content—Oven Dry Method—Part 3: Moisture in General Analysis Sample*; British Standards Institution: Bonn, Germany, 2015; pp. 1–14. (In Polish)
45. DS/EN. *Solid biofuels—Determination of Particle Size Distribution for Uncompressed Fuels—Part 2: Vibrating Screen Method Using Sieves with Aperture of 3,15 mm and Below*; Dansk standard: Copenhagen, Denmark, 2010. (In Polish)
46. Jewiarz, M.; Mudryk, K.; Wróbel, M.; Frączek, J.; Dziedzic, K. Parameters affecting RDF-based pellet quality. *Energies* **2020**, *13*, 910. [CrossRef]
47. Manouchehrinejad, M.; Yue, Y.; Lopes de Morais, R.; Souza, L.; Singh, H.; Mani, S. Densification of thermally treated energy cane and Napier grass. *BioEnergy Res.* **2018**, *11*, 538–550. [CrossRef]
48. European Committee for Standardization (CEN). *Solid Biofuels—Determination of Mechanical Durability of Pellets and Briquettes—Part 1: Pellets*; CEN: Brussels, Belgium, 2010. (In Polish)
49. ISO, BSEN. *Solid Biofuels—Fuel Specifications and Classes—Part 6: Graded Non-Woody Pellets*; British Standards Institution: Bonn, Germany, 2014. (In Polish)
50. Serrano, C.; Monedero, E.; Lapuerta, M.; Portero, H. Effect of moisture content, particle size and pine addition on quality parameters of barley straw pellets. *Fuel Process. Technol.* **2011**, *92*, 699–706. [CrossRef]
51. Tumuluru, J.S.; Wright, C.T.; Hess, J.R.; Kenney, K.L. A review of biomass densification systems to develop uniform feedstock commodities for bioenergy application. *Biofuels Bioprod. Biorefin.* **2011**, *5*, 683–707. [CrossRef]
52. Obernberger, I.; Thek, G. Physical characterisation and chemical composition of densified biomass fuels with regard to their combustion behaviour. *Biomass Bioenergy* **2004**, *27*, 653–669. [CrossRef]
53. Van Quyen, T.; Nagy, S.; Csőke, B. Effect of moisture content and particle size on beech biomass agglomeration. *Adv. Agric. Bot.* **2017**, *9*, 79–89.
54. Rhén, C.; Gref, R.; Sjöström, M.; Wästerlund, I. Effects of raw material moisture content, densification pressure and temperature on some properties of Norway. *Fuel Process. Technol.* **2005**, *87*, 11–16. [CrossRef]
55. Obidziński, S.; Joka, M.; Luto, E.; Bieńczak, A. Research of the densification process of post-harvest tobacco waste. *J. Res. Appl. Agric. Eng.* **2017**, *62*, 149–154.
56. Van Quyen, T.; Sándor, N. Agglomeration of *Acacia mangium* biomass. *Vietnam J. Sci. Technol.* **2018**, *56*, 198. [CrossRef]
57. Samuelsson, R.; Larsson, S.H.; Tyhrel, M.; Lestander, T.A. Moisture content and storage time influence the binding mechanisms in biofuel wood pellets. *Appl. Energy* **2012**, *99*, 109–115. [CrossRef]
58. Colley, Z.; Fasina, O.O.; Bransby, D.; Lee, Y.Y. Moisture effect on the physical characteristics of switchgrass pellets. *Trans. ASABE* **2006**, *49*, 1845–1851. [CrossRef]
59. Ishii, K.; Furuichi, T. Influence of moisture content, particle size and forming temperature on productivity and quality of rice straw pellets. *Waste Manag.* **2014**, *34*, 2621–2626. [CrossRef]
60. Kaliyan, N.; Vance Morey, R. Factors affecting strength and durability of densified biomass products. *Biomass Bioenergy* **2009**, *33*, 337–359. [CrossRef]

© 2020 by the authors. Licensee MDPI, Basel, Switzerland. This article is an open access article distributed under the terms and conditions of the Creative Commons Attribution (CC BY) license (http://creativecommons.org/licenses/by/4.0/).

Article

Energy Utilization of Spent Coffee Grounds in the Form of Pellets

Radovan Nosek [1],*, Maw Maw Tun [2] and Dagmar Juchelkova [2]

1. Department of Power Engineering, Faculty of Mechanical Engineering, University of Žilina, Univerzitna 1, 010-26 Žilina, Slovakia
2. VŠB-Technical University of Ostrava, 17, Listopadu 2172/15, 708 00 Ostrava-Poruba, Czech Republic; maw.maw.tun.st@vsb.cz (M.M.T.); dagmar.juchelkova@vsb.cz (D.J.)
* Correspondence: radovan.nosek@fstroj.uniza.sk; Tel.: +46-222-46-04

Received: 7 February 2020; Accepted: 4 March 2020; Published: 6 March 2020

Abstract: Nowadays it is important to limit the use and combustion of fossil fuels such as oil and coal. There is a need to create environmentally acceptable projects that can reduce or even stop greenhouse gas emissions. In this article, we dealt with the objectives of energy policy with regard to environmental protection, waste utilization, and conservation of natural resources. The main objective of the research was to assess the possibility of the use of spent coffee grounds (SCG) as fuel. As a part of the solution, the processing of coffee waste in the form of pellets, analysis of calorific value and combustion in the boiler were proposed. The experiments were done with four samples of pellets. These samples were made from a mixture of wood sawdust and spent coffee grounds with ratio 30:70 (wood sawdust: spent coffee grounds), 40:60, 50:50 and 100% of spent coffee grounds. The calorific values were compared with wood sawdust pellets (17.15 MJ.kg^{-1}) and the best lower calorific value of 21.08 MJ.kg^{-1} was measured for 100% of spent coffee grounds. This sample did not achieve the desired performance during the combustion in the boiler due to the low strength of the sample.

Keywords: spent coffee grounds; sawdust; pellets; calorific value; combustion

1. Introduction

In recent years, renewable energy has come to the forefront and biomass is the world's fourth-largest energy source. One of the main problems of biomass grown primarily for energy use is that it displaces the cultivation of conventional crops intended to provide food to the population. Thus, the use of waste biomass, for example, sawdust, straw, etc., is much more advantageous. Renewable energy sources are a means of reducing greenhouse gases and alternative forms of biomass also fall into this category. Generally, alternative biomass sources form unused residues of food processing, which often end up unused as waste in landfills. Spent coffee ground is included in the category of waste biomass. Therefore, the recycling of waste to energy and value-added products is one effective way to solve the problem where many countries face a serious challenge in dealing with the huge amount of waste generated daily due to their increasing populations, industrial growth, and human consumption [1] as well as in dealing with the loss of valuable resources from waste and environmental degradation [2]. Spent coffee ground (SCG) constitutes a large number of organic compounds (more than 1000 individual compounds) such as proteins, carbohydrates, tannins, fibers, caffeine, cellulose, non-protein nitrogenous, fatty acids, amino acids, polyphenols, minerals lignin and polysaccharides [1]. The lower heating value of wet SCG accounts for approximately 8.4 MJ.kg^{-1} [3] while the lower calorific values of dried SCG range between 19.3–24.9 MJ.kg^{-1} [1].

Nowadays, there are a variety of options for the utilization of SCG in the field of agriculture and numerous possibilities for converting SCG to biofuels. The most recent technologies for valorization of SCG include anaerobic digestion, pyrolysis, liquefaction or gasification, oil extraction, fermentation [4] and other technologies for value-added products from SCG such as composting, adsorbents, antioxidants and nutrients. Atabani et al. reported that SCG can be recycled in different ways to produce various types of biofuels such as biohydrogen, biobutanol, biodiesel, fuel pellets, bio-oil, bioethanol, biogas and hydrocarbon fuels, or value-added products such as bioactive compounds for food, pharmaceutical, cosmetic and chemical industries and antioxidants and anti-tumor activities, adsorbents, composting, co-composting, vermicomposting, nanocomposites, biopolymers, creams-scrubs, soaps and detergents, odor control, textile, facile preparation of pyrolytic carbon as anode in sodium-ion battery, inks and screen painting, yarn, and pulp and paper production [1].

The recent research studies underlined the valorization of SCG as a valuable source of phenolic compounds and bioenergy [5], the integration of chlorogenic acid recovery and bioethanol production from SCG [4], the co-production of biodiesel and bioethanol from SCG [6,7], the coffee oil extraction from SCG using four solvents and prototype-scale extraction [8] and the production of bio-oil and biochar from the defatted SCG by slow pyrolysis [6]. The high calorific value of spent coffee grounds (SCG) has the potential for producing refuse-derived fuel (RDF), however the burning of pure SCG pellets can lead to low boiler efficiency resulting in increased particle emissions, thus additional material is required to produce good quality pellets [9]. Therefore, the research studies highlight the utilization of waste paper and coffee residue for briquette production [8], production of the carbonized briquettes from Rain tree (Samanea Saman) and SCG/tea waste [10], analysis of the effect of mixing SCG and coffee silverskin (CS) on the quality of pellet fuel produced [11] and production of the eco-fuel briquettes with 32% spent coffee ground, 23% coal fines, 11% sawdust to benefit lower toxic emissions compared to fossil fuel [12].

Likewise, in the last decade, a great deal of research has been conducted to burn waste biomass in various boilers, for example, for the analysis of the combustion tests in a commercial residential wood pellet boiler with a pure SCG pellet, a blended pellet (50% SCG and 50% sawdust) and a pure pine wood pellet [13], for the fuel and combustion test in a small boiler (6.5 kW) with SCG [2], the combustion tests of wood pellet on a fixed bed reactor with various conditions [14], and the combustion test of a fluidized bed boiler with fuel of a normal sold waste and mixing with animal wastes [15]. The combustion of straw, olives, tomatoes, cocoa beans, etc., achieved a relatively high boiler efficiency, but the problem is the emerging ash with a relatively low melting point. It is necessary to limit the clogging of the heat exchanger. Mixing these waste biofuels with wood becomes very advantageous, thus avoiding this undesirable phenomenon. By burning this biofuel, we would be able to reduce waste and obtain a renewable energy source at the same time.

Therefore, the study dealt with the objectives of energy policy with regard to environmental protection, waste utilization and conservation of natural resources in order to assess the possibility for the use of spent coffee grounds (SCG) as a fuel and to propose the processing of coffee waste in the form of pellets, analysis of calorific value and combustion in the boiler. In this article, the technical feasibility of producing biofuel from SCG is evaluated. Additionally, the experimental combustion of produced samples with the content of SCG was performed and the measurement of emissions from tests is examined.

2. Materials and Methods

In this research sawdust and SCG (Figure 1a,b) with moisture contents ranging from 45%–55% were used as input material. There was a risk of degradation of SCG with such a high water content, therefore, the moisture was reduced to 6.50% by drying in a laboratory oven at 100 °C for 12 h.

Figure 1. Input material: (**a**) spent coffee grounds and (**b**) sawdust.

The SCG was mixed with an appropriate amount of sawdust and the prepared material was then used to produce pellets in a vertical pellet press. The samples were made in the following proportions: 100% of spent coffee grounds, 50/50, 40/60 and 30/70 of SCG/sawdust, respectively.

Pelleting of each sample took approximately three minutes. Due to the low strength of the pellets, the procedure was usually repeated three times. Approximately 5 kg of samples (pellets) were made from each type of material mixture.

In the case of samples composed of 100% SCG, even after six repeated attempts and several moisture changes, pellets of the desired shape were not formed. The samples were soft and disintegrating. A similar result was obtained for the sample containing 50% of SCG and 50% of sawdust, but the pellets had a better cylindrical shape compared to the sample with 100% SCG (Figure 2a,b).

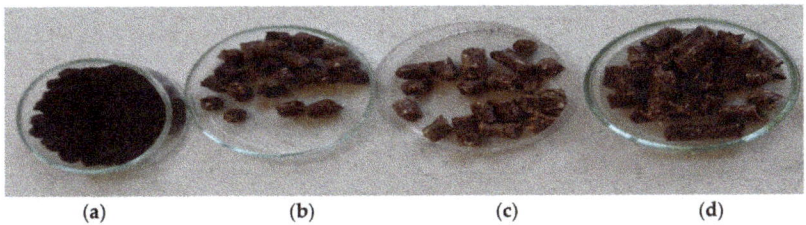

Figure 2. Samples: (**a**) 100% spent coffee grounds (SCG), (**b**) 50/50 SCG/sawdust, (**c**) 40/60 SCG/sawdust and (**d**) 30/70 SCG/sawdust.

Further, the SCG and sawdust were combined in the ratio 40/60, respectively, with an output moisture of 15%. The produced pellets had a more cylindrical shape compared to the 50/50 sample, but this sample did not have the required strength and durability. This was checked by hand compression and compared to the strength of the wood pellets. The sample with a 30/70 SCG/sawdust ratio also had the same moisture content of 15% of input material. The result showed that pellets have strength properties close to the reference sample (wood sawdust pellets). Subsequently, the combustion tests were carried out in an automatic boiler (type Lokca, model Uspor, power 18 kW).

3. Experimental Research and Results

3.1. Calorific Values

The calorific values of produced samples were measured using the LECO AC 500 device. This calorimeter is intended for the determination of the higher calorific value (HCV) of solids and liquids. The measurement of the HCV, the analysis of carbon, hydrogen, and nitrogen in the samples was performed. Its objective was to determine the hydrogen content of the produced pellets and to calculate the lower calorific value (LCV). The elemental compositions of the samples are given on a dry basis in Table 1.

Table 1. Ultimate analysis of produced samples.

Parameter	100% of SCG	50/50 SCG/Sawdust	40/60 SCG/Sawdust	30/70 SCG/Sawdust
Total Carbon (%)	54.56	52.69	52.13	51.29
Total Nitrogen (%)	17.78	46.87	50.64	61.66
Total Hydrogen (%)	7.44	6.99	6.89	6.74

The collected results of studies compare the C, H and N content of SCG (Table 2). It can be seen that the concentration of carbon (54.56%) and hydrogen (7.44%) of tested SCG are in agreement with reported ranges [1]. In the case of nitrogen, a higher content was recorded, compared to the values presented in Table 2. The fertilization of soil contributes to a higher content of nitrogen and increases its concentration in the coffee beans during growth.

Table 2. Comparison of ultimate analysis of SCG [3].

	C (%)	H (%)	N (%)
SCG	46.42–71.6	6.04–8.99	2.03–15.5

Figure 3 shows calculated lower calorific values of samples; the highest value was obtained from 100% spent coffee grounds. As already mentioned, the LCV of dried coffee grounds ranges between 19.3–24.9 MJ.kg^{-1} [1] and the LCV of pure SCG in this study had 21.08 MJ.kg^{-1}.

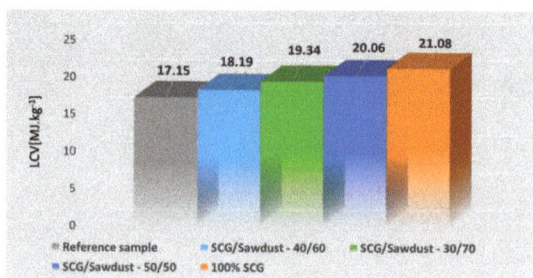

Figure 3. Lower calorific values of tested samples.

3.2. The Boiler Measurements

The boiler was placed on the scale and the supply of combustion air was provided by a ventilator. Emission probes were inserted to the outlet of the boiler, as well as a thermocouple and a draft pressure sensor. The actual boiler heat power was measured by a heat exchanger station and all data was recorded by the data logger (Figure 4).

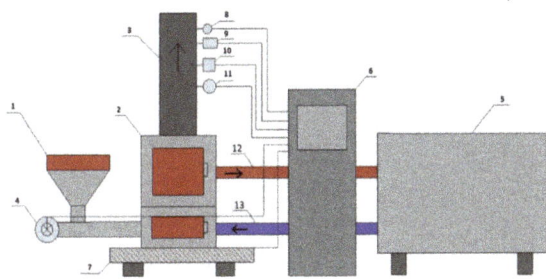

Figure 4. Scheme of experimental setup: 1—fuel tank, 2—pellet boiler, 3—chimney, 4—blower, 5—heat exchanger station, 6—flue gas analyzer and data logger, 7—scale, 8—draft pressure sensor, 9— and 10—mission probes, 11—thermocouple, 12—cold water, 13—hot water.

3.3. The Analyses of Results

Five samples were tested and each measurement lasted around 60 min. Heat power and emissions measurements were performed continuously throughout the experiment. The concentration of CO and CO_2 were measured in the flue gas and the results are shown in Figure 5a,b.

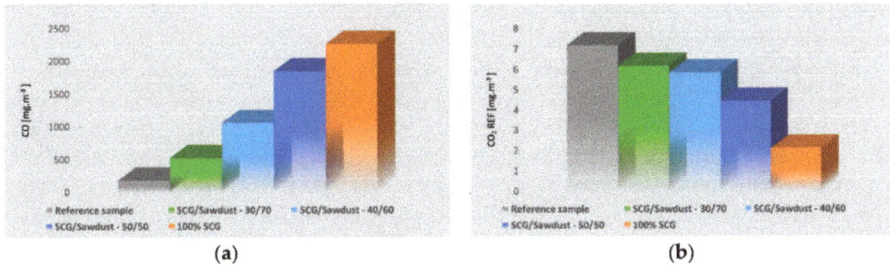

Figure 5. (a) Comparison of average CO emissions and (b) comparison of average CO_2 emissions.

The highest CO concentrations were recorded in the 100% coffee ground samples due to low pellet strength. This was checked by hand compression and compared to the strength of the wood pellets. Most of the sample volume was coffee powder, so the combustion was incomplete. In the 50/50 samples, similar results were observed. Also, these pellets did not have the required strength and therefore high CO emissions were recorded. In general, coffee pellets had incomparably higher CO emissions compared to certified wood pellets. The standard STN EN 303-5 2012 gives a maximum CO concentration of 3000 mg.m^{-3} and all of the produced samples meet this standard. In contrast to CO, CO_2 emissions were the lowest during 100% of SCG combustion, indicating precisely the incomplete combustion. Thus, most of the carbon contained in the coffee grounds has left the combustion chamber in the form of CO. Samples with ratios 40/60 and 30/70 had emissions comparable to wood pellets.

The formation of nitrogen oxides is accompanied mainly by high combustion temperature and chemical composition of the fuel. During the test of the 100% SCG sample, incomplete combustion occurred, the combustion temperature was low, and this was reflected in low NO values (Figure 6a). Samples containing SCG and sawdust show comparable NO emissions since the nitrogen content in SCG (17.78%) is higher than the concentration of nitrogen in wood pellets (0.3%).

The performance of the boiler is mainly related to the quality of the fuel and proper settings of the heat source's control system. As mentioned above, the durability of the pellets from 100% of coffee grounds was low and the boiler heat power was four times lower compared to wood pellets (Figure 6b).

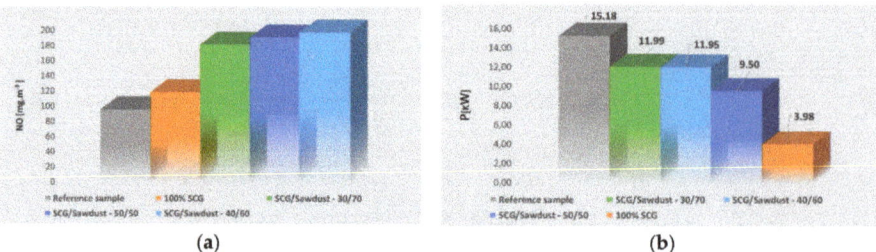

Figure 6. (a) Comparison of average NO emissions and (b) comparison of average heat power of the boiler.

4. Discussion and Conclusions

The chemical properties of coffee waste take the top position among the different types of material for wood pellet production and the short growing period for coffee beans is also an advantage of SCG compared to trees. Lisowski et al. [16] confirmed that the chemical composition of the SCG is a very good quality raw material for use as a biofuel, by highlighting that moisture had a greater and nonlinear effect on the density and strength of pellets than the height of the pelleting die.

This study showed that compared to the other samples studied here, the highest LCV was obtained from 100% SCG, due to the presence of a higher carbon content. The LCV of 100% SCG sample yielded about 21.08 MJ.kg^{-1} while the LCV of SCG/Sawdust (30/70) sample yielded 20 MJ.kg^{-1}. However, the use of pure SCG can lower boiler efficiency along with an increase in particle and gas emissions [1,15]. The SCG mixed with pine sawdust (50/50 wt%) could provide similar combustion parameters (emissions and boiler efficiency) to wood pellets and satisfy the NF agro-pellets standard (French standards), as stated by Atabani et al., 2019, Limousy et al., 2013 and Jeguirim, 2014 [1,13,17].

The study conducted by Kristanto and Wijaya [9] on the determination of the characteristics of SCG and CS and the effect of mixing SCG and CS on the quality of pellet fuel showed that the physical characteristics of CS were suitable for producing pellets with better density, durability, and combustion levels, but large amounts of CS in pellets potentially produces particulate and NOx emissions during combustion. Based on the German pellet standard DIN 51731, the sample with 75% SCG, 20% CS, and 5% artificial adhesive has the highest durability and optimum water content. This kind of sample had a constant flame and a decrease in temperature during the combustion process. The decrease in temperature caused by mixing CS with SCG was relatively constant because the pellet density affects combustion efficiency.

The aim of this research was to verify the potential of the coffee ground as an alternative fuel. The samples from 100% SCG proved the highest value of LCV, reduced durability, and therefore the results show low heat-power of the boiler and high CO emissions. The high concentration of carbon monoxide can be a result of incomplete combustion and a carbon content higher than 50% of SCG. It has been found that the combustion of pure SCG decreases boiler efficiency and increases the concentrations of emissions, especially in the case of carbon monoxide [13]. Moreover, CO emission during combustion of 100% SCG (2248 mg.Nm^{-3} at 10% O$_2$) is lower than those obtained by Limousy et al. during pure SCG combustion (3069 mg.Nm^{-3} at 10% O$_2$) in a 12 kW boiler (Pellematic PES12—PVB 2000) [13]. Kang et al. combusted dried spent coffee ground and the results also confirmed the presence of the highest concentration of CO among the measured emissions [3]. Pilusa et al. reported that carbon monoxide from the combustion of already mentioned eco-fuel briquettes dominated in the flue gas [12]. Besides, CO is the most abundant gas in the flue gas and it is necessary to cut this emission. The 30/70 (spent coffee grounds/sawdust) sample appears to be the most ideal when compared to other analyzed samples, taking into account the measured performance of the boiler and emissions. For practical use of coffee grounds, it would be necessary to produce pellets with higher quality pelletizing machines in order to improve their strength and durability [18–20]. These parameters affect the reported emissions and heat power of the boiler. The combustion of SCG could be one of the options for utilization of coffee waste, based on its high calorific value [21], but according to the Paris Agreement, one of the key targets for 2030 is to reduce greenhouse gases and therefore, there is a need to find an environmentally acceptable solution.

Author Contributions: R.N.; experimental methodology, investigation, data collection, writing—original draft, M.M.T.; preparation of the testing equipment, formal analysis, writing and editing, D.J.; analysis of the data and validation. All authors have read and agree to the published version of the manuscript.

Funding: This research was funded by KEGA No. 038ŽU-4/2019 "Piping systems in heat supply" and VEGA No. 1/0479/19 "Influence of combustion conditions on the production of solid pollutants in small heat sources" and the project: CZ.02.1.01/0.0/0.0/ 16_019/0000753.

Conflicts of Interest: The funders had no role in the design of the study; in the collection, analyses, or interpretation of data; in the writing of the manuscript, or in the decision to publish the results.

References

1. Atabani, A.E.; Ala'a, H.; Kumar, G.; Saratale, G.D.; Aslam, M.; Khan, H.A.; Said, Z.; Mahmoud, E. Valorization of spent coffee grounds into biofuels and value-added products: Pathway towards integrated bio-refinery. *Fuel* **2019**, *254*, 115640. [CrossRef]
2. Tun, M.M.; Juchelková, D.; Raclavská, H.; Sassmanová, V. Utilization of biodegradable wastes as a clean energy source in the developing countries: A case study in Myanmar. *Energies* **2018**, *11*, 3183. [CrossRef]
3. Kang, S.B.; Oh, H.Y.; Kim, J.J.; Choi, K.S. Characteristics of spent coffee ground as a fuel and combustion test in a small boiler (6.5 kW). *Renew. Energy* **2017**, *113*, 1208–1214. [CrossRef]
4. Karmee, S.K. A spent coffee grounds based biorefinery for the production of biofuels, biopolymers, antioxidants and biocomposites. *Waste Manag.* **2018**, *72*, 240–254. [CrossRef] [PubMed]
5. Zuorro, A.; Lavecchia, R. Spent coffee grounds as a valuable source of phenolic compounds and bioenergy. *J. Clean. Prod.* **2012**, *34*, 49–56. [CrossRef]
6. Burniol-Figols, A.; Cenian, K.; Skiadas, I.V.; Gavala, H.N. Integration of chlorogenic acid recovery and bioethanol production from spent coffee grounds. *Biochem. Eng. J.* **2016**, *116*, 54–64. [CrossRef]
7. Kwon, E.E.; Yi, H.; Jeon, Y.J. Sequential co-production of biodiesel and bioethanol with spent coffee grounds. *Bioresour. Technol.* **2013**, *136*, 475–480. [CrossRef] [PubMed]
8. Somnuk, K.; Eawlex, P.; Prateepchaikul, G. Optimization of coffee oil extraction from spent coffee grounds using four solvents and prototype-scale extraction using circulation process. *Agric. Nat. Resour.* **2017**, *51*, 181–189. [CrossRef]
9. Kristanto, G.A.; Wijaya, H. Assessment of spent coffee ground (SCG) and coffee silverskin (CS) as refuse derived fuel (RDF). *IOP Conf. Ser. Earth Environ. Sci.* **2018**, *195*, 012056. [CrossRef]
10. Kansai, N.; Chaisuwan, N.; Supakata, N. Carbonized Briquettes as a Tool for Adding Value to Waste from Rain tree (Samanea Saman) and Coffee Ground/Tea Waste. *Eng. J.* **2018**, *22*, 47–63. [CrossRef]
11. Patcharee, P.; Naruephat, T. A Study on How to Utilize Waste Paper and Coffee Residue for Briquettes Production. *Int. J. Environ. Sci. Dev.* **2015**, *6*, 201. [CrossRef]
12. Pilusa, T.J.; Huberts, R.; Muzenda, E. Emissions analysis from combustion of eco-fuel briquettes for domestic applications. *J. Energy South. Afr.* **2013**, *24*, 30–36. [CrossRef]
13. Limousy, L.; Jeguirim, M.; Dutournie, P.; Kraiem, N.; Lajili, M.; Said, R. Gaseous products and particulate matter emissions of biomass residential boiler fired with spent coffee grounds pellets. *Fuel* **2013**, *107*, 323–329. [CrossRef]
14. Arce, M.E.; Saavedra, Á.; Míguez, J.L.; Granada, E.; Cacabelos, A. Biomass fuel and combustion conditions selection in a fixed bed combustor. *Energies* **2013**, *6*, 5973–5989. [CrossRef]
15. Moradian, F.; Pettersson, A.; Svärd, S.H.; Richards, T. Co-combustion of animal waste in a commercial waste-to-energy BFB boiler. *Energies* **2013**, *6*, 6170–6187. [CrossRef]
16. Lisowski, A.; Olendzki, D.; Swietochowski, A.; Dabrowska, M.; Mieszkalski, L.; Ostrowska-Ligeza, E.; Stasiak, M.; Klonowski, J.; Piatek, M. Spent coffee grounds compaction process: Its effects on the strength properties of biofuel pellets. *Renew. Energy* **2019**, *142*, 173–183. [CrossRef]

17. Jeguirim, M.; Limousy, L.; Dutournie, P. Pyrolysis kinetics and physicochemical properties of agropellets produced from spent ground coffee blended with conventional biomass. *Chem. Eng. Res. Des.* **2014**, *92*, 1876–1882. [CrossRef]
18. Kantová, N.; Holubčík, M.; Jandačka, J.; Čaja, A. Comparison of Particulate Matters Properties from Combustion of Wood Biomass and Brown Coal. *Procedia Eng.* **2017**, *192*, 416–420. [CrossRef]
19. Palacka, M.; Vician, P.; Holubčík, M.; Jandačka, J. The Energy Characteristics of Different Parts of the Tree. *Procedia Eng.* **2017**, *192*, 654–658. [CrossRef]
20. Recycling Coffee Grounds into Biomass Pellets for Home Heating. Available online: https://wood-pellet-line.com/recycling-coffee-grounds-into-biomass-pellets-for-home-heating/ (accessed on 2 February 2020).
21. Sołowiej, P.; Neugebauer, M. Impact of coffee grounds addition on the calorific value of the selected biological materials. *Agric. Eng.* **2018**, *20*, 177–183. [CrossRef]

© 2020 by the authors. Licensee MDPI, Basel, Switzerland. This article is an open access article distributed under the terms and conditions of the Creative Commons Attribution (CC BY) license (http://creativecommons.org/licenses/by/4.0/).

Article

Finite Element Modeling of Geothermal Source of Heat Pump in Long-Term Operation †

Elżbieta Hałaj *, Leszek Pająk and Bartosz Papiernik

AGH University of Science and Technology, Faculty of Geology, Geophysics and Environmental Protection, 30-059 Krakow, Poland; pajakl@agh.edu.pl (L.P.); papiern@geol.agh.edu.pl (B.P.)
* Correspondence: ehalaj@agh.edu.pl
† This paper is an extended version of article presented at the 6th Scientific Conference Renewable Energy Systems, Engineering, Technology, Innovation ICoRES 2019, to be published in E3S Web of Conferences.

Received: 31 January 2020; Accepted: 11 March 2020; Published: 13 March 2020

Abstract: Model simulation allows to present the time-varying temperature distribution of the ground source for heat pumps. A system of 25 double U-shape borehole heat exchangers (BHEs) in long-term operation and three scenarios were created. In these scenarios, the difference between balanced and non-balanced energy load was considered as well as the influence of the hydrogeological factors on the temperature of the ground source. The aim of the study was to compare different thermal regimes of BHEs operation and examine the influence of small-scale and short-time thermal energy storage on ground source thermal balance. To present the performance of the system according to geological and hydrogeological factors, a Feflow® software (MIKE Powered by DHI Software) was used. The temperature for the scenarios was visualized after 10 and 30 years of the system's operation. In this paper, a case is presented in which waste thermal energy from space cooling applications during summer months was used to upgrade thermal performance of the ground (geothermal) source of a heat pump. The study shows differences in the temperature in the ground around different Borehole Heat Exchangers. The cold plume from the not-balanced energy scenario is the most developed and might influence the future installations in the vicinity. Moreover, seasonal storage can partially overcome the negative influence of the travel of a cold plume. The most exposed to freezing were BHEs located in the core of the cold plumes. Moreover, the influence of the groundwater flow on the thermal recovery of the several BHEs is visible. The proper energy load of the geothermal source heat pump installation is crucial and it can benefit from small-scale storage. After 30 years of operation, the minimum average temperature at 50 m depth in the system with waste heat from space cooling was 2.1 °C higher than in the system without storage and 1.6 °C higher than in the layered model in which storage was not applied.

Keywords: finite element modeling; thermal imbalance of the ground source; small-scale thermal energy storage; FEFLOW; ground source heat pump; Borehole Heat Exchanger (BHE)

1. Introduction

Increasing climate change awareness and the popularity of renewable energy sources are connected with the increasing attention to geothermal energy and devices such as heat pumps. According to [1], in 2050, electric heat pumps will become more common in most parts of the world. In 2017, there were 10,572,395 heat pumps units in operation in Europe [2]. This number is growing constantly. The main advantage of a heat pump solution is the possibility of providing not only heating during cold periods, but also space cooling during summer hot days. Ground source heat pumps (GSHP) are a popular solution, especially for bigger installations. Several regional studies of the shallow geothermal potential of heat pumps in different geological conditions have been conducted, for example, in Japan [3],

Germany [4], and Finland [5]. The efficiency of the system and the economic advantage should be estimated for each specific case [6].

The increase in popularity of geothermal source heat pumps may create future problems regarding the space available for new ground sources and the negative influence of neighbouring heat exchangers and their arrays on each other [7]. Therefore, evaluating of the thermal impact is crucial for high-density Ground Source Heat Pumps (GSHP) [8].

Knowing hydrogeological conditions and the performance of the ground source system is very important for its exploitation [9]. Thermal parameters and ground water flow can have a significant effect on the development of borehole heat exchangers systems [10] and the temperature recovery rate [11]. In cases in which the layout is sufficiently constructed, the ground water flow can improve the long-term operation conditions [12]. The influence of groundwater flow on storage systems has also been investigated [13].

Several studies considered the performance of the ground sources at different times of operation, —for example [14]—during 15 years and the optimization of borehole heat exchangers (BHE) for both deep [15] and shallow units [16]. Typically, the ground source heat pump itself is considered to work for about 20 years [17]. However, the heat source may operate longer, for 30 or even 50 years. Therefore, it is crucial to conduct long-term modelling [18] and measurements of such systems [19]. Ground temperature changes depend on the annual heat imbalance between injected and extracted heat rate [20]. According to [21], the soil's thermal imbalance provides problems like soil temperature decrease, heating performance deterioration and a decline of the heating reliability. Moreover, there is a risk of the influence of the remaining thermal imbalance on new installations. There is also a need to provide more efficient heating/cooling applications. In this context, a seasonal thermal energy storage is gaining attention. Several applications of thermal energy storage implement Borehole Thermal Energy Storage (BTES) [22], often associated with solar [23] or waste heat [24] or developed for industrial projects [25]. These research projects regard mostly larger installations.

This paper presents a case in which waste thermal energy from space cooling applications during the summer months was used to upgrade the thermal performance of the ground source. Thirty years of operation of the system was used for the simulation.

The aim of the study was to compare the different thermal regimes of BHEs' operation and examine the influence of small-scale and short-time thermal energy storage on ground source thermal balance. In the study, three-dimensional modelling of 25 borehole heat exchangers (ground source) was carried out in order to determine the performance of the system in long-term operation. It is assumed that ground temperature partially regenerates during the summer months due to space cooling in central Poland conditions.

The simulation allows to show the temperature distribution in the ground source for the heat pump. The temperature distribution and the cold plume are affected by the ground water flow and its nature as well as the balancing of the energy load. The central Poland study area was chosen because of recent intensive deep geothermal research in this area [26]. An added value of this research is that it provides another option to compare different geothermal applications in the area.

2. Materials and Methods

This study presents a hypothetical system of a ground source for heat pumps implemented for central Poland conditions. To present the performance of the system according to its geological and hydrogeological factors, a Feflow® software (MIKE Powered by DHI Software, Hørsholm, Denmark) was used. The software provides Finite Element Modeling of Flow, Mass and Heat Transport in Porous and Fractured Media. The nodes belong to the boreholes were coupled to the rest of the modelled area [11]. Borehole heat exchangers (BHEs) were modeled according to [27]. A BHE is considered as a specific fourth type of boundary condition (BC) [28] which applies a predefined extraction or injection of thermal energy to a model [29].

2.1. The Model

A non-uniform grid of the area was used. In the vicinity of wells, grid points were refined to obtain a better accuracy (Figure 1). Optimal mesh refinement for nodes around BHEs were calculated according to the Nillert method [27]. The total model dimensions were set to 240 × 240 × 105 m. The mesh consists of 169,477 elements and 92,940 nodes.

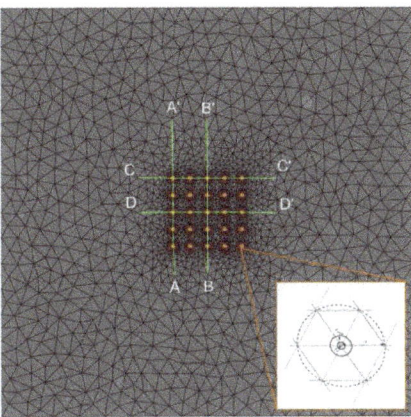

Figure 1. Non-uniform grid model of the study area and 25 borehole heat exchangers (2D), cross-section lines (green). Spatial discretization of BHE nodes (in frame) according to [27].

To minimize the influence on the result and calculation time, the size of the model domain was chosen after pre-calculations. Moreover, the second type of boundary conditions were applied to the model's northern and southern edges. A larger domain size does not affect the results. All boundary conditions applied to the models' domain are given in Figure 2. There were 3 types of boundary conditions applied in this model: the 1st boundary condition (Dirichlet) is temperature applied at the top surface; the 2nd boundary condition (Neumann) is heat-flux at the bottom surface and fluid-flux applied at the edges or in the aquifer elements (in the 3rd scenario); the 4th boundary conditions are borehole heat exchangers.

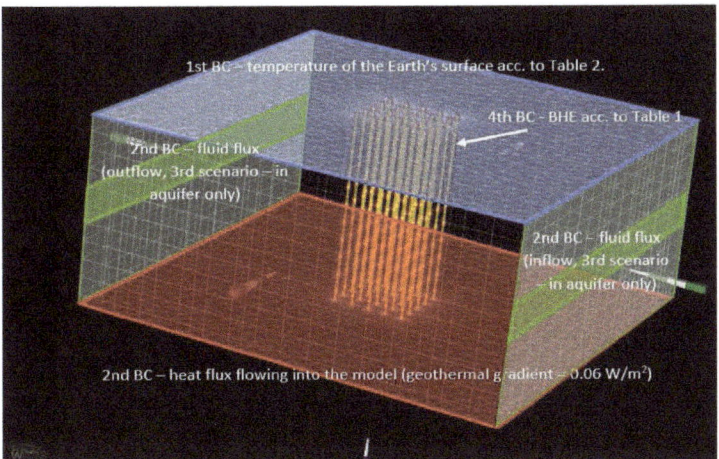

Figure 2. The scheme of boundary conditions applied to the model.

The multiple system considered consists of 25 borehole heat exchangers (BHEs). They are in 5 × 5 layout with 10-m spacing. The spacing was chosen to fulfill the heat pump's branch recommendations [30]. The BHEs are not connected in arrays, all (equal) parameters are set for single BHE. Each BHE is 100 m in length, which equates to 2500 m of heat exchangers' total length. The exchangers were of the double U-shaped type, which consist of two inlet pipes and two outlet pipes surrounded by grout (Figure 3). A single BHE setting is provided in Table 1.

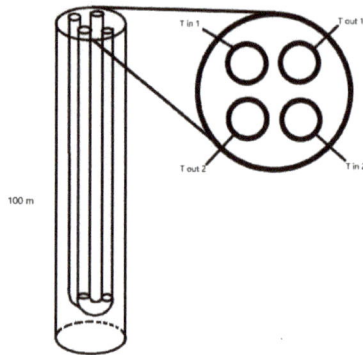

Figure 3. Double U-shaped pipe scheme.

Table 1. Borehole heat exchanger (BHE) settings.

BHE Geometry	Double U-Shape
Borehole diameter	0.15 m
Inlet Pipe diameter	0.032 m
Refrigerant volumetric heat capacity	4 MJ/m/s/K
Refrigerant thermal conductivity	0.48 J/m/s/K
Computational method	Eskilson & Claesson method
Power	Variable, acc. to monthly energy load (Table 2)
Flow rate	53 m^3/d

In the simulation, an Eskilson and Claesson [31] (quasi-stationary) method, which assumes the local thermal equilibrium between all the elements of BHE during the whole time of the simulation, was applied.

Table 2. Earth's average temperature in the study area [32] and energy load during a year.

Month	Earth's Surface Temperature [°C]	Average Energy per BHE [kWh]		
		1st scenario	2nd scenario	3rd scenario
I	−3.5	558	558	558
II	−2.2	504	504	504
III	2.3	558	558	558
IV	9.2	360	360	360
V	15.3	186	186	186
VI	18.3	0	−180 (cooling)	0
VII	20.9	0	−372 (cooling)	0
VIII	20.6	0	−186 (cooling)	0
IX	14.7	0	0	0
X	8.6	186	186	186
XI	1.8	540	540	540
XII	−2.5	558	558	558

The difference in thermal energy per time between inlet and outlet of BHE is variable. It was set according to energy demand on a given day. The BHEs did not work longer than 6 hours a day. Their power was variable, from 0 to 3 kW. Each simulation was run for 30 years (262,800 h).

2.2. Thermal and Hydrogeological Settings

The model is considered as fully saturated. The initial temperature condition across the model was set to 9 °C (referenced ground temperature in the central Poland area). To show the thermal regime, heat flux from the ground during the year was set to the top surface according to the change of the earth's surface monthly average temperature (Table 2). Earth's temperature for the central Poland area is provided. These values were set as the first type of boundary conditions. Moreover, a geothermal heat flux of 0.06 W/m^2 assessed for this area was applied to the base surface of the model.

The BHE system was operating according to the assessed monthly load, as shown in Table 2.

In order to compare different operating regimes, three scenarios were taken into consideration:

- 1st scenario: homogeneous parameters of the ground source, not balanced energy load (only heating in winter)
- 2nd scenario: homogeneous parameters of the ground source, partially balanced energy load (heating in winter and cooling in summer)
- 3rd scenario: parameters according to different layers of the model, not balanced energy (only heating in winter).

The hydrogeological and thermal parameters are shown in Table 3. The initial conditions of the models were obtain by applying the simulation in steady-state conditions first.

Table 3. Ground and water initial parameters.

Parameter	1st Scenario	2nd Scenario	3rd Scenario
Porosity	0.3	0.3	0.18–0.2; 0.3 in aquifer
Volumetric heat capacity of groundwater	4.2 10^6 J/m^3/K	4.2 10^6 J/m^3/K	4.2 10^6 J/m^3/K
Volumetric heat capacity of solid phase	2.52 10^6 J/m^3/K	2.52 10^6 J/m^3/K	2.52 10^6 J/m^3/K
Thermal conductivity of groundwater	0.65 W/m/K	0.65 W/m/K	0.65 W/m/K
Thermal conductivity of solid phase	2 W/m/K	2 W/m/K	according to layer (Figure 4)
Fluid flux	4000 mm/a	4000 mm/a	4000 mm/a in aquifer
Groundwater conductivity	2.5 10^{-4} m/s	2.5 10^{-4} m/s	1.0 10^{-4} m/s; 2.5 10^{-4} m/s in aquifer

The area was poorly recognized in the field of shallow geothermal energy and this study reflects some assumptions according to thermal and hydrogeological parameters in the area. The model is considered as the confined one. Aquifers have good flow conditions. The fluid flow direction is from the south to the north, as the fluid flux speed is constant, being limited by groundwater conductivity. Other borders of the model are defined as impermeable. The geological settings of the model are based on variability observed in boreholes located in the vicinity of Poddebice town, central Poland [33].

In the 1st and 2nd scenarios, the whole model had good porosity and permeability conditions and the groundwater flowed throughout the model. In the 3rd scenario, the permeability conditions worsened and the water flow was mostly limited to an aquifer of 10 m. The aquifer layer has better conditions of thermal and fluid conductivity (Figure 4).

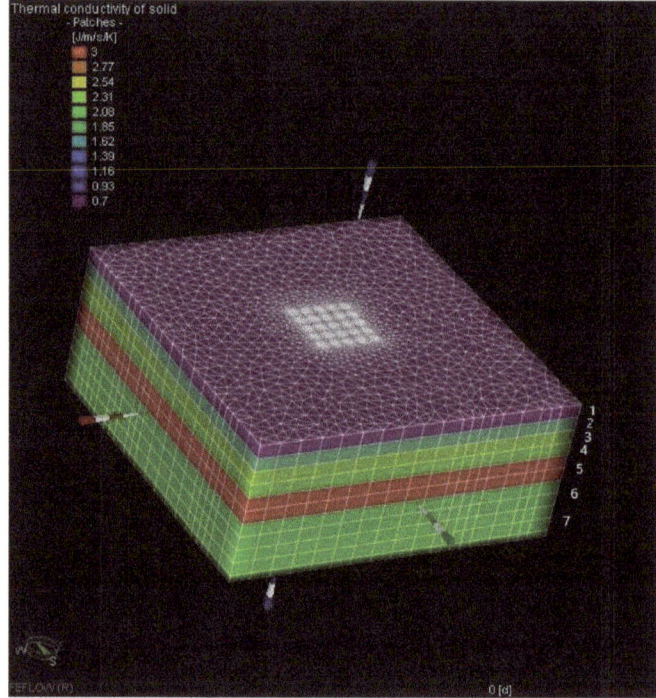

Figure 4. Thermal conductivity in the 3rd scenario model. The numbers refer to lithological layers (1: medium sands, 2: brown-gray marl, 3: brown-gray marly mudstone, 4: light gray marly mudstone, 5: sandstone, 6: beige mudstone, 7: marly claystone).

3. Results and Discussion

In this project, three sets of simulations were run to verify different scenarios for the performance of heat pump geothermal heat exchangers. For each simulations, two check-points were selected: after 10 and 30 years of the system operation, at 87,600 and 262,800 h, respectively.

3.1. First Scenario: Homogeneous Parameters of the Ground Source, Not Balanced Energy Load

According to the simulation, the temperature of the ground decreased with time. At the beginning stage, the drop of temperature was most visible in the centre between BHEs. The difference in temperature between the points of the minimum temperature and the surrounding domain after the first year of operation is ca. 4.5 °C. Later, the influence of the water flow was more visible in the model and since the second year of the system operation, the points with minimum temperature shifted towards the north. At the same time, different BHEs started to work on different ground parameters. In Figure 5, the temperature model is presented for a depth of 50 m (the mid-section of the BHE). After 10 years of system operation (A), an increase in the cold plume can be observed. The range of the model zone where the temperature dropped to 5 °C was ca. 45 m from the centre of the BHE system and ca. 24 m from the BHE on the edge. During this time, the minimum ground temperature dropped to 0.1 °C.

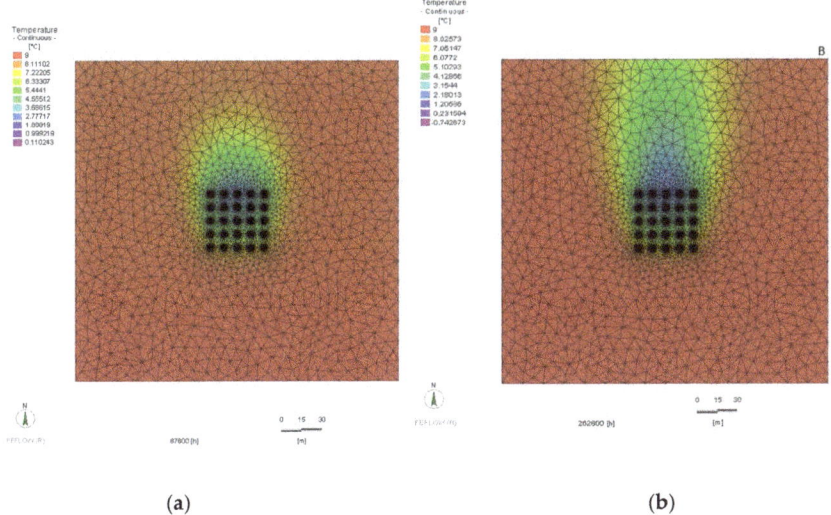

Figure 5. Results of the simulation of ground temperature after 10 years (**a**) and 30 years (**b**) of heat pump operation within the 1st scenario (homogeneous model). Two-dimensional view of the surface at a depth of 50 m.

After 30 years of system operation (B), the decrease of temperature was even more visible and the cold plume travelled further north due to the fluid flow conditions. The minimum average temperature at the same surface of 50 m depth was −0.7 °C, which implies the risk of ground freezing. Particularly, the refrigeration fluid temperature in several BHEs can even drop to −3.5 °C. After 30 years of system operation (B), the range of the model zone in which the temperature dropped to 5 °C was ca. 100 m from the centre of the BHE system and ca. 75 m from the BHE on the north edge. The distribution of temperature in 3D view is shown in Figure 6.

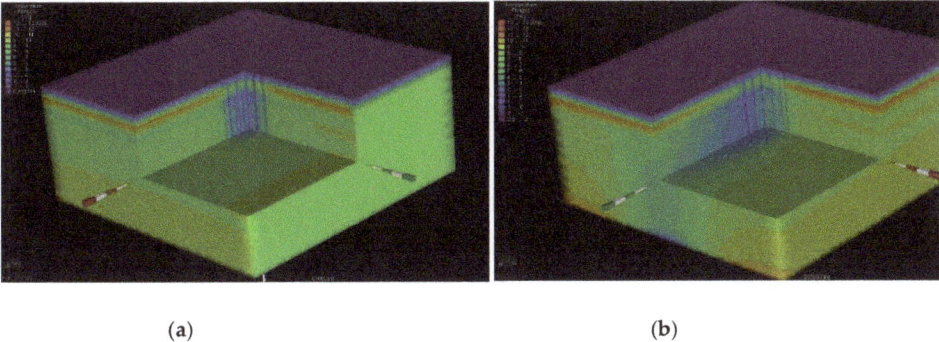

Figure 6. Results of 3D simulation of ground temperature after 10 years (**a**) and 30 years (**b**) of heat pump operation within the 1st scenario (homogeneous model); the view from the north-west.

In the most affected BHEs, the average working fluid's temperature decreased significantly during the first 15 years of operation. Then, the drop of the temperature was rather small as time passed (Figure 7). The difference between the warmest and the coldest BHE's average temperature in this layout is 4.5 °C after 30 years of operation.

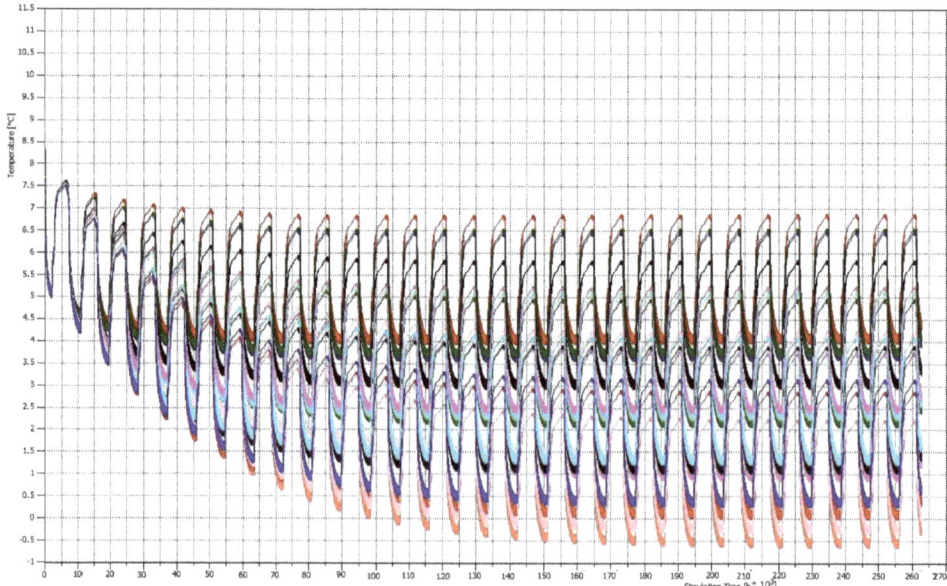

Figure 7. Average temperature of BHEs within the 1st scenario (system without space cooling) during 30 years of operation.

The ground water flow provides the differences in temperature. The temperature in some BHEs (in the centre and north) dropped due the cold plume travelling. However, water of a higher temperature moved from the south to the BHE system. The southern edges of BHE system, which recover faster, have a higher temperature. In fact, the average temperature in the southern BHEs was almost constant after the initial first few years. For example, the minimal average temperature of the BHE on the southern edge stayed at level of ca. 4 °C after the fifth year.

The vertical distribution of temperature is given in Figure 8. In cross-sections A-A' and B-B', it is clearly visible that the northern side of the BHE system has a significantly lower temperature than the southern one. The BHEs' had an influence on each other and there was a decrease in temperature in the central BHEs.

3.2. Second Scenario: Homogeneous Parameters of the Ground Source, Balanced Energy Load (Heating in Winter and Cooling in Summer)

A balanced energy load affects the drop in the system's temperature less. The average temperature of the system did not drop below 1 °C during the 30 years of operation.

At the beginning stage, the drop in temperature is the most visible in the centre, between BHEs. The difference in temperature between the points of the minimum temperature and the surrounding domain after the first year of operation is ca. 4.4 °C.

Starting from the second year, different BHEs started to work on different ground parameters due to the interaction between the BHEs and water flow. In Figure 9, the temperature model is presented for a depth of 50 m (the mid-section of the BHE). After 10 years of system operation (A), the zone with a temperature of 5 °C was concentrated near BHEs and was not present outside of the BHEs layout.

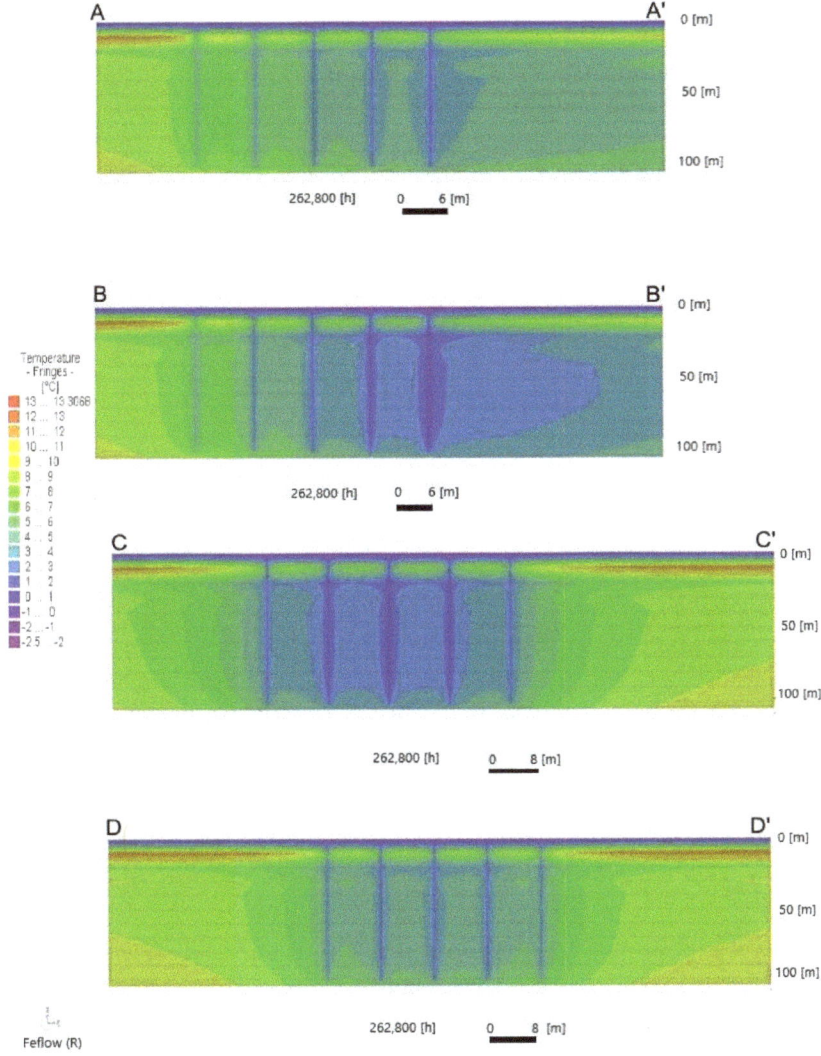

Figure 8. The temperature distribution as cross-section views for the 1st scenario. The cross-section lines are the same as those in Figure 1.

The distribution and layout of the temperature in the 3D model is provided in Figure 9. After 10 years of operation, the minimum ground temperature dropped to 1.9 °C at a 50-m depth.

After 30 years of system operation (Figure 10), despite the partly balanced energy load, there is a cold plume observed to be travelling north, according to the fluid flow direction. However, its shape is more slender than those observed in the first scenario. The range of the model zone in which the temperature dropped to 5 °C is ca. 50 m from the centre of the BHE system and ca. 30 m from the BHE on the north edge. The minimum average temperature at the surface of a 50 m depth was 1.4 °C, which dismiss the risk of ground freezing.

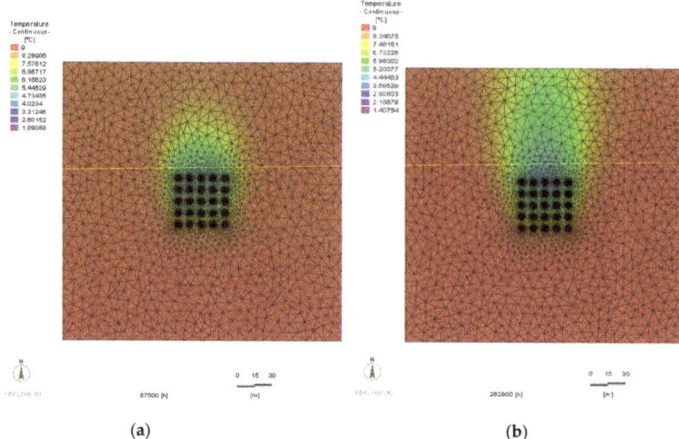

Figure 9. Results of the simulation of ground temperature after 10 years (**a**) and 30 years (**b**) of heat pump operation within the 2nd scenario (homogeneous model). Two-dimensional view of the surface at a depth of 50 m.

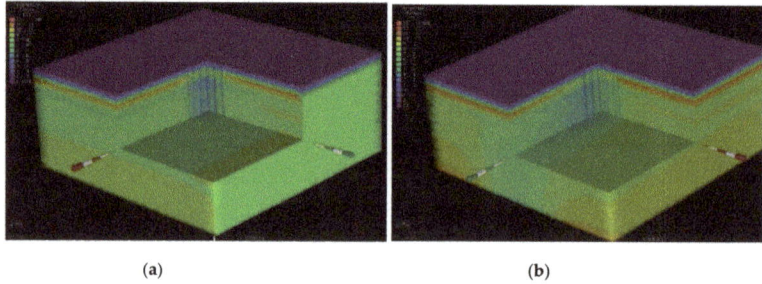

Figure 10. Results of the 3D simulation of ground temperature after 10 years (**a**) and 30 years (**b**) of heat pump operation within the 2nd scenario (homogeneous model); the view from the north-west.

In the most affected BHEs, the temperature decreased significantly during the first 15 years of operation. Then, the drop in temperature was rather small over time (Figure 11). The difference between the warmest and the coldest BHE's average temperature was ca. 3.1 °C after 30 years of operation.

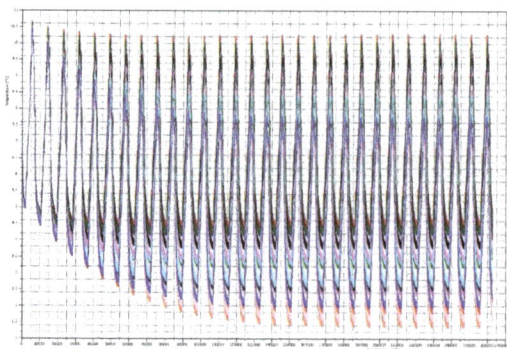

Figure 11. Average temperature of BHEs within the 2nd scenario (system with space cooling) during 30 years of operation.

The vertical distribution of the temperature is given in Figure 12. In cross-sections A-A′ and B-B′, smaller difference between the northern and southern site can be observed compared to the first scenario. The influence of the system into the neighboring ground is lower than in the first scenario (C-C′, D-D′). As the groundwater flow was set as homogenous, the BHEs in the centre were the most affected by the heat withdrawal.

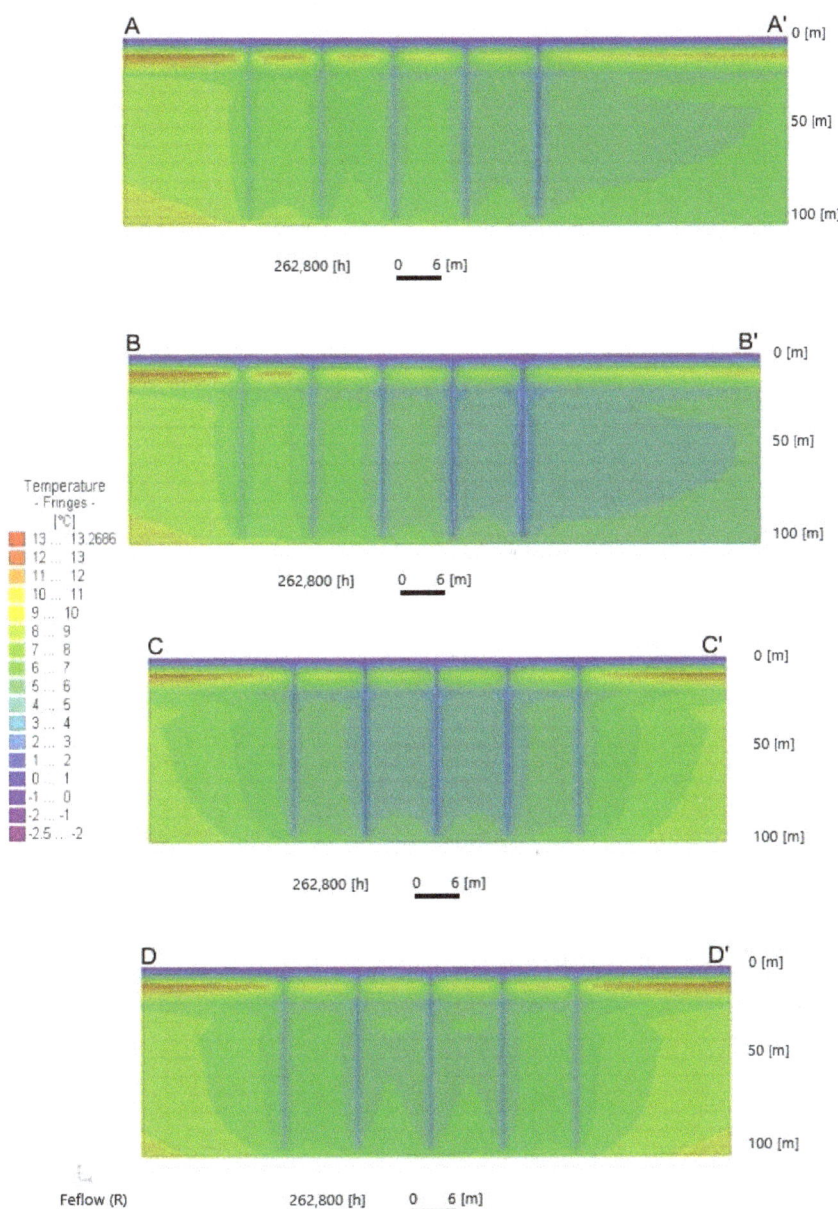

Figure 12. The temperature distribution as cross-section views for the 2nd scenario. The cross-section lines are the same as those used in Figure 1.

3.3. Third Scenario (Layered Model): Parameters According to Different Layers of the Model, Not Balanced Energy (Only Heating in Winter)

In Figure 13, the temperature of the ground source is shown after 10 and 30 years of the simulation at a 50-m depth. After 10 years of system operation, the temperature at a depth of 50 m was 2.7 °C. The cold plume is rather round-shaped and did not shift significantly towards the north (it was not affected by the ground water flow).

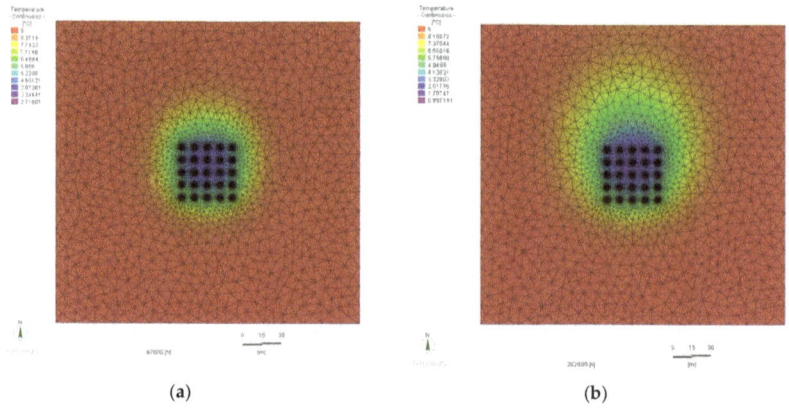

Figure 13. Results of the simulation of ground temperature after 10 years (**a**) and 30 years (**b**) of heat pump operation within the 3rd scenario (layered model). Two-dimensional view of the surface at a depth of 50 m.

After 30 years of operation, the cold plume travelled slightly north and was slightly more visible in the permeable layers. At a depth of 50 m, the range of the model zone in which the temperature dropped to 5 °C was ca. 45 m from the centre of the BHE system and ca. 25 m from the BHE on the northern edge. The most visible decrease of the temperature is observed inside the BHEs' system layout (Figure 14).

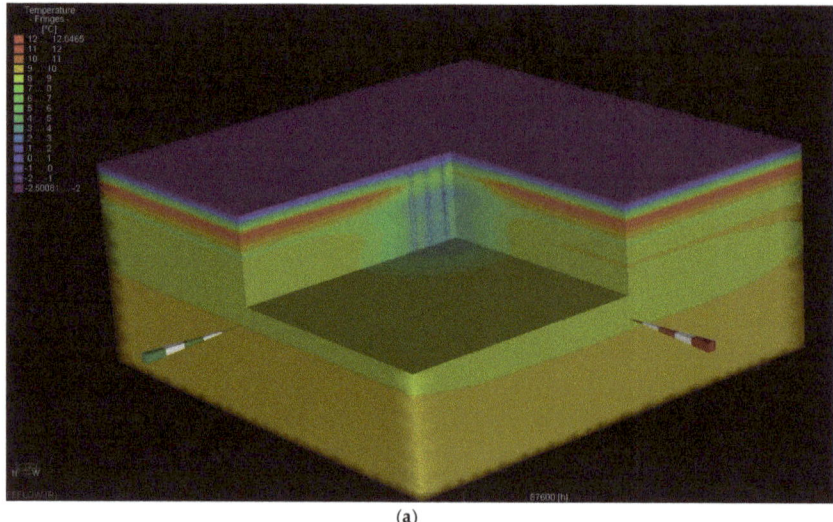

(**a**)

Figure 14. *Cont.*

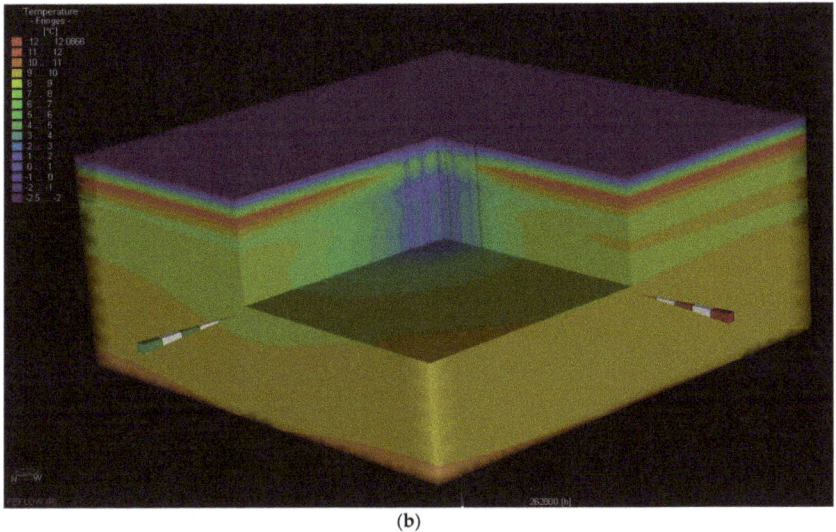

(b)

Figure 14. Results of 3D simulation of ground temperature after 10 years (**a**) and 30 years (**b**) of heat pump operation within the 3rd scenario (layered model); view from the north-west.

In the northern BHEs, there is a risk of ground freezing, while the ground temperature can drop to −0.2 °C (layers with the worst thermal conditions). At the same time, the average temperature of the refrigerant can be as low as −2 °C at the coldest time of the season (Figure 15). The figure shows the average temperature of BHE's working fluid temperature within the 3rd scenario (system without cooling) within a single year. This was for the last and 30th year considered in the simulation.

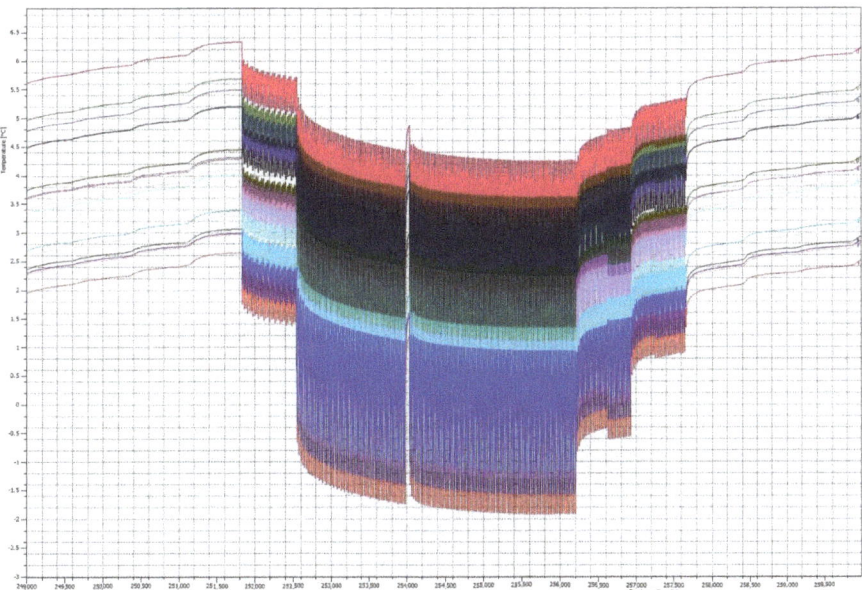

Figure 15. Average temperature of BHEs within the 3rd scenario (system without cooling) during the 30th year of operation.

Figure 16 depicts differences in the vertical distribution of temperature around BHEs. Some sections or layers interfered more strongly than the others due to the different hydrogeological and thermal parameters of the layers. The influence of the groundwater flow in the aquifer is less significant to the system.

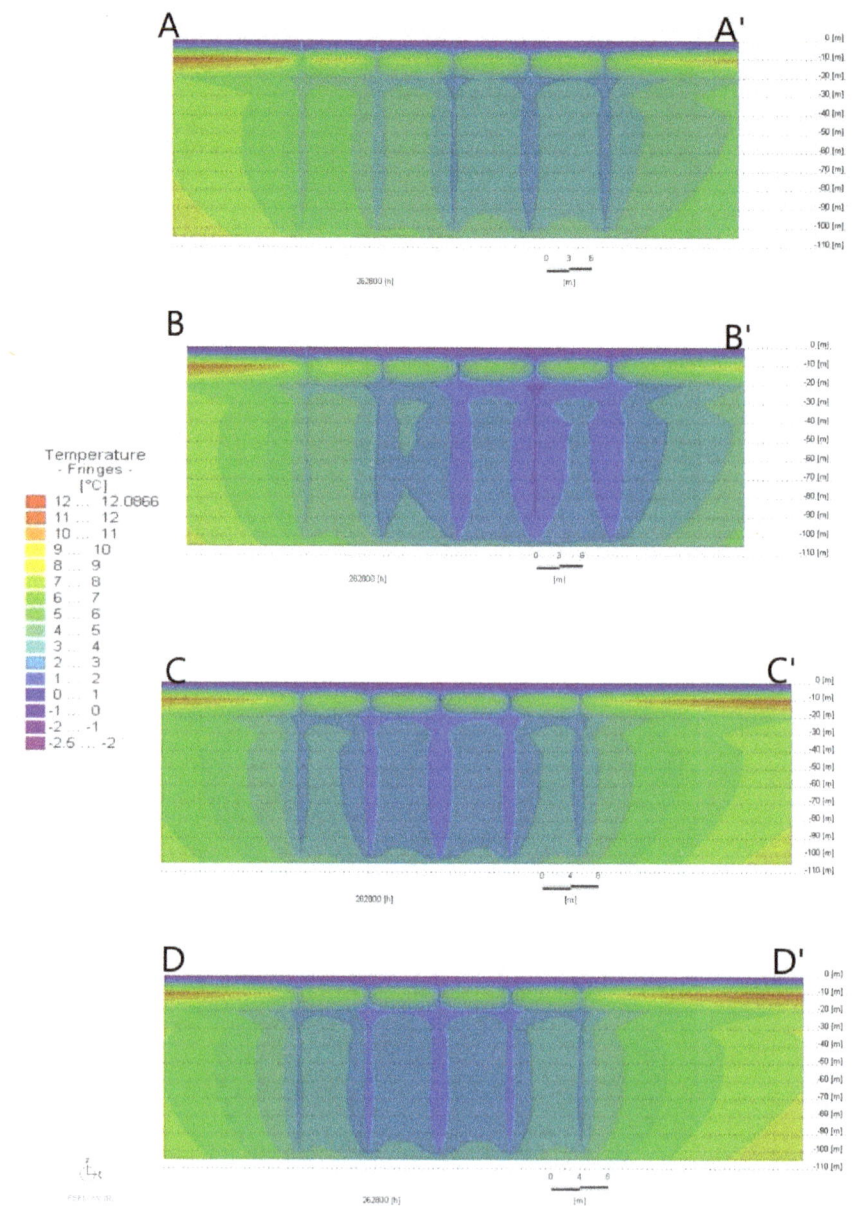

Figure 16. The temperature distribution as a cross-section view for the 3rd scenario. The cross-section lines are the same as those used in Figure 1.

Figure 17 shows the comparison between the cold plumes that appeared after 30 years of operation for the different scenarios. A temperature value of 5 °C was arbitrary chosen as the cold plume boundary. Normally, heat pumps can work on temperatures below 5 °C; however, lower temperatures are no clearly visible in these 3D models.

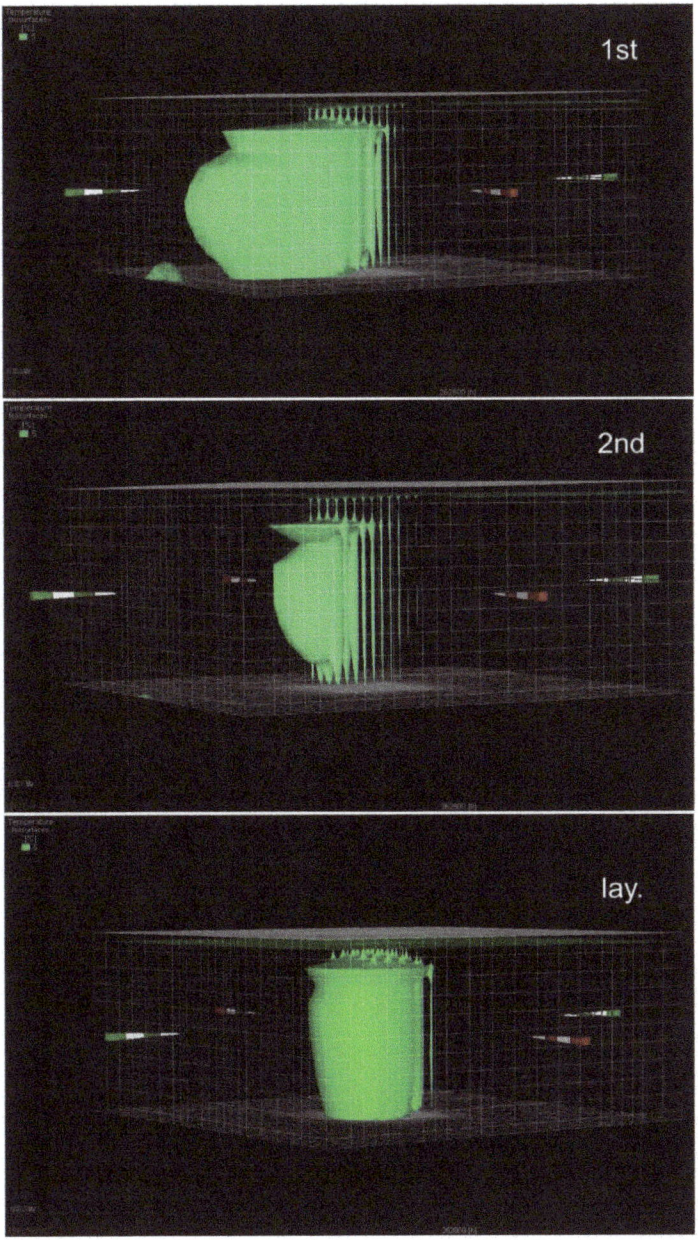

Figure 17. Cold plumes in 3D models for 3 different scenarios (1st scenario, 2nd scenario and layered model (3rd scenario)) after 30 years of system operation.

The cold plume from the 1st scenario is the most developed in length and width. It travelled further to the north according to the ground flow direction. This BHE system might threats future installations in the vicinity.

In the 2nd scenario, due to the balanced energy load and energy storage, the range of the cold plume is definitely smaller, even also directed to the north (groundwater flow). Moreover, thermal energy storage (understood here as cooling of space in summer) can partially overcome the negative influence of traveling of a cold plume.

The most exposed to freezing are BHEs from the north edge of the system located at the core of the plumes. In the 1st and 2nd scenario, the influence of the groundwater flow on the thermal recovery of the southern BHEs is clearly visible. They worked with a higher temperature regime the whole time.

In the 3rd scenario, where the best groundwater conditions are limited to the aquifer, the heat conductions have a greater influence on the system (more rounded shape of the plume). There was no significant thermal recovery in the south. The cold plume travelled much more slowly to the north (similarly to the 2nd scenario with a balanced energy load) and the cold plume shift was slight.

The model allows to simulate the time-varying temperature of the ground source for heat pumps. The temperature regime is important for proper heat pump operation. Even the ground source heat pumps can work with a temperature below 0 °C; this temperature is associated with the freezing of the ground and water in the BHE vicinity, which can have a negative impact on the BHE itself (cracks, air presence, etc.) [30]. It was observed that after 15 years, the system's temperature stabilized and the system's elements worked with similar parameters. However, after that long period of exploitation, there is a risk of ground freezing in several locations. It is important to determine how the cold plume location changes, how it propagates and how BHEs interfere with each other and with the cold plume.

In the layered model, the influence of the groundwater flow was limited to the aquifer and provided lower temperatures in the centre BHEs. The thermal parameters' diversity in each section of the BHE have a lower influence on the thermal layout of the whole system than the balanced energy load. This balance can be obtained by small-scale and short-time thermal energy storage, which requires the use of waste heat from space cooling during the summer months. In the system with small-scale energy storage, the zone of a temperature of 5 °C is two times smaller and the cold plume does not travel as far as in the system without storage.

The proper energy load of the geothermal source heat pump installation is crucial, and it can benefit from small-scale storage. After 30 years of operation, the minimum average temperature at a 50-m of depth in the system with waste heat from space cooling was 2.1 °C higher than in the system without storage and 1.6 °C higher than in the layered model where storage was not applied.

The groundwater flow influences the improvement of parameters of BHEs located in the south (in the model). At the same time, it shifts the cold plume to the north, which negatively influences the northern BHEs. BHEs that are the most affected by the cold plume or by interference with other BHEs can be found thanks to the simulation.

Balanced energy load (including e.g., summer cooling) can increase system efficiency by providing higher temperatures. The study shows differences in the temperature in the ground around BHEs. Therefore a proper layout of the BHE system can be adopted prior to the system construction.

4. Conclusions

Geological and hydrogeological conditions have a great impact on BHEs' system performance. Thus, good recognition of the parameters and suitable long-term simulation are necessary prior to installing a sustainable ground source for heat pumps. Installations with seasonal storage (if possible) are preferred for long-time operation due to their better ground thermal recovery which causes a higher effectiveness of heat pumps' installation and prevents ground freezing. FEM modelling of such installations would answer the question of whether a given heat pump and BHE's installation can work efficiently in specific conditions and locations of installation.

Author Contributions: Methodology E.H.; validation, E.H. and L.P.; writing—original draft preparation, E.H.; writing—review and editing, B.P. and L.P.; supervision, L.P. All authors have read and agreed to the published version of the manuscript.

Funding: This research was funded by AGH University of Science and Technology statutory work No. 16.16.140.315/05.

Acknowledgments: This research was supported by DHI by making the mentioned MIKE Powered by DHI Software available for the research. I would thank DHI for support. Although, they may not agree with all of the interpretations/conclusions of this paper.

Conflicts of Interest: The authors declare no conflict of interest. The funders had no role in the design of the study; in the collection, analyses, or interpretation of data; in the writing of the manuscript, or in the decision to publish the results.

References

1. IRENA. *Global Energy Transformation. A Roadmap to 2050*; International Renewable Energy Agency (IRENA): Abu Dhabi, UAE, 2018; ISBN 978-92-9260-059-4.
2. EHPA. *The European Heat Pump Market and Statistics Report 2018*; European Heat Pump Association: Brussels, Belgium, 2018.
3. Ishihara, T.; Shrestha, G.; Kaneko, S.; Uchida, Y. Analysis of Shallow Subsurface Geological Structures and Ground Effective Thermal Conductivity for the Evaluation of Ground-Source Heat Pump System Installation in the Aizu Basin, Northeast Japan. *Energies* **2018**, *11*, 2098. [CrossRef]
4. Agemar, T.; Schellschmidt, R.; Schulz, R. Subsurface temperature distribution in Germany. *Geothermics* **2012**, *44*, 65–77. [CrossRef]
5. Janiszewski, M.; Caballero Hernández, E.; Siren, T.; Uotinen, L.; Kukkonen, I.; Rinne, M. In Situ Experiment and Numerical Model Validation of a Borehole Heat Exchanger in Shallow Hard Crystalline Rock. *Energies* **2018**, *11*, 963. [CrossRef]
6. Christodoulides, P.; Aresti, L.; Florides, G. Air-conditioning of a typical house in moderate climates with Ground Source Heat Pumps and cost comparison with Air Source Heat Pumps. *Appl. Therm. Eng.* **2019**, *158*, 113722. [CrossRef]
7. Tolooiyan, A.; Hemmingway, P. A preliminary study of the effect of groundwater flow on the thermal front created by borehole heat exchangers. *Int. J. Low-Carbon Technol.* **2014**, *9*, 284–295. [CrossRef]
8. Meng, B.; Vienken, T.; Kolditz, O.; Shao, H. Evaluating the thermal impacts and sustainability of intensive shallow geothermal utilization on a neighborhood scale: Lessons learned from a case study. *Energy Convers. Manag.* **2019**, *199*, 111913. [CrossRef]
9. Li, Y.; Geng, S.; Han, X.; Zhang, H.; Peng, F. Performance Evaluation of Borehole Heat Exchanger in Multilayered Subsurface. *Sustainability* **2017**, *9*, 356. [CrossRef]
10. Li, M.; Mao, J.; Xing, Z.; Zhou, J.; Li, Y. Analysis of geo-temperature restoration performance under intermittent operation of borehole heat exchanger fields. *Sustainability* **2016**, *8*, 35. [CrossRef]
11. Emad Dehkordi, S.; Schincariol, R.A.; Olofsson, B. Impact of groundwater flow and energy load on multiple borehole heat exchangers. *Groundwater* **2015**, *53*, 4. [CrossRef]
12. Capozza, A.; De Carli, M.; Zarrella, A. Investigations on the influence of aquifers on the ground temperature in ground-source heat pump operation. *Appl. Energy* **2013**, *107*, 350–363. [CrossRef]
13. Mielke, P.; Bauer, D.; Homuth, S.; Götz, A.E.; Sass, I. Thermal effect of a borehole thermal energy store on the subsurface. *Geotherm. Energy* **2014**, *2*, 5. [CrossRef]
14. Chan Choi, J.; Park, J.; Rae Lee, S. Numerical evaluation of the effects of groundwater flow on borehole heat exchanger arrays. *Renew. Energy* **2013**, *52*, 230–240. [CrossRef]
15. Pan, S.; Kong, Y.; Chen, C.; Pang, Z.; Wang, J. Optimization of the utilization of deep borehole heat exchangers. *Geotherm. Energy* **2020**, *8*, 6. [CrossRef]
16. Kumar, S.; Murugesan, K. Optimization of geothermal interaction of a double U-tube borehole heat exchanger for space heating and cooling applications using Taguchi method and utility concept. *Geothermics* **2020**, *83*, 101723. [CrossRef]
17. Ma, W.; Kim, M.K.; Hao, J. Numerical Simulation Modeling of a GSHP and WSHP System for an Office Building in the Hot Summer and Cold Winter Region of China: A Case Study in Suzhou. *Sustainability* **2019**, *11*, 3282. [CrossRef]

18. Bae, S.M.; Nam, Y.; Shim, B.O. Feasibility Study of Ground Source Heat Pump System Considering Underground Thermal Properties. *Energies* **2018**, *11*, 1786. [CrossRef]
19. Bockelmann, F.; Fisch, M.N. It Works—Long-Term Performance Measurement and Optimization of Six Ground Source Heat Pump Systems in Germany. *Energies* **2019**, *12*, 4691. [CrossRef]
20. Capozza, A.; Zarrella, A.; De Carli, M. Long-term analysis of two GSHP systems using validated numerical models and proposals to optimize the operating parameters. *Energy Build.* **2015**, *93*, 50–64. [CrossRef]
21. You, T.; Wu, W.; Shi, W.; Wang, B.; Li, X. An overview of the problems and solutions of soil thermal imbalance of ground-coupled heat pumps in cold regions. *Appl. Energy* **2016**, *177*, 515–536. [CrossRef]
22. Janiszewski, M.; Siren, T.; Uotinen, L.; Oosterbaan, H.; Rinne, M. Effective Modelling of Borehole Solar Thermal Energy Storage Systems in High Latitudes. *Geomech. Eng.* **2018**, *16*, 503–512.
23. Naranjo-Mendoza, C.; Oyinlola, M.A.; Wright, A.J.; Greenough, R.M. Experimental study of a domestic solar-assisted ground source heat pump with seasonal underground thermal energy storage through shallow boreholes. *Appl. Therm. Eng.* **2019**, *162*, 114218. [CrossRef]
24. Skarphagen, H.; Banks, D.; Frengstad, B.S.; Gether, H. Design Considerations for Borehole Thermal Energy Storage (BTES): A Review with Emphasis on Convective Heat Transfer. *Geofluids* **2019**, 4961781. [CrossRef]
25. Nilsson, E.; Rohdin, P. Performance evaluation of an industrial borehole thermal energy storage (BTES) project e Experiences from the first seven years of operation. *Renew. Energy* **2019**, *143*, 1022–1034. [CrossRef]
26. Sowiżdżał, A.; Hajto, M.; Hałaj, E. Thermal waters of central Poland: A case study from Mogilno–Łódź Trough, Poland. *Environ. Earth Sci.* **2020**, *79*, 112. [CrossRef]
27. Diersch, H.J.G. *FEFLOW*; Springer: Berlin/Heidelberg, Germany, 2014.
28. Diersch, H.J.G.; Bauer, D.; Heidemann, W.; Ruhaak, W.; Schatzl, P. *Finite Element Formulation for Borehole Heat Exchangers in Modeling Geothermal Heating Systems by FEFLOW*; White Papers; DHI-WASY GmbH: Berlin, Germany, 2010; Volume V.
29. Diersch, H.J.G.; Bauer, D. Analysis, modeling and simulation of underground thermal energy storage (UTES) systems. In *Advances in Thermal Energy Storage System*; Elsevier Ltd.: Amsterdam, The Netherlands, 2015.
30. Lachman, P. *Guidelines for Projecting and Constructing Heat Pumps Installations*; PORT PC: Krakow, Poland, 2013. (In Polish)
31. Eskilson, P.; Claesson, J. Simulation model for thermally interacting heat extraction boreholes. *Numer. Heat Transf.* **1988**, *13*, 149–165. [CrossRef]
32. NASA. *NASA's Satellite Provided for Use with the RETScreen Software via the NASA Prediction of Worldwide Energy Resource (POWER) Project*; NASA: Washington, DC, USA, 2012.
33. Central Geological Database–Portal. Available online: http://geoportal.pgi.gov.pl/ (accessed on 28 February 2020).

© 2020 by the authors. Licensee MDPI, Basel, Switzerland. This article is an open access article distributed under the terms and conditions of the Creative Commons Attribution (CC BY) license (http://creativecommons.org/licenses/by/4.0/).

Article

Influence of Raw Material Drying Temperature on the Scots Pine (*Pinus sylvestris* L.) Biomass Agglomeration Process—A Preliminary Study

Marek Wróbel *, Marcin Jewiarz, Krzysztof Mudryk and Adrian Knapczyk

Department of Mechanical Engineering and Agrophysics, University of Agriculture in Krakow, Balicka 120, 30-149 Kraków, Poland; marcin.jewiarz@urk.edu.pl (M.J.); krzysztof.mudryk@urk.edu.pl (K.M.); adrian.knapczyk@urk.edu.pl (A.K.)
* Correspondence: marek.wrobel@urk.edu.pl; Tel.: +48-12-6624645

Received: 14 February 2020; Accepted: 26 March 2020; Published: 9 April 2020

Abstract: For biomass compaction, it is important to determine all aspects of the process that will affect the quality of pellets and briquettes. The low bulk density of biomass leads to many problems in transportation and storage, necessitating the use of a compaction process to ensure a solid density of at least 1000 kg·m^{-3} and bulk density of at least 600 kg·m^{-3}. These parameters should be achieved at a relatively low compaction pressure that can be achieved through the proper preparation of the raw material. As the compaction process includes a drying stage, the aim of this work is to determine the influence of the drying temperature of pine biomass in the range of 60–140 °C on the compaction process. To determine whether this effect is compensated by the moisture, compaction was carried out on the material in a dry state and on the materials with moisture contents of 5% and 10% and for compacting pressures in the 130.8–457.8 MPa range. It was shown that drying temperature affects the specific density and mechanical durability of the pellets obtained from the raw material in the dry state, while an increase in the moisture content of the raw material neutralizes this effect.

Keywords: drying; compaction; biomass; pellets; mechanical durability; specific density; Scots pine

1. Introduction

Cell walls are the main structural elements of plant biomass. The composition and structure of cell walls vary depending on the plant species [1], cell function and response to environmental conditions [2]. However, from a chemical point of view, the cell walls of most plants consist of three main components, namely cellulose, hemicellulose and lignin. The biomass of this composition is called lignocellulose biomass [3–5]. Additionally, this natural composite may also contain extractives and a mineral fraction. The most common extractives are simple sugars, fats, waxes, proteins, phenols, gums, terpenes, saponins, pectin, resins, fatty acids and essential oils that mainly fulfill a protective function for the plants [6].

Chemically, lignocellulose biomass contains mainly about 50% of carbon and hydrogen—on average 8% [7,8]. This is why it is mainly used as a biofuel. Besides this, there are investigations into the use of biomass as a raw material for hydrogen production from biomass gasification [9–12].

From the point of view of energy use of biomass, most of the energy accumulated in a given biomass is found in cellulose, hemicellulose and lignin. An elementary analysis of several types of biomass has shown [7] that the differences between cellulose and hemicellulose are insignificant, so that their energy parameters are similar. The cellulosic heat of combustion is 17.6 MJ·kg^{-1} and that for hemicellulose is 17.9 MJ·kg^{-1}, whereas lignin has higher energy parameter values. The lignin heat of combustion depends on the type of biomass from which it was extracted. For example, the heats of combustion for lignins from hazelnut shells, olive pomace, walnut shells, spruce and beech wood are in the 27–28 MJ·kg^{-1} range [13,14]. Additionally, much higher heats of combustion of 35 MJ·kg^{-1} are obtained for the extractives [6]. Thus, the biomass heat of combustion depends mainly on the relative contents of cellulose and hemicellulose on the one hand and lignin on the other hand. While the highest heat of combustion is found for extractives, particularly those containing large amounts of resins and terpenes, their small content in biomass (not exceeding 9.5%) [15–17] does not give rise to a significant increase in the heat of the combustion value of the raw material. The highest concentration of the extractives is found in the wood of coniferous plants, making the heat of the combustion values of these materials slightly higher compared to the wood of deciduous plants or herbaceous biomass [15,18,19]. Nevertheless, the main factor determining the high value of heat of combustion is the proportion of lignin in relation to holocellulose. It has been shown that, for certain biomass groups, this relationship is linear [7,17] (Figure 1). Therefore, despite the different levels of lignin, cellulose and hemicellulose in different types of biomass [15,20–23], their heat of combustion values lie within a narrow range of 18–22 MJ·kg^{-1} [24–27].

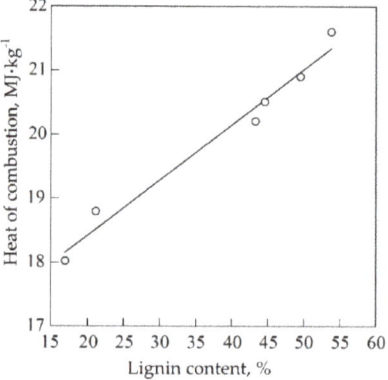

Figure 1. Heat of combustion of dry, ash-free and extractive-free biomass of different plant species seed coatings as a function of the lignin content [17].

This relatively energy-homogeneous raw material has a wide range of sources such as coniferous and deciduous wood, grasses, dicotyledonous perennials, shells, and stones, and forms such as log wood, saw dust, chips, straw, and pomace. Therefore, biomass materials exhibit a wide range of specific and bulk density values. Most typical biomass raw materials used for pellet production have low specific densities in the approximately 200–450 kg·m^{-3} range for the herbaceous type of biomass materials, and in the 350–720 kg·m^{-3} range for the woody type of biomass materials [3,28,29]. According to EN ISO 17225-1:2014 standard [30], the main types of biomass for solid biofuel production are woody, herbaceous and fruity biomass (less common is so-called water biomass). This means that a very wide range of species are already used for this purpose and the range of the different plant species is being continuously extended as a result of research conducted at many scientific institutions worldwide [27,31–39].

This variability of biomass materials gives rise to many inconveniences related to the logistics and combustion of biomass. To fully exploit the energy potential of biomass and avoid these problems, biomass materials are densified, obtaining a higher energy density, and lower transportation and storage costs. Standardization of solid biofuels (pellets and briquettes) allows the automatization of feeding and burning processes in domestic and industrial-size boilers that is impossible for low-processed biomass. For pellets that represent the most compacted biofuels, quality standards EN ISO 17225-2:2014 standard [40], require the specific density (DE) to be equal to or greater than 1000 kg·m^{-3} and mechanical durability (DU) to be equal to or greater than 97.5% (A1 quality class). To obtain these parameters, it is necessary to properly prepare the raw material and properly set the parameters of the pressure compaction process to the requirements of the processed material. Many factors influence the compaction process. Pressure and compaction temperature, type, degree of fragmentation, friction coefficient and moisture content of the raw material are the most important factors [41–47].

The moisture content of the raw material is one of the most important factors that affect the compaction process, and is in the range of 6%–17% for most biomass materials [43]. In most cases, the biomass raw material has a moisture content higher than this level, so that a drying stage is almost always used in the granulation process. This process is carried out in a wide temperature range 40–200 °C or more [48–53]. According to the literature, the main components of the biomass such as lignin, cellulose, and hemicellulose can be transformed as a result of the temperature effect [49,54,55], which may affect the compaction process. During the drying of coniferous tree biomass, terpenic compounds that act as binders and energy carriers are also released [51]. It is known that the high temperature in the 150–300 °C range used in the torrefaction process also affects the compaction. This leads to an improvement in the raw material grindability but complicates the compactions process, mainly due to the decomposition of lignin, which acts as the binder during the densification [56–62].

Lignin plasticizes at temperatures of about 150 °C [15,63]; plasticization can occur at lower temperatures of 75–90 °C with the presence of moisture in the material [42,64]. A promising way of activating the properties of lignin as a binder is steam explosion [65,66]. It causes structural changes in biomass fiber and, under some pretreatment conditions, lignin in biomass cells comes out of the fiber. During the agglomeration process, lignin melts both on the particle surface and between particles. This results in the creation of a cover layer around pellets and bonds inside them [67]. The durability of such pellets is significantly higher than pellets from conventional materials [68].

The aim of this study was to investigate the effect of drying temperature on the process of Scots pine biomass compaction. Scots pine was chosen in this study because it is a typical biomass material that is widely used for the production of fuel pellets.

2. Materials and Methods

The research in this work was carried out according to the procedure shown in Figure 2.

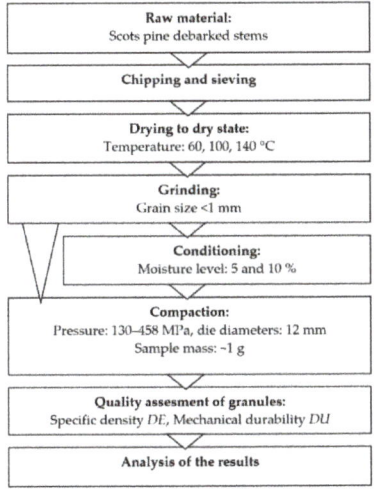

Figure 2. Flowchart of the procedure used for the investigations in this work.

2.1. Materials

Because it is the most popular raw material for fuel pellet production, Scots pine (*Pinus sylvestris* L.) biomass was used for the present study. According to the biomass classification [30], this material represents woody biomass. After the growth season, Scots pine stems were collected, debarked and chipped.

A set of analytical sieves with a diameter of 400 mm and hole sizes of 16 and 8 mm (Morek Multiserw, Marcyporęba, Poland) and a sieve shaker (LPzE-4e, Morek Multiserw, Marcyporęba, Poland) were used to separate the particle size fraction that passes through the sieve with the hole size of 16 mm and remains on the sieve with the hole size of 8 mm in order to ensure that the material is geometrically uniform prior to the drying stage. Pine chips were divided into three samples and were dried in a laboratory dryer (SLW 115, Pol-Eko, Wodzisław Śląski, Poland) to the dry state at the temperatures of 60, 100 and 140 °C.

After drying, all three samples were ground to obtain particles with a size of <1 mm using a hammer mill (PX-MFC 90D Polymix, Kinematika, Luzern, Switzerland). According to the literature [66,69–73], grain size and grain size distribution has a significant impact on the biomass agglomeration process. Therefore, in order to exclude this effect, the grain size distribution of all samples was standardized to the grain size distribution of the biomass sample dried in 60 °C. Samples were divided into four dimensional fractions, (sieve classes): C_1: 0.1—grain diameter d ≤ 0.1 mm, C_2: 0.25—grain diameter between 0.1 < d ≤ 0.25 mm, C_3: 0.5–0.25 < d ≤ 0.5 mm and C_4: 1–0.5 < d ≤ 1 mm. The standardized grain size distribution of all samples is contained in Table 1.

Table 1. Standardized particle size distribution of samples.

Sieve Classes (mm)	C_1: 0.1	C_2: 0.25	C_3: 0.5	C_4: 1
Share [%]	8.2	16.1	38.2	37.5

The sample dried at 60 °C was also analyzed in terms of energy properties. The values of determined parameters are shown in Table 2.

Table 2. Characterization of raw material—all values refer to the dry state of sample.

Parameter	Unit	Value	Standard
Heat of combustion	MJ/g	20.9	EN ISO 18125 [74]
Ash content	%	0.2	EN ISO 18122 [75]
Volatile matter	%	79.8	EN ISO 18123 [76]

Each sample of dry, ground and size standardized material was divided into three subsamples, giving a final total of nine test samples. Three of the samples (dried at the examined temperatures) were left in the dry state, and from the remaining six samples, three samples were moisturized to 5% and three samples were moisturized to 10%. Based on the literature, these moisturization values were considered to be most suitable for investigating the effects of the moisture content on the pine pellet quality parameters [46,47,77]. The samples in the dry state were treated as controls (Figure 3). The materials were moisturized and conditioned in a climate chamber with controlled temperature and humidity (KBF-S 115, Binder, Tuttlingen, Germany). This allowed us to maintain a stable moisture content during the experiments, which was highly important for the accuracy of the investigations.

Figure 3. Samples ready for densification: from left to right, the samples of the raw material dried at 60, 100, and 140 °C, respectively, in the dry state are shown.

2.2. Samples Densification

The biomass compacts (test pellets) were produced using a compaction stand (hydraulic press P400, Sirio, Meldola, Italy) with a special appliance. The components of this stand are shown in Figure 4. The main parts of the stand are the compaction unit that consisted of hardened steel and a cylindrical die (with an internal die channel diameter of 12 mm and length of 110 mm). The unit also contains a piston and bottom (diameter of both was 11.9 mm). The compaction unit is placed between the press plate and pressure is applied by the pistons.

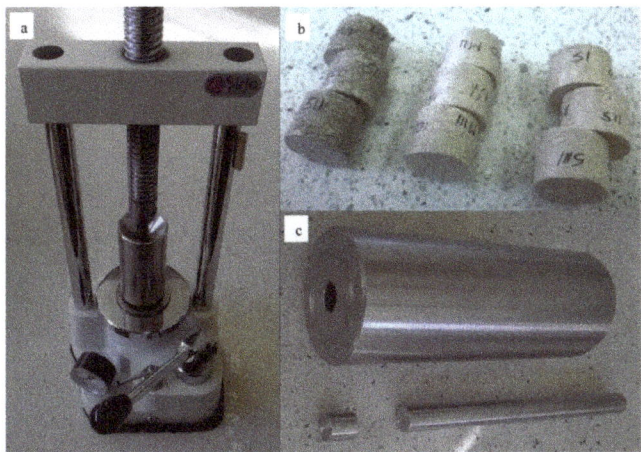

Figure 4. Lab-scale compaction stand for test pellets production: (**a**) hydraulic press, (**b**) samples of compacts, and (**c**) die, piston and bottom.

Using the compaction stand, we obtained precise information regarding the pressure used for sample densification. Based on the previous studies reported in the literature [3,78,79], six pressure values were selected, namely 130.8, 196.2, 216.6, 327, 392.4, and 457.8 MPa (14.5, 21.7, 29, 36.2, 43.5, 50.7 kN, respectively),

The mass of the sample (approximately 1 g) allowed us to obtain compacts with heights that were smaller than their diameters (approximately 8–10 mm). A total of 54 combinations of the different dry temperature, moisture content and pressure values were investigated in this work. To ensure the quality of the obtained results, for each set of conditions we examined three test pellets, even though the standards for DU and DE require the use of only two samples [80,81]. Thus, a total of 162 test pellets were produced. After compaction, the samples were placed into individual labeled compartments to prevent moisture change and human error during further measurements.

2.3. Pellets Quality Parameters Analysis

After waiting for 24 h after the compaction, the height, diameter and mass of the test pellets were measured using a caliper with an accuracy of 0.01 mm and a scale (AS 160.R2, Radwag, Radom, Poland) with an accuracy of 0.1 mg. The DE of each pellet was calculated according to the values of the specific densities, as shown in Equation (1)

$$DE = \frac{m_{24}}{V_{24}} \tag{1}$$

where DE is the specific density, m_{24} is the mass of the sample 24 h after compaction and V_{24} is the volume of the sample 24 h after compaction.

The mechanical durability (DU) was also determined for these samples. Hardness and durability are very important parameters for logistics. While many standards refer to DU [57], the requirements for energetic pellets are described in the EN ISO 17831-1:2015 standard [81] dedicated standard. In this case, where only a few granules were produced, some modification of this method was necessary. This is because the standard procedure requires the use of approximately 500 g of pellets per test. In this research, only three granules were made, and the equipment described in the EN ISO 17831-1:2016-02 standard was used to simulate the damages incurred by the particles during logistics operations. The main experimental apparatus is a cuboidal chamber with two baffles inside that rotates at a rate of 50 rpm for 10 min.

The requirement for the 500 g mass of the sample was met by using ballast material in which three tested pellets were placed (Figure 5). For the volume of the ballast sample to be comparable to that of the classic sample, the ballast material should have a specific density similar to that of a typical pellet (approximately 1000 kg·m^{-3}). Therefore, polystyrene with a specific density of 1070 kg·m^{-3} was used. The grains forming the ballast material had a sphere-like shape with a diameter of 5 mm (Figure 5). Some of the samples were made from the biomass in the dry state, which means that the strength of some compacts was low and, therefore, in order not to cause their complete disintegration during the test, it was decided to reduce the time of rotation. Based on previous research, the time of rotations was reduced to 5 min [3].

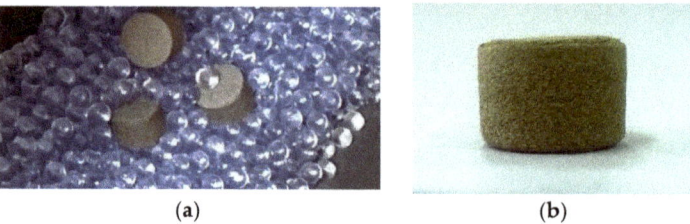

(a) (b)

Figure 5. Test pellet sample: (a) the ballast material placed in the testing chamber, (b) after test with the visible crumbling zones.

After the test, the compacts were removed from the box, and weighed (scale AS 160.R2, Radwag, Radom, Poland with accuracy of 0.1 mg), and using Formula 2, the DU values of each sample were obtained. The obtained values are only compared for the materials used in this research, but further research on this method is in progress in order to compare the results obtained using this method for a variety of materials.

$$DU = \frac{m_A}{m_E} \cdot 100 \tag{2}$$

where DU is the mechanical durability, m_A is the mass of the sample after the test and m_E is the mass before the test.

The last modification to the basic procedure was the mode used for the calculation of the compact material's mechanical durability (DU). In the standard, the DU result is related to the mass of the sample as a whole, while in the case of the conducted tests, the durability of each individual granule was determined, so that the result had to be related to its mass. The cylindrical granule was crushed only along the circumference of its base, while the central part of the cylinder side remained undamaged (Figure 5b). Thus, although the mass of the fines for two granules with the same diameter but different heights is comparable, when using the classic formula for DU, different calculated durability values are obtained for these two granules because the mass of the longer granule is greater than that of the shorter granule. To make the calculation of DU independent of the granule height, the mechanical durability of the equivalent DU_{10} granules with a constant height of 10 mm h_{10} was calculated as

$$DU_{10} = \frac{m_{A10}}{m_{E10}} \cdot 100 \tag{3}$$

where DU_{10} is the mechanical durability of the substitute compact, m_{A10} is the mass of the substitute compact after the test and m_{E10} is the mass of the substitute compact prior to the test.

Due to the use of equivalent granules, the mechanical durability value does not depend on the height of the granules, enabling comparison of the DU_{10} results regardless of the height of the granules. The height of the granules was between 8 and 10 mm, and a more detailed description of the procedure use in this work can be found in a previous report [3].

3. Results

Figure 6 shows the changes in the pellets' specific density (DE) due to the effects of the raw material drying temperature and the increase in the densification pressure. Figure 6a shows the changes for the material in the dry state and Figure 6b,c show the changes for the materials with moisture contents of 5% and 10%, respectively. For the dry materials, the applied pressure results in the pellets with DE values in the 794–1100 kg·m^{-3} range. The threshold value of 1000 kg·m^{-3} (the dividing line between the red and green part of the density variation area) was reached most rapidly for the material that was dried at 60 °C. This was achieved at the pressure of 261 MPa, whereas for the materials dried at 100 and 140 °C, the threshold values were achieved at pressures of 319 and 333 MPa, respectively.

For a test material moisturized to 5%, the area of the specific density variation over the entire range of the test pressures was shifted up and the DE values of the resulting pellets were in the 930–1120 kg·m^{-3} range. The density threshold value was obtained in this case at 169 MPa, showing a lower value than that of the material in the dry state. As the drying temperature increases, this value increases slightly to 187 MPa, suggesting that the drying temperature of the raw material is less important in this case, while the increase in humidity has resulted in an increase in the pellet DE. The increase in the moisture content to 10% caused the whole pressure range to reach specific density values above the threshold value for all pellets. Similar evolutions of the density values were observed for all drying temperatures, showing that the density changes do not depend strongly on the raw material drying temperature. At first pressure level (130.8 MPa), the specific density was 1070 kg·m^{-3}. This values increased with pressure, and was 1110 kg·m^{-3} for 196.2 MPa and 1130 kg·m^{-3} for 261.6 MPa. A further increase in the compaction pressure did not lead to an increase in the pellet specific density.

Changes in mechanical durability (DU) for the same pellets are shown in Figure 7. The DU threshold value was set at 97.5% (this value is required for wood pellets of class A1 according to standard EN ISO 17225-2:2014 [40]). This is a theoretical value because the DU of the tested compacts was determined by the modified method. For the pellets obtained from raw material in the dry state (Figure 7a), the DU threshold level of the obtained pellets was obtained at a pressure of 392.4 MPa, and compacts fabricated at 130.8 MPa had a DU of only 65. No effect of the drying temperature was observed.

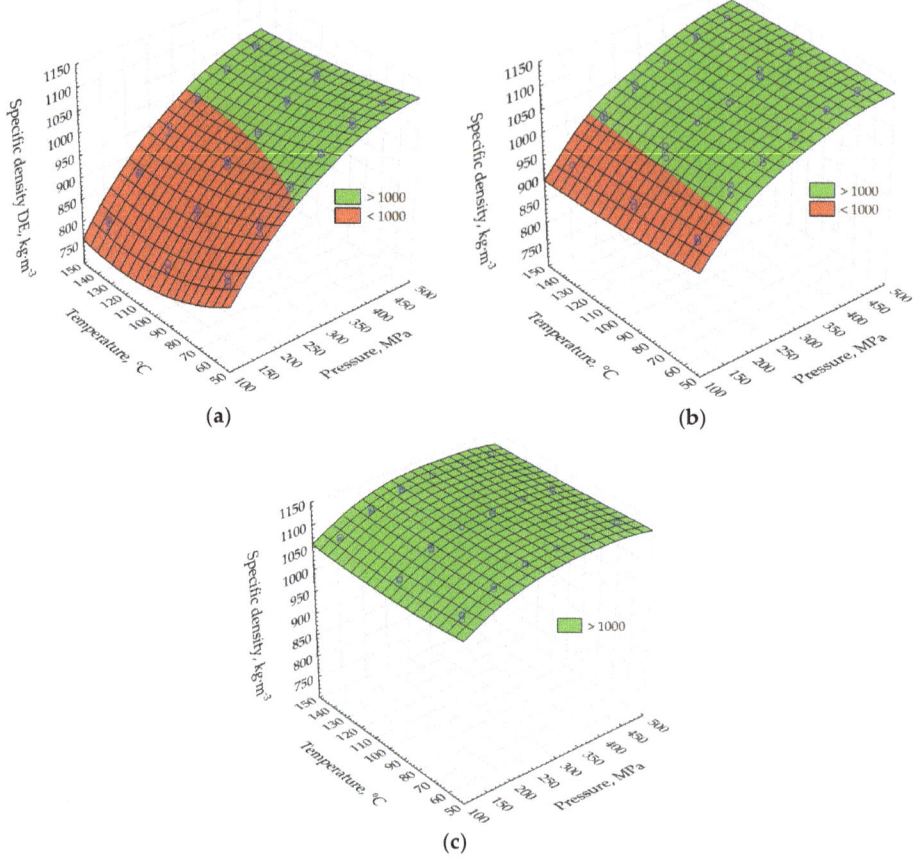

Figure 6. Pellet specific density (*DE*) changes as a function of the raw material drying temperature and densification pressure: (**a**) raw material in the dry state, (**b**) with 5% moisture content, and (**c**) with 10% moisture content.

An increase in the raw material moisture content up to 5% causes the *DU* threshold value to be reached for pressures greater than 196.2 MPa, regardless of the raw material drying temperature (Figure 7b). At the lowest pressure, the *DU* of pellets was approximately 91%, which is much higher than that of the dry pellets. A further increase in the moisture content of the raw material to 10% led to an increase in the mechanical durability of the compacts above the threshold value, regardless of the applied pressure. (Figure 7c).

To confirm the significance of the discussed evolution of *DE* and *DU*, a statistical analysis was carried out for the obtained results. A detailed statistical analysis was performed for two hypotheses:

- The drying temperature of the raw material significantly affects the *DE* and *DU* values of the pellets obtained at the tested levels of pressure and raw material moisture content;
- The increase in the raw material moisture content compensated the influence of the drying temperature on the quality parameters (*DE* and *DU*) for the tested pressure levels.

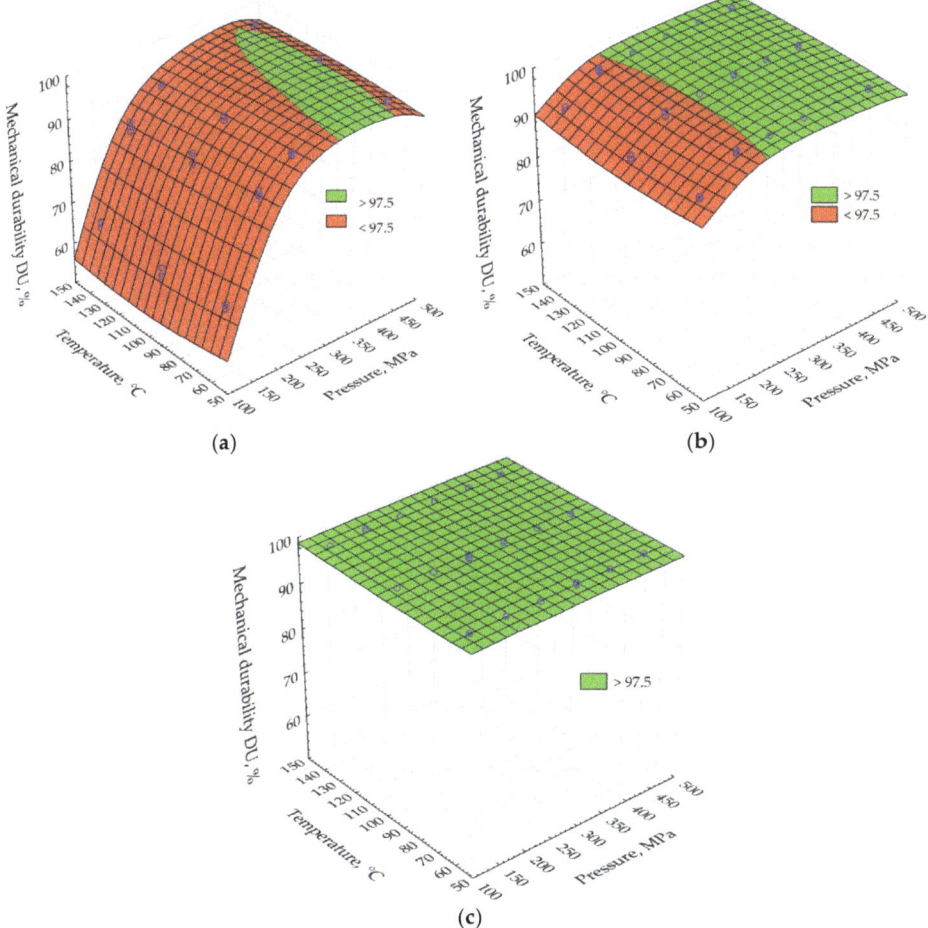

Figure 7. Pellet mechanical durability (*DU*) changes as a function of the raw material drying temperature and densification pressure: (**a**) raw material in dry state, (**b**) 5% moisture content, (**c**) 10% moisture content.

In this work, ANOVA analysis was carried out, and the normality of the decomposition was checked using the Shapiro–Wilk test. It was found that the distribution was normal for all cases. Then, the assumption of the equality of variance was also examined using the Brown–Forsythe test. For all cases, this equality was met. Then, ANOVA and post-hoc analysis (Scheffé's test) were carried out to determine whether there were statistically significant differences between the groups.

Figure 8 shows the results of the three-factor ANOVA examining the effects of the drying temperature, moisture content and pressure on the specific density. The graphs show the compaction profiles derived from the mean values together with the standard deviations.

Post-hoc analyses have shown that for the dry material, there are significant differences between the specific density values of the pellets produced from the material dried at 60 °C and those produced from the material dried at 100 and 140 °C. These differences exist for each of the examined pressures with the exception of 261.6 MPa and 457.8 MPa, for which the change in the drying temperature from 60 and 100 °C does not have a significant effect on the specific density of compacts. For each tested pressure, the differences in the specific density values between the pellets obtained from the materials dried at 100 and 140 °C were insignificant. As mentioned above, the specific densities of the pellets decreased as the drying temperature increased from 60 to 100 °C. This means that the drying of raw material at 100 and 140 °C leads to a decrease in the densification ability of the raw material.

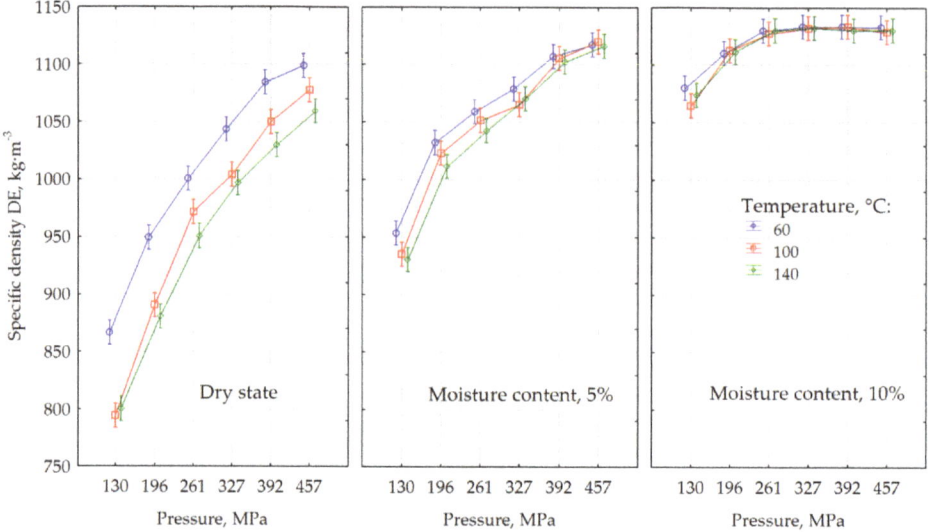

Figure 8. Changes in the specific density (DE) of the pellets as a function of the compaction process parameters.

For the materials moistened to 5% and 10% moisture content, the effect of the drying temperature on the obtained DE values was insignificant in the entire examined pressure range. This means that moistening the material even to the level of only 5% eliminates the influence of the drying temperature and at the same time allows us to obtain pellets with higher DE compared to those obtained from the material in the dry state.

A similar analysis was performed for DU. Figure 9 shows the result of the three-factor ANOVA examining the effects of the drying temperature, moisture content and pressure on the DU values. The graphs show the compactibility profiles derived from the average values together with the standard deviations.

Post-hoc analyses showed that, for the dry material, there are significant differences between the DU values of the pellets produced from the material dried at 60 °C and those produced from the material dried at 100 and 140 °C. These differences are observed only for pressures lower than 261.6 MPa, while for higher pressures, the differences are insignificant. At a pressure of 130.8 MPa, the significant difference occurs between the drying temperature of 60 and 100 °C, at 196.2 and 261.6 MPa a significant difference is observed between the samples obtained using drying temperatures of 60 and 140 °C. The differences in mechanical durability between the pellets made of the materials dried at 100 and 140 °C were insignificant in the entire pressure range. Similar to the specific density results, DU decreased with increasing drying temperature for the dry materials, while for the materials moistened to the 5% and 10% moisture content, the influence of the drying temperature on the obtained mechanical durability values was insignificant in the entire range of tested pressures.

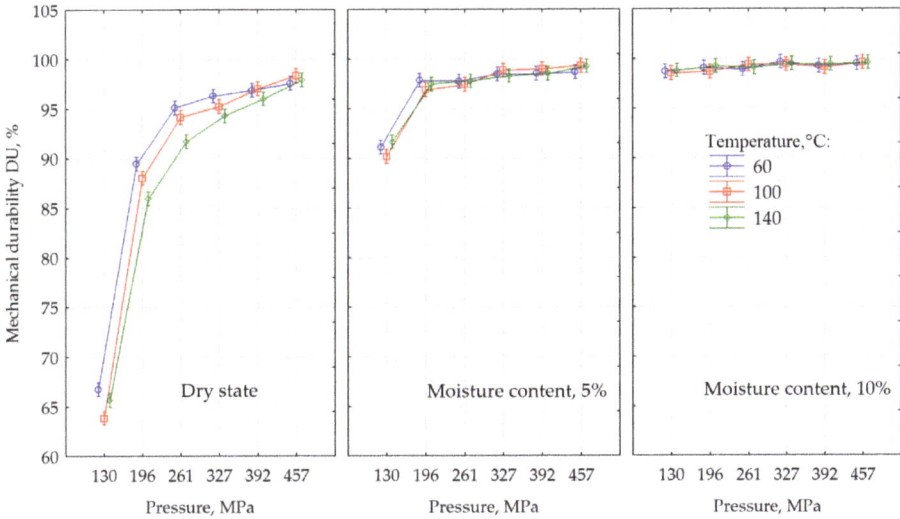

Figure 9. Changes in the mechanical durability (*DU*) of the pellets for different compaction process parameters.

Thus, the conducted tests show that the drying temperature significantly influences the obtained values of the specific density of the pellets produced from the material in a dry state in the whole pressure range, while for *DU*, the changes were significant only in the 130.8–261.6 MPa range. Significant changes occur between the drying temperature of 60 °C on the one hand, and the drying temperature of 100 and 140 °C on the other hand, while the differences between 100 and 140 °C are not significant. An increase in the moisture content to 5% and 10% eliminates the influence of the drying temperature and leads to an increase in the values of the tested quality parameters. What is remarkable is that this range of moisture content is optimal for the pelletization process and meets the quality standard requirements [40,82]. In order to obtain high parameters of *DE* and *DU*, it is not necessary use the maximum tested pressure values and that these pressure are lower than the pressure occurring in the real process of pelletization up to 750 MPa [83].

4. Discussion

Higher values of the quality parameters obtained for the pellets produced from the material dried at 60 °C compared to those produced from the materials dried at 100 and 140 °C (compacted in dry state) may be explained by the fact that, at this temperature, only water is removed from the material and other components such as terpenes, resins, and essential oils (generally called volatile substances) remain in the material, while when the drying temperature exceeds 100 °C, these substances are also gradually released. This phenomenon is confirmed by studies performed by Stahl et al. [84] comparing the drying process on industrial systems. They have shown that the highest terpene emissions occur in dryers with high drying temperatures (above 100 °C) and long material residence times in the dryer. The type of dryer also affects the emission of terpenes – steam dryers cause less emission compared to rotary drum dryers. Moreover, the Englund and Nussbaum research [85] confirms that the emission of terpenes at a drying temperature of 60 °C is lower compared to emission at 110 °C.

Generally, these substances act as binders, and also play the role of binder in the material dried at 60 °C. For the pellets produced from the biomass dried at 100 and 140 °C, the removal of the binder substances from the material results in a decrease in the *DE* and *DU* values. For the material with 5% and 10% moisture content, the water contained in the material plays the role of a binder and its higher agglomeration capacity compared to the volatile substances leads to the better connection between the material particles, resulting in an increase in the DE and DU values. At the same time, water eliminates the weaker impact of volatile substances, as evident from the absence of the significant effect of the drying temperature on the obtained values of DE and DU at this raw material moisture content. The increase in the water content in the raw material to the level of 10% results in an increase in the number of bonds between the particles (so-called interparticle bridges) [3,86], which further increases the DE and DU values. This is probably caused by the increase in the moisture content of the raw material. This has a significant effect on the plasticization of lignin. This is probably the main reason for the mentioned increase in the number of bonds between the particles.

A different course of DE and DU changes was observed by Filbak et al. [83] during a study on the influence of drying methods for Scots pine raw material on pellets' DE and DU. Pellets made of raw material dried at 75 °C had a higher DU value compared to those made of material dried at 450 °C. An opposite relationship was observed for DE. In these studies, the pellets were made on an industrial line where, apart from moisture content, the main factor influencing the agglomeration process is temperature changes occurring during the process [19].

Similar to most lignocellulose materials, pine wood has a spatial porous structure. Therefore, the specific density of dry pine wood is approximately 490 kg·m^{-3}, while its absolute density is approximately 1470 kg·m^{-3} (lignocellulose material without any pores) and its bulk density is approximately 180 kg·m^{-3} [3]. During compaction, the applied pressure is responsible for reducing the external (between material particles) and internal (inside material particles) pores. Drying at 60 °C removes the water from the inner pores of the material, while an increase in the temperature to 100 and 140 °C also leads to the removal of the volatile components, increasing the volume of the pores. The pressure reduces the volume of the pores (both external and internal), and the degree of pore reduction is greater for the material dried at 60 °C. Upon the moistening of the material to 5% water content, the water infiltrates the wood and the material becomes plasticized, making a lower pressure sufficient for compacting it. Then, a further moisturization leads to an increase in the plasticity and a further increase in the density at lower pressures. The maximum obtained specific density values oscillate at 1130 kg·m^{-3}. A value of DE equal to the absolute density (1470 kg·m^{-3}) cannot be achieved due to the expansion of the material when the pressure stops. The value of this parameter depends on the elasticity of the material and the amount and strength of the joints between the particles formed during the compaction. In this case, the best results were achieved by moisturizing the material to the 10% water content and compacting at the pressure 261.6 MPa, and a further increase in the pressure did not increase the achieved DE. For DU, 130.8 MPa and 10% moisture content is sufficient to obtain the maximum value. Although these studies have shown that the influence of the drying temperature on the compaction process is important for the material in a dry state, and while moistening to 5% and 10% water content eliminates this influence, it is an open question as to whether this dependence is also found for other materials used in solid biofuel production. While a classic material that can be easily agglomerated was tested in this work, materials that are difficult to compact, e.g., straw, miscanthus, and reed canary grass, may be more sensitive to the influence of the drying temperature. Our investigations showed that the drying temperature changes the properties of the raw material. This means that these differences can affect other stages of raw material processing, such as grinding. The fact that the moisture content reduces this effect during pelletization is important because it means that, for the compaction stage, it is possible to use raw material dried over a wide temperature range without affecting the main quality parameters. If further investigations confirm that the drying temperature affects the grinding process, then the drying temperature can be selected so that grinding can be carried out with minimum effort and without fear of losing the quality of the final product. The drying temperature may affect the energy parameters of the raw material. High temperature causes a loss of extractives, which may decrease the pellets' combustion heat and change the combustion process [51,85]. The study of these relationships will be taken into account in further research.

5. Conclusions

The aim of this study was to determine the effect of the drying temperature of the raw material on the compacting process of pine biomass. This effect was determined for three different material moisture contents and six different compaction pressures. Based on the presented data, the following key results were obtained:

- The drying temperature significantly affects the specific density of the compacts over the entire pressure range when the compacted material is in the dry state;
- For DU, the influence of the drying temperature is significant in the pressure range of 130.8–261.6 MPa for the materials in the dry state;
- An increase in the drying temperature leads to a decrease in the DE and DU values (dry material);
- Significant differences are observed between the material dried at 60 °C and the materials dried at 100 and 140 °C, while the differences between the materials treated at the drying temperatures of 100 and 140 °C are not significant;
- An increase in the moisture content of the raw material to 5% eliminates the influence of the drying temperature on the obtained values of DE and DU and increases the obtained DE and DU values relative to those of the dry material;
- An increase in the moisture content to 10% results in a further increase in DE and DU;
- Assumed threshold values of specific density and mechanical durability were obtained for each drying temperature and moisture content level. As the moisture content of the raw material increases, the pressure necessary to obtain the threshold DE and DU decreases.

These results reveal an effect that has not been described in previous reports in the literature. Further research on the effects of the drying temperature on the other biomass raw materials and the stages of the biomass compaction process should be carried out to improve the understanding of this topic.

Author Contributions: Conceptualization, M.W.; methodology, K.M., M.W., M.J.; investigation, M.W., K.M. and M.J.; resources, M.J. and K.M.; data curation, M.W. and M.J.; writing—original draft preparation, M.W. and M.J.; writing—review and editing, M.W. and M.J.; visualization, M.W.; supervision, K.M., M.J. and A.K.; statistical analysis, A.K. All authors have read and agreed to the published version of the manuscript.

Funding: This research was supported by Ministry of Science and Higher Education of the Republic of Poland (Faculty of Production and Power Engineering, University of Agriculture in Krakow).

Conflicts of Interest: The authors declare no conflict of interest.

References

1. Popper, Z.A.; Michel, G.; Hervé, C.; Domozych, D.S.; Willats, W.G.T.; Tuohy, M.G.; Kloareg, B.; Stengel, D.B. Evolution and Diversity of Plant Cell Walls: From Algae to Flowering Plants. *Annu. Rev. Plant Biol.* **2011**, *62*, 567–590. [CrossRef] [PubMed]
2. Knox, J.P. Revealing the structural and functional diversity of plant cell walls. *Curr. Opin. Plant Biol.* **2008**, *11*, 308–313. [CrossRef] [PubMed]
3. Wróbel, M. *Zagęszczalność i Kompaktowalność Biomasy Lignocelulozowej*; PTIR: Kraków, Poland, 2019; ISBN 978-83-64377-35-8.
4. Welker, C.; Balasubramanian, V.; Petti, C.; Rai, K.; DeBolt, S.; Mendu, V.; Welker, C.M.; Balasubramanian, V.K.; Petti, C.; Rai, K.M.; et al. Engineering Plant Biomass Lignin Content and Composition for Biofuels and Bioproducts. *Energies* **2015**, *8*, 7654–7676. [CrossRef]
5. Kumar, P.; Barrett, D.M.; Delwiche, M.J.; Stroeve, P. Methods for Pretreatment of Lignocellulosic Biomass for Efficient Hydrolysis and Biofuel Production. *Ind. Eng. Chem. Res.* **2009**, *48*, 3713–3729. [CrossRef]
6. Zethræus, B. The Bioenergy System Planners Handbook—BISYPLAN Chapter 04. Available online: http://bisyplan.bioenarea.eu/html-files-en/02-02.html (accessed on 12 December 2018).
7. Demirbaş, A. Estimating of structural composition of wood and non-wood biomass samples. *Energy Sour.* **2005**, *27*, 761–767. [CrossRef]

8. Lamlom, S.H.; Savidge, R.A. A reassessment of carbon content in wood: Variation within and between 41 North American species. *Biomass Bioenergy* **2003**, *25*, 381–388. [CrossRef]
9. Fajín, J.L.C.; Cordeiro, M.N.D.S. Probing the efficiency of platinum nanotubes for the H2 production by water gas shift reaction: A DFT study. *Appl. Catal. B Environ.* **2020**, *263*, 118301.
10. García-Moncada, N.; González-Castaño, M.; Ivanova, S.; Centeno, M.Á.; Romero-Sarria, F.; Odriozola, J.A. New concept for old reaction: Novel WGS catalyst design. *Appl. Catal. B Environ.* **2018**, *238*, 1–5. [CrossRef]
11. Chianese, S.; Fail, S.; Binder, M.; Rauch, R.; Hofbauer, H.; Molino, A.; Blasi, A.; Musmarra, D. Experimental investigations of hydrogen production from CO catalytic conversion of tar rich syngas by biomass gasification. *Catal. Today* **2016**, *277*, 182–191. [CrossRef]
12. Li, W.; Li, Q.; Chen, R.; Wu, Y.; Zhang, Y. Investigation of hydrogen production using wood pellets gasification with steam at high temperature over 800 °C to 1435 °C. *Int. J. Hydrogen Energy* **2014**, *39*, 5580–5588. [CrossRef]
13. Demirbas, A. Higher heating values of lignin types from wood and non-wood lignocellulosic biomasses. *Energy Sources Part A Recover. Util. Environ. Eff.* **2017**, *39*, 592–598. [CrossRef]
14. Dziurzyński, A. *Zależności Między Wybranymi Właściwościami Fizycznymi Drewna Sosny i Buka a Zawartością jego Strukturalnych Składników Chemicznych*; Wydawnictwo Akademii Rolniczej im. Augusta Cieszkowskiego: Poznań, Poland, 2003.
15. Kaliyan, N.; Vance Morey, R. Factors affecting strength and durability of densified biomass products. *Biomass Bioenergy* **2009**, *33*, 337–359. [CrossRef]
16. Li, W.; Amos, K.; Li, M.; Pu, Y.; Debolt, S.; Ragauskas, A.J.; Shi, J. Fractionation and characterization of lignin streams from unique high-lignin content endocarp feedstocks. *Biotechnol. Biofuels* **2018**, *11*, 304. [CrossRef] [PubMed]
17. Demirbaş, A. Relationships between heating value and lignin, fixed carbon, and volatile material contents of shells from biomass products. *Energy Sources* **2003**, *25*, 629–635. [CrossRef]
18. White, R.H. Effect of lignin content and extractives on the higher heating value of wood. *Wood Fiber Sci.* **1987**, *19*, 446–452.
19. Križan, P. *The Densification Process of Wood Waste*; De Gruyter Open Ltd.: Warsaw, Poland; Berlin, Germany, 2015; ISBN 978-3-11-044001-0.
20. Robbins, M.P.; Evans, G.; Valentine, J.; Donnison, I.S.; Allison, G.G. New opportunities for the exploitation of energy crops by thermochemical conversion in northern Europe and the UK. *Prog. Energy Combust. Sci.* **2012**, *38*, 138–155. [CrossRef]
21. Bajpai, P. *Pretreatment of Lignocellulosic Biomass for Biofuel Production*; Springer: Singapore, 2016; ISBN 978-981-10-0686-9.
22. Mendu, V.; Harman-Ware, A.E.; Crocker, M.; Jae, J.; Stork, J.; Morton, S.; Placido, A.; Huber, G.; Debolt, S. Identification and thermochemical analysis of high-lignin feedstocks for biofuel and biochemical production. *Biotechnol. Biofuels* **2011**, *4*, 43. [CrossRef]
23. van Dam, J.E.G.; van den Oever, M.J.A.; Teunissen, W.; Keijsers, E.R.P.; Peralta, A.G. Process for production of high density/high performance binderless boards from whole coconut husk: Part 1: Lignin as intrinsic thermosetting binder resin. *Ind. Crops Prod.* **2004**, *19*, 207–216. [CrossRef]
24. Niemczyk, M.; Kaliszewski, A.; Jewiarz, M.; Wróbel, M.; Mudryk, K. Productivity and biomass characteristics of selected poplar (Populus spp.) cultivars under the climatic conditions of northern Poland. *Biomass Bioenergy* **2018**, *111*, 46–51. [CrossRef]
25. Bryś, A.; Bryś, J.; Głowacki, S.; Tulej, W.; Zajkowski, P.; Sojak, M. *Analysis of Potential Related to Grass-Derived Biomass for Energetic Purposes*; Springer: Cham, Switzerland, 2018; pp. 443–449.
26. Knapczyk, A.; Francik, S.; Wójcik, A.; Bednarz, G. Influence of Storing Miscanthus x Giganteus on Its Mechanical and Energetic Properties. In *Renewable Energy Sources: Engineering, Technology, Innovation*; Springer: Cham, Switzerland, 2018; pp. 651–660.
27. Wróbel, M.; Mudryk, K.; Jewiarz, M.; Głowacki, S.; Tulej, W. Characterization of Selected Plant Species in Terms of Energetic Use. In *Proceedings of the Renewable Energy Sources: Engineering, Technology, Innovation. Springer Proceedings in Energy ICORES 2017*; Mudryk, K., Werle, S., Eds.; Springer Proceedings in Energy: Cham, Switzerland, 2018; pp. 671–681.
28. Stelte, W.; Sanadi, A.R.; Shang, L.; Holm, J.K.; Ahrenfeldt, J.; Henriksen, U.B. Recent developments in biomass pelletization—A review. *BioResources* **2012**, *7*, 4451–4490.

29. Bilbao, R.; Lezaun, J.; Abanades, J.C. Fluidization velocities of sand/straw binary mixtures. *Powder Technol.* **1987**, *52*, 1–6. [CrossRef]
30. EN ISO 17225-1:2014. *Solid Biofuels—Fuel Specifications and Classes—Part 1: General Requirements*; International Organization for Standardization: Geneva, Switzerland, 2014.
31. Brzychczyk, B.; Hebda, T.; Giełżecki, J. Physical and Chemical Properties of Pellets Produced from the Stabilized Fraction of Municipal Sewage Sludge. In *Renewable Energy Sources: Engineering, Technology, Innovation*; Springer: Cham, Switzerland, 2018; pp. 613–622.
32. Brzychczyk, B.; Hebda, T.; Giełżecki, J. Energy Characteristics of Compacted Biofuel with Stabilized Fraction of Municipal Waste. In *Renewable Energy Sources: Engineering, Technology, Innovation*; Springer: Cham, Switzerland, 2018; pp. 451–462.
33. Hebda, T.; Brzychczyk, B.; Francik, S.; Pedryc, N. Evaluation of suitability of hazelnut shell energy for production of biofuels. In Proceedings of the Engineering for Rural Development, Jelgava, Latvia, 23–25 May 2018; Volume 17, pp. 1860–1865.
34. Knapczyk, A.; Francik, S.; Fraczek, J.; Slipek, Z. Analysis of research trends in production of solid biofuels. In Proceedings of the Engineering for Rural Development, Jelgava, Latvia, 22–24 May 2019; Volume 18, pp. 1503–1509.
35. Karbowniczak, A.; Hamerska, J.; Wróbel, M.; Jewiarz, M.; Nęcka, K. Evaluation of Selected Species of Woody Plants in Terms of Suitability for Energy Production. In *Renewable Energy Sources: Engineering, Technology, Innovation. Springer Proceedings in Energy ICORES 2017*; Mudryk, K., Werle, S., Eds.; Springer Proceedings in Energy: Cham, Switzerland, 2018; pp. 735–742.
36. Telmo, C.; Lousada, J. Heating values of wood pellets from different species. *Biomass Bioenergy* **2011**, *35*, 2634–2639. [CrossRef]
37. Viana, H.; Vega-Nieva, D.J.; Ortiz Torres, L.; Lousada, J.; Aranha, J. Fuel characterization and biomass combustion properties of selected native woody shrub species from central Portugal and NW Spain. *Fuel* **2012**, *102*, 737–745. [CrossRef]
38. Luca, C.; Pilu, R.; Tambone, F.; Scaglia, B.; Adani, F. New energy crop giant cane (Arundo donax L.) can substitute traditional energy crops increasing biogas yield and reducing costs. *Bioresour. Technol.* **2015**, *191*, 197–204. [PubMed]
39. Zegada-Lizarazu, W.; Monti, A. Energy crops in rotation. A review. *Biomass Bioenergy* **2011**, *35*, 12–25. [CrossRef]
40. EN ISO 17225-2:2014. *Solid Biofuels—Fuel Specifications and Classes—Part 2: Graded Wood Pellets*; International Organization for Standardization: Geneva, Switzerland, 2014.
41. Sitzmann, W.; Buschhart, A. *Pelleting as Prerequisite for the Energetic Utilization of Byproducts of the Wood-Processing Industry*; Amandus Kahl GmBH & Co.: Oslo, Norway, 2009.
42. Kaliyan, N.; Morey, R.V. Natural binders and solid bridge type binding mechanisms in briquettes and pellets made from corn stover and switchgrass. *Bioresour. Technol.* **2010**, *101*, 1082–1090. [CrossRef]
43. Dyjakon, A.; Noszczyk, T. The Influence of Freezing Temperature Storage on the Mechanical Durability of Commercial Pellets from Biomass. *Energies* **2019**, *12*, 2627. [CrossRef]
44. Wójcik, A.; Frączek, J.; Wota, A.K. The methodical aspects of the friction modeling of plant granular materials. *Powder Technol.* **2019**, *344*, 504–513. [CrossRef]
45. Wójcik, A.; Frączek, J.; Niemczewska-Wójcik, M. The relationship between static and kinetic friction for plant granular materials. *Powder Technol.* **2019**, *361*, 739–747. [CrossRef]
46. Nielsen, N.P.K.; Gardner, D.J.; Poulsen, T.; Felby, C. Importance of temperature, moisture content, and species for the conversion process of wood residues into fuel pellets. *Wood Fiber Sci.* **2009**, *41*, 414–425.
47. Arshadi, M.; Gref, R.; Geladi, P.; Dahlqvist, S.A.; Lestander, T. The influence of raw material characteristics on the industrial pelletizing process and pellet quality. *Fuel Process. Technol.* **2008**, *89*, 1442–1447. [CrossRef]
48. Amos, W.A. *Report on Biomass Drying Technology*; National Renewable Energy Laboratory: Golden, CO, USA, 1999.
49. Fagernäs, L.; Brammer, J.; Wilén, C.; Lauer, M.; Verhoeff, F. Drying of biomass for second generation synfuel production. *Biomass Bioenergy* **2010**, *34*, 1267–1277. [CrossRef]
50. Gebreegziabher, T.; Oyedun, A.O.; Hui, C.W. Optimum biomass drying for combustion—A modeling approach. *Energy* **2013**, *53*, 67–73. [CrossRef]

51. Ståhl, M.; Granström, K.; Berghel, J.; Renström, R. Industrial processes for biomass drying and their effects on the quality properties of wood pellets. *Biomass Bioenergy* **2004**, *27*, 621–628. [CrossRef]
52. Francik, S.; Łapczyńska-Kordon, B.; Francik, R.; Wójcik, A. Modeling and Simulation of Biomass Drying Using Artificial Neural Networks. In *Renewable Energy Sources: Engineering, Technology, Innovation*; Springer: Cham, Switzerland, 2018; pp. 571–581.
53. Głowacki, S.; Tulej, W.; Jaros, M.; Sojak, M.; Bryś, A.; Kędziora, R. Kinetics of Drying Silver Birch (Betula pendula Roth) as an Alternative Source of Energy. In *Renewable Energy Sources: Engineering, Technology, Innovation*; Springer: Cham, Switzerland, 2018; pp. 433–442.
54. Kumar, A.; Jones, D.; Hanna, M. Thermochemical Biomass Gasification: A Review of the Current Status of the Technology. *Energies* **2009**, *2*, 556–581. [CrossRef]
55. Jahirul, M.I.; Rasul, M.G.; Chowdhury, A.A.; Ashwath, N. Biofuels production through biomass pyrolysis—A technological review. *Energies* **2012**, *5*, 4952–5001. [CrossRef]
56. Peng, J.H.; Bi, X.T.; Sokhansanj, S.; Lim, C.J. Torrefaction and densification of different species of softwood residues. *Fuel* **2013**, *111*, 411–421. [CrossRef]
57. Li, H.; Liu, X.; Legros, R.; Bi, X.T.; Jim Lim, C.; Sokhansanj, S. Pelletization of torrefied sawdust and properties of torrefied pellets. *Appl. Energy* **2012**, *93*, 680–685. [CrossRef]
58. Wróbel, M.; Hamerska, J.; Jewiarz, M.; Mudryk, K.; Niemczyk, M. Influence of Parameters of the Torrefaction Process on the Selected Parameters of Torrefied Woody Biomass. In *Renewable Energy Sources: Engineering, Technology, Innovation. Springer Proceedings in Energy ICORES 2017*; Mudryk, K., Werle, S., Eds.; Springer Proceedings in Energy: Cham, Switzerland, 2018; pp. 691–700.
59. Chandler, C.; Cheney, P.; Thomas, P.; Traband, L.D.W. *Fire in Forestry—Vol.I. Forest Fire Behavior and Effects*; John Wiley & Sons: New York, NY, USA, 1983.
60. Chen, D.; Gao, A.; Cen, K.; Zhang, J.; Cao, X.; Ma, Z. Investigation of biomass torrefaction based on three major components: Hemicellulose, cellulose, and lignin. *Energy Convers. Manag.* **2018**, *169*, 228–237. [CrossRef]
61. Chen, W.H.; Kuo, P.C. Torrefaction and co-torrefaction characterization of hemicellulose, cellulose and lignin as well as torrefaction of some basic constituents in biomass. *Energy* **2011**, *36*, 803–811. [CrossRef]
62. Mierzwa-Hersztek, M.; Gondek, K.; Jewiarz, M.; Dziedzic, K. Assessment of energy parameters of biomass and biochars, leachability of heavy metals and phytotoxicity of their ashes. *J. Mater. Cycles Waste Manag.* **2019**, *21*, 786–800. [CrossRef]
63. Jankowska, A.; Kozakiewicz, P. Determination of fibre saturation point of selected tropical wood species using different methods. *Drewno* **2016**, *59*, 89–97.
64. Gilbert, P.; Ryu, C.; Sharifi, V.; Swithenbank, J. Effect of process parameters on pelletisation of herbaceous crops. *Fuel* **2009**, *88*, 1491–1497. [CrossRef]
65. Lam, P.S.; Sokhansanj, S.; Bi, X.; Lim, C.J.; Melin, S. Energy input and quality of pellets made from steam-exploded douglas fir (Pseudotsuga menziesii). *Energy Fuels* **2011**, *25*, 1521–1528. [CrossRef]
66. Shaw, M.D.; Karunakaran, C.; Tabil, L.G. Physicochemical characteristics of densified untreated and steam exploded poplar wood and wheat straw grinds. *Biosyst. Eng.* **2009**, *103*, 198–207. [CrossRef]
67. Anglès, M.N.; Ferrando, F.; Farriol, X.; Salvadó, J. Suitability of steam exploded residual softwood for the production of binderless panels. Effect of the pre-treatment severity and lignin addition. *Biomass Bioenergy* **2001**, *21*, 211–224. [CrossRef]
68. Biswas, A.K.; Yang, W.; Blasiak, W. Steam pretreatment of Salix to upgrade biomass fuel for wood pellet production. *Fuel Process. Technol.* **2011**, *92*, 1711–1717. [CrossRef]
69. Carone, M.T.; Pantaleo, A.; Pellerano, A. Influence of process parameters and biomass characteristics on the durability of pellets from the pruning residues of Olea europaea L. *Biomass Bioenergy* **2011**, *35*, 402–410. [CrossRef]
70. Kirsten, C.; Lenz, V.; Schröder, H.W.; Repke, J.U. Hay pellets—The influence of particle size reduction on their physical-mechanical quality and energy demand during production. *Fuel Process. Technol.* **2016**, *148*, 163–174. [CrossRef]
71. Tumuluru, J. Effect of Moisture Content and Hammer Mill Screen Size on the Briquetting Characteristics of Woody and Herbaceous Biomass. *KONA Powder Part.* **2019**. [CrossRef]
72. Payne, J.D. Troubleshooting the Pelleting Process. In *Feed Technology Technical Report Series*; American Soybean Association: Singapore, 2006.

73. Muramatsu, K.; Massuquetto, A.; Dahlke, F.; Maiorka, A. Factors that Affect Pellet Quality: A Review. *J. Agric. Sci. Technol. A* **2015**, *5*, 717–722. [CrossRef]
74. EN ISO 18125:2017. *Solid Biofuels—Determination of Calorific Value*; International Organization for Standardization: Geneva, Switzerland, 2017.
75. EN ISO 18122:2015. *Solid Biofuels—Determination of Ash Content*; International Organization for Standardization: Geneva, Switzerland, 2015.
76. EN ISO 18123:2015. *Solid Biofuels—Determination of the Content of Volatile Matter*; International Organization for Standardization: Geneva, Switzerland, 2015.
77. Stelte, W.; Holm, J.K.; Sanadi, A.R.; Barsberg, S.; Ahrenfeldt, J.; Henriksen, U.B. A study of bonding and failure mechanisms in fuel pellets from different biomass resources. *Biomass Bioenergy* **2011**, *35*, 910–918. [CrossRef]
78. Stelte, W.; Holm, J.K.; Sanadi, A.R.; Barsberg, S.; Ahrenfeldt, J.; Henriksen, U.B. Fuel pellets from biomass: The importance of the pelletizing pressure and its dependency on the processing conditions. *Fuel* **2011**, *90*, 3285–3290. [CrossRef]
79. Wróbel, M. Assessment of Agglomeration Properties of Biomass—Preliminary Study. In *Renewable Energy Sources: Engineering, Technology, Innovation. Springer Proceedings in Energy ICORES 2018*; Wróbel, M., Jewiarz, M., Szlęk, A., Eds.; Springer Proceedings in Energy: Cham, Switzerland, 2020; pp. 411–418.
80. EN ISO 18847:2016. *Solid Biofuels—Determination of Particle Density of Pellets and Briquettes*; International Organization for Standardization: Geneva, Switzerland, 2016.
81. EN ISO 17831-1:2015. *Solid Biofuels—Determination of Mechanical Durability of Pellets and Briquettes—Part 1: Pellets*; International Organization for Standardization: Geneva, Switzerland, 2015.
82. EN ISO 17225-6:2014. *Solid Biofuels—Fuel Specifications and Classes—Part 6: Graded Non-Woody Pellets*; International Organization for Standardization: Geneva, Switzerland.
83. Filbakk, T.; Skjevrak, G.; Høibø, O.; Dibdiakova, J.; Jirjis, R. The influence of storage and drying methods for Scots pine raw material on mechanical pellet properties and production parameters. *Fuel Process. Technol.* **2011**, *92*, 871–878. [CrossRef]
84. Ståhl, M.; Berghel, J. Energy efficient pilot-scale production of wood fuel pellets made from a raw material mix including sawdust and rapeseed cake. *Biomass Bioenergy* **2011**, *35*, 4849–4854. [CrossRef]
85. Englund, F.; Nussbaum, R.M. Monoterpenes in Scots pine and Norway spruce and their emission during kiln drying. *Holzforschung* **2000**, *54*, 449–456. [CrossRef]
86. Pietsch, W. *Agglomeration Processes. Phenomena, Technologies, Equipment*; Wiley-VCH Verlag GmBH.: Weinheim, Germany, 2002; ISBN 3-527-30369-3.

© 2020 by the authors. Licensee MDPI, Basel, Switzerland. This article is an open access article distributed under the terms and conditions of the Creative Commons Attribution (CC BY) license (http://creativecommons.org/licenses/by/4.0/).

Article

Parameters Affecting RDF-Based Pellet Quality

Marcin Jewiarz [1],*, Krzysztof Mudryk [1], Marek Wróbel [1], Jarosław Frączek [1] and Krzysztof Dziedzic [2],*

[1] Department of Mechanical Engineering and Agrophysics, University of Agriculture in Kraków, Balicka 120, 30-149 Kraków, Poland; krzysztof.mudryk@urk.edu.pl (K.M.); marek.wrobel@urk.edu.pl (M.W.); jaroslaw.fraczek@urk.edu.pl (J.F.)
[2] Beskid Żywiec Sp. z o.o., Department of Research and Development, Kabaty 2, 34-300 Żywiec, Poland
* Correspondence: marcin.jewiarz@urk.edu.pl (M.J.); dziedzickrzysiek@poczta.fm (K.D.); Tel.: +48-12-6624643 (M.J.)

Received: 31 January 2020; Accepted: 14 February 2020; Published: 18 February 2020

Abstract: Increasing production of waste has compelled the development of modern technologies for waste management. Certain fractions of municipal solid wastes are not suitable for recycling and must be utilised in other ways. Materials such as refuse-derived fuel (RDF) fractions are used as fuel in cement or CHP (combined heat and power) plants. The low bulk density leads to many problems pertaining to transportation and storage. In the case of biomass, these problems cause reduction in pelletisation. This paper therefore presents a comprehensive study on RDF pellet production, which is divided into three major areas. The first describes laboratory-scale tests and provides information on key factors that affect pellet quality (e.g., density and durability). Based on this, the second part presents a design of modified RDF dies to form RDF pellets, which are then tested via a semi-professional line test. The results show that RDF fraction can be compacted to form pellets using conventional devices. Given that temperature plays a key role, a special die must be used, and this ensures that the produced pellets exhibit high durability and bulk density, similar to biomass pellets.

Keywords: refuse-derived fuel; refuse-derived fuel (RDF); pellets; durability; density; die; design

1. Introduction

The current model for waste management in Poland assumes that waste is pre-sorted at, or near, its place of production. Polish and European laws regulate the possible application of waste-derived materials. Figure 1 shows the waste hierarchy where prevention (i.e., reduction) takes the highest level of priority. Re-use and recycling of waste as well as energy recovery are ranked next [1]. Finally, landfilling ranks last, being the least favoured method of waste processing. Energy recovery is realised by incineration plants, cement plants, and CHP (combined heat and power) plants [2–7]. Incineration plants are usually equipped for the waste stream produced by communities. By contrast, cement and CHP plants are more challenging because they require high-quality fuel. In particular, they require conditions of high HHV (higher heating value) as well as low moisture, chlorine, and sulphur content. In Poland, approximately 30% of material, classified as refuse-derived fuel (RDF) or solid refused fuel (SRF), is applied to cement plants [8,9]. The remaining stream is incinerated. Solutions for recovery from municipal solid waste (MSW) streams with energetic potential and processing are needed. It is important that the resulting fuel is attractive to other sectors of the energy economy, especially specialised CHP plants [10–12].

Mechanical–biological treatment (MBT) plants produce most of the Polish RDF fraction from municipal solid waste (MSW) [13]. These units use a combination of separation, drying, and biological processes for the treatment. The first process involves sieving of the particles on drum screens, with various mesh sizes (from 50 to 100 mm), which enables separation of fine particles. The oversized fraction (particles remaining on the drum screen) can be dried (if necessary) and processed. The undersized fraction contains a significant amount of organic matter. In most cases, this fraction is composted or undergoes other forms of biological treatment, while the oversized fraction is sent for energy recovery [14]. Previous studies have shown that the morphological composition of this fraction can vary. Even in a single region, it can vary based on the time of year [2]. MSW comprises organic matter, minerals (including glass), plastics, textiles, paper, and cardboard. The last four materials are mainly present in the oversized fraction. These are materials with high energy parameters (low ash content and high calorific value) favourable for waste-based fuel.

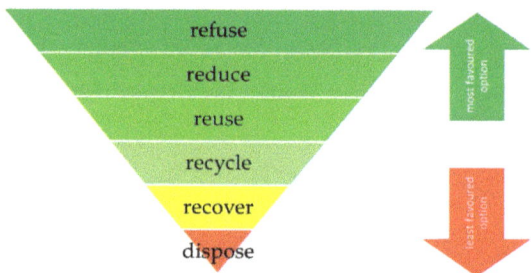

Figure 1. Waste hierarchy schematic [15,16].

According to literature [17–20], the mixed MSW fraction in Poland is very inhomogeneous. This material is dependent upon regional characteristics and residents' customs. Urban areas, characterised by highly dense social and economic infrastructure, account for the highest production of RDFs/SRFs. From 1000 g of MSW, it is possible to achieve approximately 500 g of RDF/SRF. Areas with a high proportion of single-family houses can generate up to 380 g, whereas rural areas can produce only 270 g per kg of MSW.

The growing interest in using MSW as an energy source is limited by homogenisation and low bulk density. A comparison between the bulk densities of RDF and other popular materials is shown in Figure 2. These issues, in the case of biomass, are overcome by densification processes [21,22]. These processes increase the bulk density by producing pellets or briquettes. Such fuel is more favourable for transport and storage. Literature describes that the usage of pellet instead of loose RDF would improve the production of cement [23]. It also appears to be a sought-after fuel for CHP plants [24]. Due to the higher bulk density, such fuel will be more beneficial in other thermal treatment processes such as gasification [25], pyrolysis [26,27], or even torrefaction [28]. Pellet production also leads to quality uniformity of the produced fuel. Like biomass, RDF fuel can also be compacted, but it requires special attention. The morphology of the MSW-based material is different from that of biomass. Biomass is composed of natural polymers: cellulose, hemicellulose, and lignin, and lignin acts as a natural binder during the densification of biomass [29–33]. However, such binders do not occur in RDF, so its potential for densification must be checked. The introduction of artificial binders is not desired, and may moreover be impossible to achieve. Such substances can change the fuel quality parameters and increase the cost of the fuel [34,35].

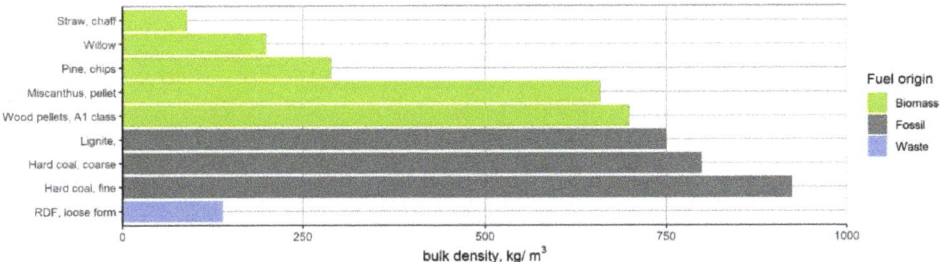

Figure 2. Comparison of bulk densities of fuel materials [32,36–38].

Other important factors for densification are grindability and grain size distribution (after milling) [39]. An oversized fraction is troublesome during conventional milling. Owing to the high share of elastic and fibrous materials, dedicated devices are needed, a requirement that is often realised upon using several knife mills working in a cascade mode. However, the achieved material is still not as fine as biomass grist, as illustrated in Figure 3. The grain size distribution is a major parameter that affects pellets production [40]. During biomass pelletisation, the rollers that move across the die also grind the particles. Given that biomass is more brittle than RDF, this process should not pose a significant issue. Another factor is moisture content, which is a major parameter in the densification of biomass. In both cases, drying is used. Based on literature, the levels of moisture range from 10% to 20%, with respect to mass [41–43]. Many studies have covered the agglomeration of biomass [44–46]. However, only a few have covered the problems connected with MSW-based pellet production [47]. The die temperature is often above 100 °C and is a result of dissipated energy from friction between the die, rollers, and material [48]. High friction results in higher temperatures, up to a limit. When the temperature becomes too high, carbonisation occurs on the pellet surface, which is specific to the material used. Given that the temperature is a result of die geometry, in general, it is not a point of interest for manufacturers. Friction increases with the die thickness or channel entry angle, which both affect the resistance generated during the pressing of material through the matrix channels.

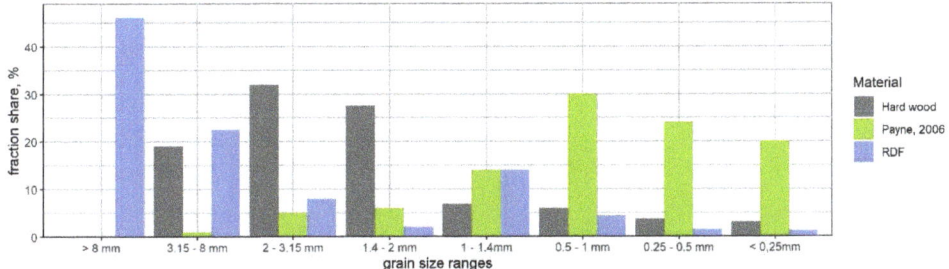

Figure 3. Grain size distribution comparison of biomass and refuse-derived fuel (RDF) grists from a knife mill [49]. Payne, 2006 is optimal grain size distribution to produce high quality biomass pellets [40].

Based on this, the main aim of this study was to develop an effective technology to produce pellets from RDF. We assumed that the majority of this product will go to CHP plants. Thus, larger pellet diameters are acceptable, such that the final product will be pellets with diameter greater than 6 mm (typical size of biomass pellets).

As in the case of biomass, it is important to determine all aspects of the process that affects product quality. The main challenge is to reduce the importance of feedstock patchiness, so the difference in morphology, grain size distribution, or even moisture content must be considered.

Thus, rigorous research is needed. First, we conducted our analysis based on the selected process parameters: durability and density. Next, we carried out a small-scale test on a modified die. Then a semi-professional line test was carried out, to verify all aspects of this study.

2. Materials and Methods

For all research phases, the oversized MSW fraction was obtained from a local MBT plant (Kraków, Poland). This material was collected in spring, 2018. This was a high-quality fraction, which is used as RDF for a cement plant after milling. The morphological structure of this material is detailed in [2]. Given that MSW is composed of bulky materials, such as plastic elements or textile cloths, it must be shredded. For this, we used a knife mill (MF-4, Protechnika, Poland) with a mesh size of 10 mm, which is typical for RDF. Before grinding was performed, all hard materials, such as rocks, metal elements, or glass, were removed. The produced grist was then sieved to determine grain size distribution. This was assessed using a sieve shaker (LPzE-4e, Multiserw-Morek, Poland) equipped with 6, 3.15, 2, 1.4, 1, and 0.5 mm sieves (Multiserw-Morek, Poland). Next, the material was conditioned in the constant climate chamber (KBF-S 115, Binder, Germany) under controlled temperature and humidity conditions. This allows us to acquire consistent moisture content throughout the study. According to literature [42], we chose 18% as the most suitable moisture content value.

2.1. Laboratory-Scale Densification

For the laboratory-scale densification, we used the universal testing machine (EDZ 20 Type, VEB "Fritz-Heckert", Germany) with special appliance. This appliance consists of several elements, as shown in Figure 4. The main component consists of the die and piston, which can be changed for different pellet diameters. In this study, we employed three diameters: 16, 18, and 20 mm. With the testing machine, we obtained precise information about the force used to compress the material. This value is converted to pressure, using the piston base area. In the laboratory test, according to the literature, we selected five pressure levels: 80, 100, 120, 140, and 160 MPa. Table 1 presents the calculated force values needed to achieve the desired pressure levels. The heating band, mounted on the outer sleeve, heats up both the sleeve and die. The high mass of this assembly (approximately 5 kg), results in high thermal inertia, which stabilises the temperature during tests. A proportional-integral-derivative system controls the temperature, using two thermocouples. Based on literature and our assumptions, we selected three temperature levels: 80, 100, and 120 °C. The upper temperature value is limited by both the melting point of the plastics and the die temperature achieved during normal pellet production [32,50].

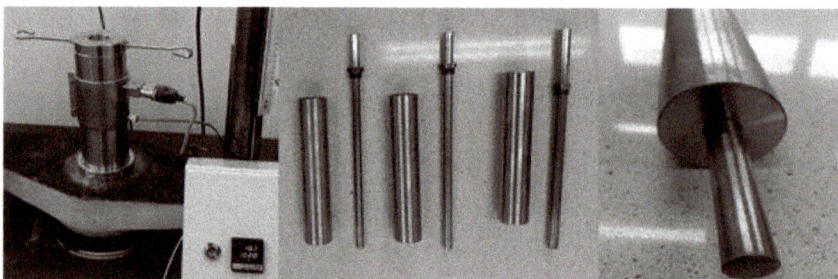

Figure 4. Laboratory-scale compaction stand for pellet production.

Table 1. Compaction pressure levels with forces for different die diameters.

Pressure	Die Diameter		
	16 mm	18 mm	20 mm
	Force, kN		
80 MPa	16.08	20.36	25.13
100 MPa	20.11	25.45	31.42
120 MPa	24.13	30.54	37.7
140 MPa	28.15	35.63	43.98
160 MPa	32.17	40.72	50.27

A total of 45 combinations of die diameter, pressure, and temperature were tested. To ensure quality of our results, we produced five sets of granules for each test. This resulted in the production of 225 pellets in total. All samples were produced from a similar mass of 4 g of RDF grist.

2.2. Laboratory-Scale Analysis

After densification, we measured the height, diameter, and mass of the individual granules. After this, we placed the samples into individually labelled compartments to prevent any misidentification during later measurements.

All samples were left for 24 h, to determine the relaxation (expansion) factor R, which is a popular quantity for biomass pellets. It provides information about the height changes after 24 h as it is normal for biomass granules to expand after densification [51]. It is given by

$$R = \frac{h_{24} - h_0}{h_0} \quad (1)$$

where R is the expansion coefficient, h_0 is the pellet height immediately after the production process, and h_{24} is the pellet height after 24 h. This quantity gives information how strong bonding forces between particles are. For elastic materials, R will have higher values than rigid ones.

Mechanical durability is a crucial parameter concerning transport and logistics. Many standards refer to this quantity [52]. In general, as described in methods, we required about 500 g of pellets per test. In laboratory scale we produced only five granules (20 g) in each condition. The small number of samples was dictated by their time-consuming production. Assuming that two people worked at this stage, it took about 2–4 min to produce one pellet. This gives an average of 12 man-hours for one full series of samples (500 g), which in our case was a total of 45 combinations. That is why we could not use such methods without modifying them. We used equipment described in the ISO 17831-1:2016-02 [53] standard, as shown in Figure 5, which simulates the damage to particles during transportation. The main component is a cuboidal chamber with two baffles inside. The box rotates at 50 rpm for 10 min. Given that we used only five granules, we filled the box with ABS (acrylonitrile butadiene styrene) balls 5 mm in diameter, to meet the 500 g noted in the standard [54]. ABS balls were chosen, as they are durable, and have similar specific density to RDF pellets. The principle of operation of the balls should be similar to that used for smoothing the edges of metal castings. After the process, we removed and weighed the granules. The mechanical durability DU is given by

$$DU = \frac{m_A}{m_E} \cdot 100 \quad (2)$$

where m_E and m_A is mass of the sample before and after testing, respectively [53]. Please note that this value is only for comparison within this study, so reference with other pellets is not possible. Further research on this method is in progress.

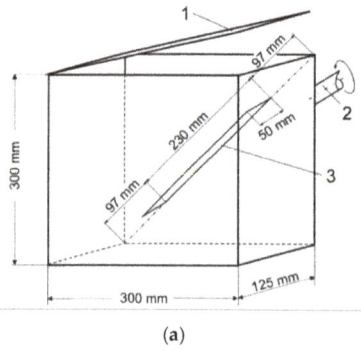 

Figure 5. Mechanical durability machine: (**a**) measurement box schematic [53] and (**b**) image of samples immersed in ABS (acrylonitrile butadiene styrene) balls.

The specific density of the obtained agglomerates was measured using a quasi-fluid pycnometer (GeoPyc 1360, Micromeritics, USA). The device measures the volume of a sample placed in the measuring chamber, using the DryFlo (Micromeritics, USA) quasi-fluid agent. Using a dry agent, instead of a liquid, eliminates the wetting phenomenon (i.e., soaking) of the tested material. This makes it possible to test very absorptive materials. The pycnometer is shown in Figure 6a. First, the device determines the volume of agent placed in the chamber (Figure 6b) of known diameter, based on the piston travel measurement. Then, the sample is added, and the total volume is determined. The volume of the sample is determined by the difference between the total volume and the agent volume. The device also automatically calculates the density, given that the mass and volume are known. It is possible to set the precise force that the piston will apply to press the material inside the chamber, which enables determination of the volume of substances with low compressive strength.

Figure 6. GeoPyc 1360 pycnometer: (**a**) general view and (**b**) measuring chamber.

2.3. Die Design

As RDF differs from biomass, a special die geometry must be proposed. The laboratory-scale tests confirmed that temperature plays a key role in the densification process. To determine the temperature of the standard biomass die, when RDF is used as the feedstock, we used an infrared (IR) camera (T600 Series, FLIR, USA). As the die moves at high rpm, temperature measurement techniques are limited. Thus, a camera was used to visualise the temperature of the die. In this application, we only need information about the range of temperature. A picture of the element is taken after several minutes of stable operation, to ensure that all parts reach their working temperature. We carried out a similar test for the modified die. From the acquired thermographs, we redefined the die dimensions.

2.4. Test on Professional Machinery

Data taken during the laboratory-scale test were used for the new die design, which should enable the production of high-quality RDF pellets. The redesigned matrix was then tested on a small-scale flat die pelletiser (MGL 200, KovoNovak, Czech Republik). For reference, we used a normal die for biomass pelletisation. We analysed the bulk density and mechanical durability, using the relevant standards. The determination of bulk density (BD) was carried out following the ISO 17828:2015 standard [55]. We used containers of volume 0.005 m^3 for pellets, and 0.05 m^3 for RDF grist. These results were used to determine how the bulk density changes during the process. For a better understanding of the process, we also took tomographic pictures using the IR camera. The temperature of the die's surface and pellets were monitored during operation. Data from the thermographs were used to confirm the effects of the design process.

The final tests were on a semi-professional line, based on ring-die pelletiser, which was equipped with all unit operations needed for pellet production. This includes drying, milling, storage, densification, and separation processes. The semi-professional line was equipped with our new die design. Due to strength issues, we were not able to produce dies with the same channel diameter. The small part width reduced the largest possible diameter to 8 mm. All other concepts, such as channel length to diameter ratio, were maintained. In this test, we were also able to acquire bulk density and durability information.

3. Results

3.1. Raw Material Analysis

Figure 7 presents the morphology of the waste stream we used in this study. The high share of plastics is normal because the RDF is formed of the oversized fraction. Before any further analysis, all visible elements of glass, metal, and minerals that could damage the mill elements were removed. Grain size analysis, before and after milling, is shown in Figure 8. It confirms what has been suggested in the literature; that RDF is a difficult material for milling. The dominant fraction is over 6 mm. These were thin and flexible plastic foils and textiles, which are very hard to cut in the knife mill.

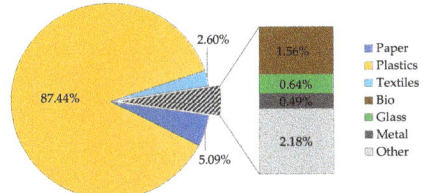

Figure 7. Morphology of the material used in this study.

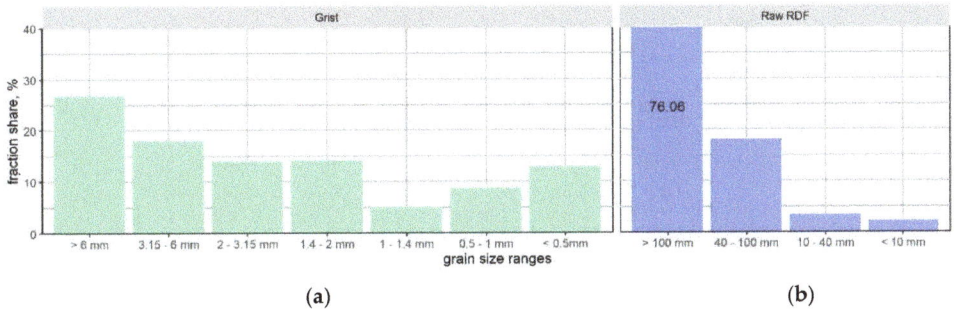

Figure 8. Comparison of RDF grain size distribution after milling (**a**), and before (**b**).

3.2. Laboratory-Scale Densification

Owing to the small number of pellets produced from each series, all granules were treated with care. Figure 9 shows the specific density for all pellets grouped by diameter and temperature. Each box represents an average value of the five samples grouped by pressure. The average value for the group is also indicated in each plot. The best pellets should have the highest specific density. The results show that it is hard to achieve granules with stable density value, even in recurring laboratory conditions. Analysis of the average values provides initial information about the trends.

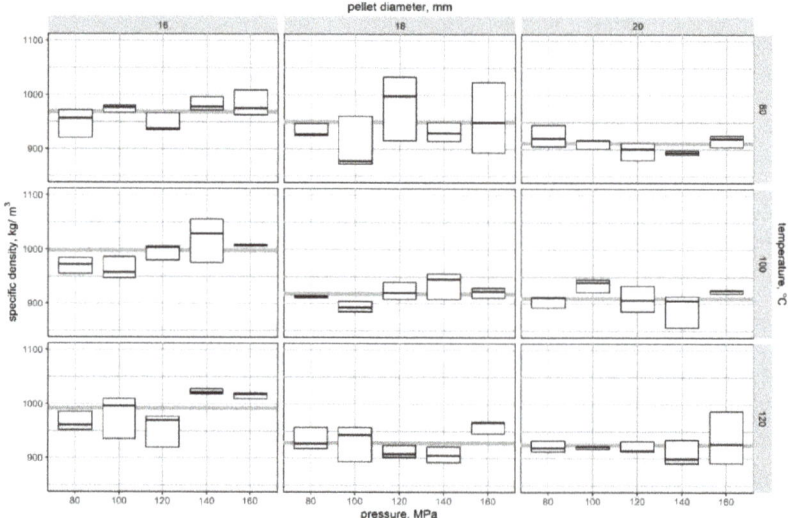

Figure 9. Variation of specific density with compaction pressure, temperature, and pellet diameter. The grey line in each plot represents the average value of each group. Top and bottom of each box represents range of values deviation.

After 24 h, the expansion coefficient of the samples was considered according to height change. Exemplary pellets with our quality criterion are shown in Figure 10. We have attributed high-quality pellets as having an expansion coefficient less than or equal to 5%. The appearance of the samples and the ranges used in this test are shown in Figure 10. The cumulative results are shown in Figure 11. As for Figure 9, average values are also indicated by grey lines. As for the density analysis, values for the same pressure group vary.

Figure 10. Pellet quality criterion based on the expansion coefficient.

The mechanical durability results, based on our modified method, are shown in Figure 12. According to biomass pellets standards, all samples with durability higher than 96% were considered to have high quality.

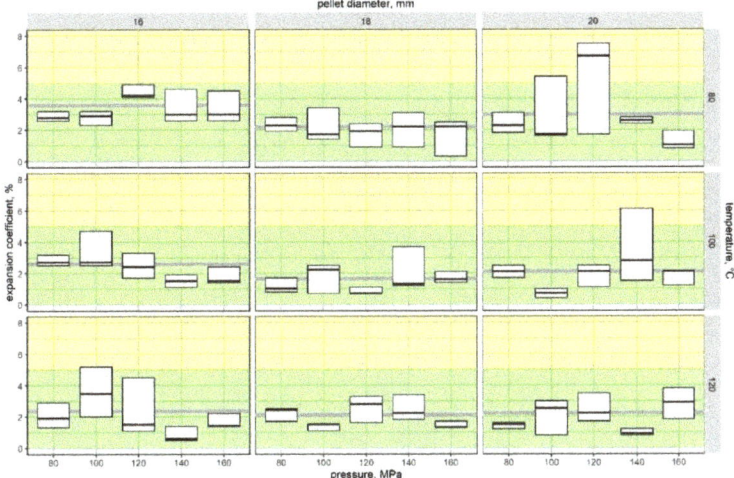

Figure 11. Variation of the expansion coefficient with compaction pressure, temperature, and pellet diameter. The grey line in each plot represents the average value of each group. Top and bottom of each box represents range of values deviation.

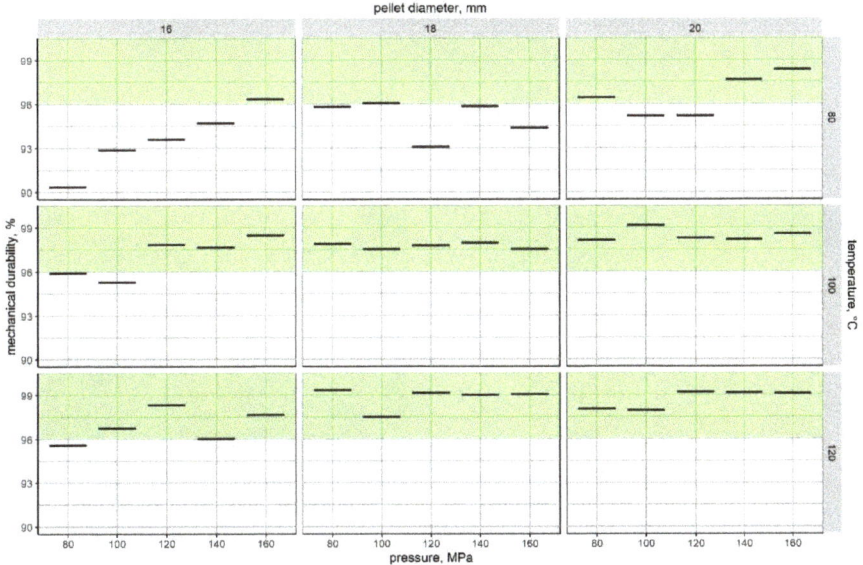

Figure 12. Variation of the mechanical durability with compaction pressure, temperature, and pellet diameter.

3.3. RDF Die Design

Before designing the new die, we tested the standard biomass die, to determine die temperature and pellet quality. We used a standard biomass die, with a channel diameter of 8 mm, a thickness of 35 mm, and an entry angle of 19.6°. The results, shown in Table 2, show that this type of channel geometry is inappropriate for RDF. Figure 13 shows images of pellets formed using this die. They are short and there are large amounts of fine particles. Thermographic analysis of the die and pellets temperatures are shown in Figure 14. For biomass material, the die temperature should be nearby 100°C [56,57]. For RDF, these temperatures are lower, reaching only 80 °C. This results in pellet temperatures at the 60 °C level. According to the laboratory-scale results, these values are too low. This was considered for the new die design.

Table 2. Pellet quality and raw materials.

ID	BD, kg/m^3	DU, %
Raw RDF	143	-
Pellets BM	411	85.1
Pellets RDFM$_{FLAT}$	665	98.6
Pellets RDFM$_{RING}$	669	98.9

BM: biomass die; RDFM$_{FLAT}$: flat RDF die; RDFM$_{RING}$: ring die; BD: bulk density; DU: mechanical durability.

Figure 13. (a) Images of the raw RDF grist, (b) pellets from biomass die, and (c) from modified RDF die.

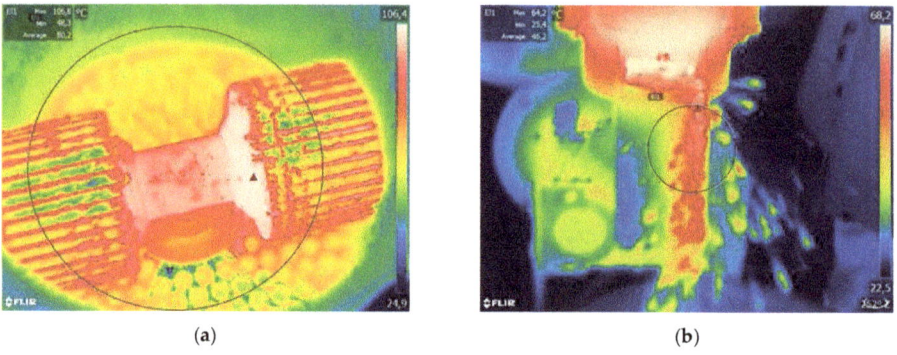

Figure 14. Thermographs from the infrared (IR) camera using the biomass die: (a) view of die and rollers, and (b) view of pellet output.

Details of the geometry, in comparison to biomass die, are shown in Figure 15a. Due to construction issues, we reduced the channel diameter to 12 mm (for details see the discussion section). The modified construction is also up to two times thicker (Figure 15b).

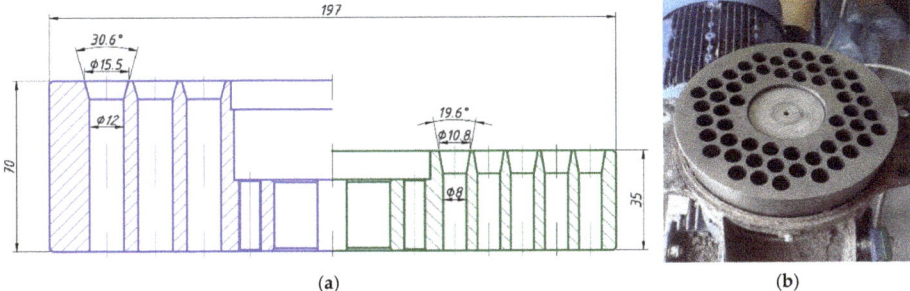

Figure 15. (**a**) Modified die (blue) compared to standard biomass die (green), and (**b**) modified die mounted into the pelletiser.

3.4. Flat Die and Semi-Professional Line Results

The standard, used for biomass densification and redesigned for the RDF dies, was compared. From this test, we acquired information about bulk density and mechanical durability of the pellets. The results are presented in Table 1.

Thermographic analysis (with the IR camera) is shown Figure 16. It is clear that the die and pellet temperatures are higher. These increases meet assumptions made during the design process. In addition, Table 2 shows that based on a comparison of BD and DU for modified and standard dies to biomass, pellets from modified dies are of higher quality.

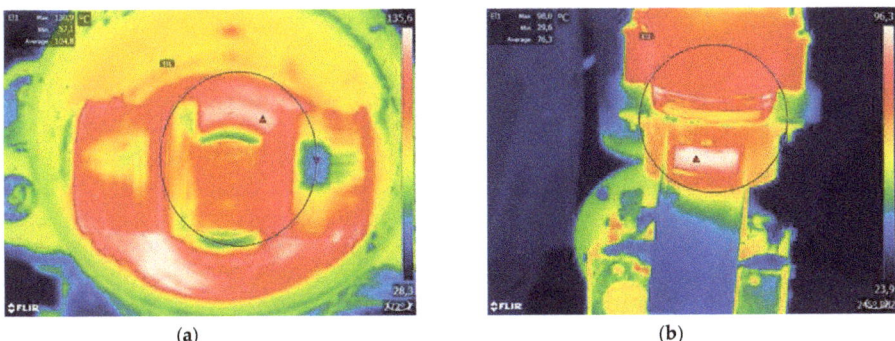

Figure 16. Thermographs from IR camera while using the modified RDF die: (**a**) view of die and rollers, and (**b**) view of pellet output.

The final stage was verification of our design incorporated in other types of pelletisers (i.e., ring die), which are the most popular in high yield installations. The ring die was modified based on previous assumptions. Table 2 presents results from the semi-professional line equipped with a modified ring die. The bulk density and durability values confirm that produced pellets are of high quality.

4. Discussion

The results show how difficult RDF is, especially in terms of agglomeration. However, several aspects must be considered. First, there is the morphology of the waste stream. Composition can vary within the year, and it is dependent on many factors. This problem is described in many studies [2,17]. Based on the average composition, we could define parameters that can be tailored to materials like RDF. Low amount of binders in a pre-densified mixture in connection with high content of plastics requires special geometry. This will ensure a sufficiently high die temperature, exceeding 100°C. Such temperature for small scale production of pellets (laboratory scale) was not found in the literature [42,47].

A comparison of the specific density values leads to some issues connected with this material. In the laboratory-scale testing, it was difficult to form each sample with the same composition. This is the main reason there are such noticeable differences within a single series. As can be seen in Figure 10, some samples have many cracks, which reduces the measured density. In addition, there were some sections of hard and dense plastics (e.g., polyamide), so the granule density would be higher. Despite this, we were able to evaluate some general trends. The highest granule density was obtained with the smallest channel diameter: 16 mm. At this stage, it is difficult to determine why this dependency has occurred. The literature describes many density and pellet diameter analyses. In most cases, specific density depends on the type of material, although there is no clear dependency with pellet diameter [57].

The expansion factor provides information about the shape stability of pellets. Here, we specified that good quality pellets should not exceed an expansion coefficient greater than 10% (Figure 11). In this case, temperature plays a major role, as only a few granules exceeded the value of 5%. The averages for all diameters at this temperature were at a similar level (from 2.1% to 2.4%). The lowest expansion coefficient, occurring at the highest temperature could be explained by the melting of plastics on the outer surface of the granule. The produced shell then stabilises the shape. Slightly higher values, in the case of the 16 mm die, can be explained by the fact that all granules have similar mass. Thus, a small die diameter will result in higher quality pellets, and compressed material will act like a spring.

Given that the density of RDF-based pellets depends on composition, it is difficult to correlate this value with durability. For higher specific densities, higher durability is generally expected. For inhomogeneous material, such as MSW, this rule may not apply. Figure 17 shows that, in our case, these parameters do not correlate at all. Thus, this needs further investigation. More important is the earlier assumption that temperature plays a key role in pellet quality. This is confirmed in Figure 12. The durability of the pellets increases with temperature. For pellets of diameter 16 mm, we also observe a dependency with pressure. At 80 °C, durability is low, compared to other diameters. A similar problem occurs at higher temperatures. Higher temperature pellets have a tendency to break during the test, which should explain this phenomenon. Almost all samples at the temperature of 120 °C have a durability higher than 96%. This value is noted as a limit value for good quality biomass pellets [32]. It is hard to compare laboratory test results with pellets from real production, as we used a modification of the standardized method [53]. Nevertheless, this type of test seems to be more representative than the compressive strength analyses which can be often found in the literature [42,58–60]. The lower durability of the longer pellets can be explained in a different way. The closed chamber compaction has some limitations. The material, during densification, does not move with respect to the channel walls. The friction between the walls and materials is very low, and heat transfer is not very intense. Lower heating rates will result in less meltdown formation. Meltdowns form the plastic shell, which maintains the shape of the pellet. More heat transferred to pellet will result in a thicker shell. For long pellets, with a thin meltdown layer, the overall strength of the structure will be weaker. Thus, a thick meltdown layer or short pellets seems to be crucial for springy material like RDF.

Considering all aspects from the laboratory-scale test, the RDF die should generate high temperatures and high compression. Higher expansion of the pellets, with smaller diameter, provides information on how to modify the geometry of the entry channel. For the biomass die, the entrance angle is usually about 20°. The conical section height is equal to, or higher than, the pellet diameter. This ensures good compression and arrangement of the particles. As a result, durable and long pellets will be produced. Different materials require modifications to the channel geometry. For RDF, this angle would be greater, as longer pellets will have greater expansion. To avoid forming long pellets, the conical section should be shorter than the channel diameter. This would result in the compaction of small portions of the material. Given that energy to heat up the die is generated from frictional forces, it was important to optimise this factor. The rollers moving on the die, and material squeezing through a channel, generates the most heat. Table 2 shows that under-heated die results in low-quality pellets. In addition, we do not observe meltdowns on the outer surface. This was one of the key conclusions from the laboratory-scale test. Therefore, in the prototype design, we decided to double the thickness of the die, increasing it to 70 mm. This would ensure that the material is exposed to high temperatures so that the meltdown of plastics occurs. It is clear the produced pellets show a different outer surface. For the biomass die, meltdowns do not occur, while the modified die produces a well-melted shell. This also improves the durability and bulk density of the pellets. This result confirms that temperature plays a major role in the pelletisation of the MSW-based materials. However, there are other crucial parameters to consider. Die geometry, especially the entry channel angle, is also important. Increasing this dimension generates good quality pellets. The die thickness also had a clear impact on durability. A longer residence time results in a thicker layer of molten plastics on the outer surface. This principle occurs with both the flat and ring die.

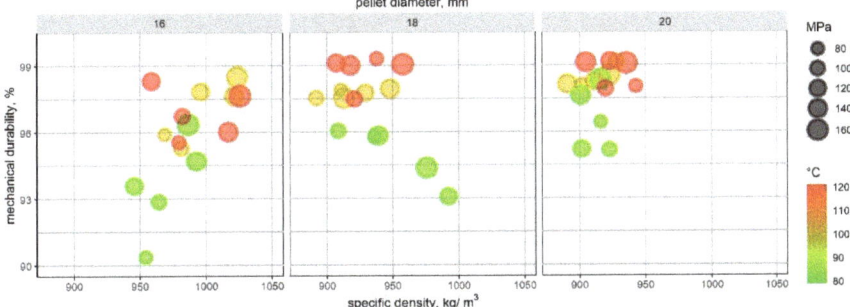

Figure 17. The relationship between mechanical durability and specific density. Average values for series (five granules). Size represents compaction pressure and colours are in connection to temperature.

The comparison of bulk densities of the raw RDF and pellets gives clear information how this process affects fuel quality. Figure 18 shows that in case of RDF the BD of pellets can be up to five times higher in relation to raw material. At shown in literature [32,36,50,61,62] these values are achieved for soft biomass materials, with low bulk density. This will improve economics of transportation and logistics [36,50].

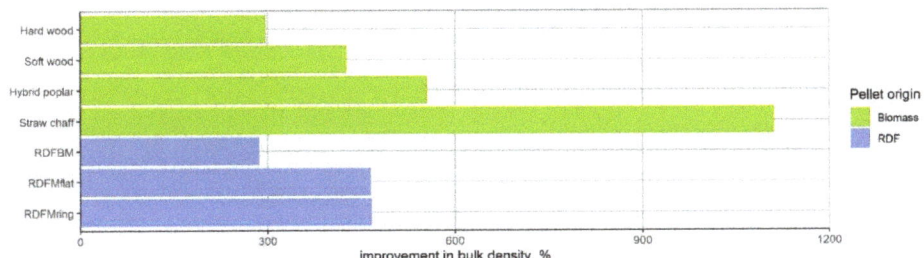

Figure 18. Improvement in bulk densities (related to raw material) of RFD samples, in connection to literature, respectively [32,36,61].

5. Conclusions

This study aimed to determine the major factors that affect RDF-based pellet production. Our results show that RDF is a difficult material to work with. The key conclusions are as follows:

- Low bulk density of the raw RDF provides high potential in densification operations.

- Negligible content of natural binders enforces redesign of die.
- It was proven that temperature significantly affects pellet quality.
- The temperature condition of 120 °C results in the most durable pellets for all variations of compaction pressure and channel diameter. This is a result of the melting of plastics.
- Dies dedicated for RDF pelletisers should be thicker (double the thickness or more), compared to those for biomass pelletisers. This will ensure the formation of the outer shell from the melted plastics.
- Improvement in bulk density (by more up to five times), results in the reduction of transportation and storage costs.

Our results are relevant to many fields, which are not described in the literature. The majority of the tests undertaken on a small scale (when several pellets are formed), present compressive strength results. Such a test does not give information on how such material will behave during handling and other logistics processes. The modified method of determining mechanical durability proposed in the paper seems to be more representative than the compression strength test used in many analyses. However, it still needs to be analysed in detail, and this will be done in the near future. Additionally, further research on channel dimensions may be useful for improving this topic. This will require the design of a new type of laboratory-scale attachment, which should enable analysis on the influence of both the entry angle and channel length on pellet quality. A similar device is presented in the literature [63].

Author Contributions: Conceptualization, K.M., M.W., and J.F.; methodology, K.M., M.W., J.F.; investigation, K.M., M.W., M.J., and K.D.; resources, M.J. and K.D; data curation, M.J. and M.W.; writing—original draft preparation, M.J. and K.D.; writing—review and editing, M.J.; visualization, M.J.; supervision, J.F., K.M., and M.W. All authors have read and agreed to the published version of the manuscript.

Funding: This study was founded as part of grant "EkoRDF—an innovative manufacturing technology of alternative fuel from municipal waste for power and heating plants—a key component of the Polish waste management system" financed by the Polish Centre for Research and Development (GEKON programme), which was undertaken at the University of Agriculture in Kraków. Grant no GEKON2/05/268002/17/2015.

Acknowledgments: The authors express many thanks to Wiktor Głowa, MSc, who performed part of the laboratory tests.

Conflicts of Interest: The authors declare no conflict of interest.

References

1. European Commission. *Life Cycle Thinking and Assessment for Waste Management*; European Commission: Brussels, Belgium, 2016.
2. Jewiarz, M.; Frączek, J.; Mudryk, K.; Wróbel, M.; Dziedzic, K. Analysis of MSW Potential in Terms of Processing into Granulated Fuels for Power Generation. In *Renewable Energy Sources: Engineering, Technology, Innovation*; Springer: Cham, Switzerland, 2018; pp. 661–670.
3. Caputo, A.C.; Palumbo, M.; Scacchia, F. Perspectives of RDF use in decentralized areas: Comparing power and co-generation solutions. *Appl. Therm. Eng.* **2004**, *24*, 2171–2187. [CrossRef]
4. Caputo, A.C.; Pelagagge, P.M. RDF production plants: II. Economics and profitability. *Appl. Therm. Eng.* **2002**, *22*, 439–448. [CrossRef]
5. Białowiec, A.; Micuda, M.; Koziel, J.A. Waste to Carbon: Densification of Torrefied Refuse-Derived Fuel. *Energies* **2018**, *11*, 3233. [CrossRef]
6. Rahman, A.; Rasul, M.G.; Khan, M.M.K.; Sharma, S.C. Assessment of energy performance and emission control using alternative fuels in cement industry through a process model. *Energies* **2017**, *10*, 1996. [CrossRef]
7. Hemidat, S.; Saidan, M.; Al-Zu'bi, S.; Irshidat, M.; Nassour, A.; Nelles, M. Potential utilization of RDF as an alternative fuel to be used in cement industry in Jordan. *Sustainability* **2019**, *11*, 5819. [CrossRef]
8. Grzesik, K.; Malinowski, M. Life cycle assessment of refuse-derived fuel production from mixed municipal waste. *Energy Sources Part A Recover. Util. Environ. Eff.* **2016**, *38*, 3150–3157. [CrossRef]
9. Nickolaos, C. Use of waste derived fuels in cement industry: A review. *Manag. Environ. Qual. Int. J.* **2016**, *27*, 178–193.

10. Ryu, C.; Shin, D. Combined heat and power from municipal solid waste: Current status and issues in South Korea. *Energies* **2013**, *6*, 45–57. [CrossRef]
11. Lupa, C.J.; Ricketts, L.J.; Sweetman, A.; Herbert, B.M.J. The use of commercial and industrial waste in energy recovery systems—A UK preliminary study. *Waste Manag.* **2011**, *31*, 1759–1764. [CrossRef]
12. Sever Akdağ, A.; Atimtay, A.; Sanin, F.D. Comparison of fuel value and combustion characteristics of two different RDF samples. *Waste Manag.* **2016**, *47*, 217–224. [CrossRef]
13. Dziedzic, K.; Łapczyńska-Kordon, B.; Malinowski, M.; Niemiec, M.; Sikora, J. Impact of Aerobic Biostabilisation and Biodrying Process of Municipal Solid Waste on Minimisation of Waste Deposited in Landfills. *Chem. Process Eng.* **2015**, *36*, 381–394. [CrossRef]
14. Debicka, M.; Zygadło, M.; Latosinska, J. Investigations of bio-drying process of municipal solid waste. *Ecol. Chem. Eng. A* **2013**, *20*, 1461–1470.
15. Waste Hierarchy. Available online: https://en.wikipedia.org/wiki/Waste_hierarchy (accessed on 28 January 2020).
16. EWIT. The Waste Hierarchy—Hazardous Waste in a Circular Economy. Available online: http://ewit.site/kb/waste-hierarchy-hazardous-waste-circular-economy/ (accessed on 28 January 2020).
17. Młodnicki, K.; Malinowski, M.; Łukasiewicz, M.; Religa, A. The share of impurities in selectively collected waste in one- and multi-sack systems. *Ecol. Chem. Eng. A* **2018**, *25*, 187.
18. Petryk, A.; Malinowski, M.; Dziewulska, M.; Guzdek, S. The impact of the amount of fees for the collection and management of municipal waste on the percentage of selectively collected waste. *J. Ecol. Eng.* **2019**, *20*, 46–53. [CrossRef]
19. Malinowski, M.; Łukasiewicz, M.; Famielec, S.; Nowińska, K. Analysis of changes in fees for the collection and management of municipal waste as regards the efficiency of waste segregation. *Ekon. Sr.* **2019**, *69*, 24–41.
20. Wróbel, M.; Frączek, J.; Mudryk, K.; Jewiarz, M.; Dziedzic, K. Conceptual Design of the RDF Granulation Line. In *Renewable Energy Sources: Engineering, Technology, Innovation*; Springer: Cham, Switzerland, 2018; pp. 813–821.
21. Wrobel, M.; Mudryk, K.; Jewiarz, M.; Knapczyk, A. Impact of raw material properties and agglomeration pressure on selected parmeters of granulates obtained from willow and black locust biomass. *Eng. Rural Dev.* **2018**, *17*, 1933–1938.
22. Knapczyk, A.; Francik, S.; Wrobel, M.; Mudryk, K. Comparison of selected agriculture universities in Europe on basis of analysis of academic output. In Proceedings of the 17th International Scientific Conference Engineering for Rural Development, Jelgava, Latvia, 23–25 May 2018; pp. 727–732.
23. Lockwood, F.C.; Ou, J.J. Review: Burning Refuse-Derived Fuel in a Rotary Cement Kiln. *Proc. Inst. Mech. Eng. Part A J. Power Energy* **1993**, *207*, 65–70. [CrossRef]
24. Del Zotto, L.; Tallini, A.; Di Simone, G.; Molinari, G.; Cedola, L. Energy enhancement of Solid Recovered Fuel within systems of conventional thermal power generation. *Energy Procedia* **2015**, *81*, 319–338. [CrossRef]
25. Nixon, J.D.; Dey, P.K.; Ghosh, S.K.; Davies, P.A. Evaluation of options for energy recovery from municipal solid waste in India using the hierarchical analytical network process. *Energy* **2013**, *59*, 215–223. [CrossRef]
26. Zhou, L.; Luo, T.; Huang, Q. Co-pyrolysis characteristics and kinetics of coal and plastic blends. *Energy Convers. Manag.* **2009**, *50*, 705–710. [CrossRef]
27. Chen, D.; Yin, L.; Wang, H.; He, P. Pyrolysis technologies for municipal solid waste: A review. *Waste Manag.* **2014**, *34*, 2466–2486. [CrossRef] [PubMed]
28. Białowiec, A.; Pulka, J.; Stępień, P.; Manczarski, P.; Gołaszewski, J.; Adefeso, I.B.; Ikhu-Omoregbe, D.I.O.; Rabiu, A.M. Sustainable co-generation plant: Refuse-derived fuel gasification integrated with high temperature PEM fuel cell system. *Waste Manag.* **2013**, *70*, 91–100. [CrossRef]
29. Whittaker, C.; Shield, I. Factors affecting wood, energy grass and straw pellet durability—A review. *Renew. Sustain. Energy Rev.* **2017**, *71*, 1–11. [CrossRef]
30. Antczak, A.; Michaluszko, A.; Klosinska, T.; Drozdzek, M. Determination of the structural substances content in the field maple wood (*Acer campestre* L.)—Comparsion of the classical methods with instrumental. *Ann. Warsaw Univ. Life Sci.-SGGW. For. Wood Technol.* **2013**, *82*, 11–17.
31. Tishler, Y.; Samach, A.; Rogachev, I.; Elbaum, R.; Levy, A.A. Analysis of Wheat Straw Biodiversity for Use as a Feedstock for Biofuel Production. *Bioenergy Res.* **2015**, *8*, 1831–1839. [CrossRef]
32. Peter, K. *The Densification Process of Wood Waste*; De Gruyter: Berlin, Germany, 2015; ISBN 978-3-11-044142-0.

33. Stelte, W.; Holm, J.K.; Sanadi, A.R.; Barsberg, S.; Ahrenfeldt, J.; Henriksen, U.B. A study of bonding and failure mechanisms in fuel pellets from different biomass resources. *Biomass Bioenergy* **2011**, *35*, 910–918. [CrossRef]
34. Finney, K.N.; Sharifi, V.N.; Swithenbank, J. Fuel pelletization with a binder: Part I—Identification of a suitable binder for spent mushroom compost—Coal tailing pellets. *Energy Fuels* **2009**, *23*, 3195–3202. [CrossRef]
35. Jamradloedluk, J.; Lertsatitthanakorn, C. Influences of Mixing Ratios and Binder Types on Properties of Biomass Pellets. *Energy Procedia* **2017**, *138*, 1147–1152. [CrossRef]
36. Van Loo, S.; Koppejan, J. *The Handbook of Biomass Combustion and Co-firing*; Taylor & Francis Group: Abingdon, UK, 2012; ISBN 9781849773041.
37. Wrobel, M.; Fraczek, J.; Francik, S.; Slipek, Z.; Krzysztof, M. Influence of degree of fragmentation on chosen quality parameters of briquette made from biomass of cup plant *Silphium perfoliatum* L. In Proceedings of the 12th International Scientific Conference on Engineering for Rural Development, Jelgava, Latvia, 23–24 May 2013; 2013; pp. 653–657.
38. Mudryk, K.; Fraczek, J.; Slipek, Z.; Francik, S.; Wrobel, M. Chosen physico-mechanical properties of cutleaf coneflower (Rudbeckia Laciniata l.) shoots. In Proceedings of the 12th International Scientific Conference on Engineering for Rural Development, Jelgava, Latvia, 23–24 May 2013; pp. 658–662.
39. Wróbel, M. *Zagęszczalność i Kompaktowalność Biomasy Lignocelulozowej*; PTIR: Kraków, Poland, 2019; ISBN 978-83-64377-35-8.
40. Payne, J.D. *Troubleshooting the Pelleting Process*; American Soybean Association: St. Louis, MO, USA, 1997.
41. Holm, J.K.; Stelte, W.; Posselt, D.; Ahrenfeldt, J.; Henriksen, U.B. Optimization of a Multiparameter Model for Biomass Pelletization to Investigate Temperature Dependence and to Facilitate Fast Testing of Pelletization Behavior. *Energy Fuels* **2011**, *25*, 3706–3711. [CrossRef]
42. Sprenger, C.J.; Tabil, L.G.; Soleimani, M.; Agnew, J.; Harrison, A. Pelletization of Refuse-Derived Fuel Fluff to Produce High Quality Feedstock. *J. Energy Resour. Technol.* **2018**, *140*, 042003. [CrossRef]
43. Sarc, R.; Lorber, K.E. Production, quality and quality assurance of Refuse Derived Fuels (RDFs). *Waste Manag.* **2013**, *33*, 1825–1834. [CrossRef] [PubMed]
44. Knapczyk, A.; Francik, S.; Fraczek, J.; Slipek, Z. Analysis of research trends in production of solid biofuels. *Eng. Rural Dev.* **2019**, *18*, 1503–1509.
45. Dyjakon, A.; Noszczyk, T. The influence of freezing temperature storage on the mechanical durability of commercial pellets from biomass. *Energies* **2019**, *12*, 2627. [CrossRef]
46. Civitarese, V.; Acampora, A.; Sperandio, G.; Assirelli, A.; Picchio, R. Production of wood pellets from poplar trees managed as coppices with Different harvesting cycles. *Energies* **2019**, *12*, 2973. [CrossRef]
47. Sprenger, C.; Tabil, L.G.; Soleimani, M. Compression and relaxation properties of municipal solid waste refuse-derived fuel fluff. *KONA Powder Part. J.* **2018**, *2018*, 200–208. [CrossRef]
48. Nielsen, N.P.K.; Holm, J.K.; Felby, C. Effect of Fiber Orientation on Compression and Frictional Properties of Sawdust Particles in Fuel Pellet Production. *Energy & Fuels* **2009**, *23*, 3211–3216.
49. Karbowniczak, A.; Hamerska, J.; Wrobel, M.; Jewiarz, M.; Necka, K. Evaluation of Selected Species of Woody Plants in Terms of Suitability for Energy Production. In *Springer Proceedings in Energy*; Springer International Publishing: Cham, Switzerland, 2018; pp. 735–742.
50. Thek, G.; Obernberger, I. *The Pellet Handbook: The Production and Thermal Utilization of Biomass Pellets*; Taylor & Francis: Abingdon, UK, 2012; ISBN 9781136539916.
51. Shaw, M.D.; Tabil, L.G. Compression and Relaxation Characteristics of Selected Biomassgrinds. In *2007 ASAE Annual Meeting*; American Society of Agricultural and Biological Engineers: St. Joseph, MI, USA, 2007.
52. Kachel, M.; Kraszkiewicz, A.; Subr, A.; Parafiniuk, S.; Przywara, A.; Koszel, M.; Zając, G. Impact of the Type of Fertilization and the Addition of Glycerol on the Quality of Spring Rape Straw Pellets. *Energies* **2020**, *13*, 819. [CrossRef]
53. International Organization for Standardization. *ISO 17831-1:2015 Solid Biofuels—Determination of Mechanical Durability of Pellets and Briquettes—Part 1: Pellets*; International Organization for Standardization: Geneva, Switzerland, 2015.
54. Larsson, S.H.; Samuelsson, R. Prediction of ISO 17831-1:2015 mechanical biofuel pellet durability from single pellet characterization. *Fuel Process. Technol.* **2017**, *163*, 8–15. [CrossRef]
55. International Organization for Standardization. *ISO 17828:2015 Solid Biofuels—Determination of Bulk Density*; International Organization for Standardization: Geneva, Switzerland, 2015.

56. Mostafa, M.E.; Hu, S.; Wang, Y.; Su, S.; Hu, X.; Elsayed, S.A.; Xiang, J. The significance of pelletization operating conditions: An analysis of physical and mechanical characteristics as well as energy consumption of biomass pellets. *Renew. Sustain. Energy Rev.* **2019**, *105*, 332–348. [CrossRef]
57. Tumuluru, J.S. Effect of process variables on the density and durability of the pellets made from high moisture corn stover. *Biosyst. Eng.* **2014**, *119*, 44–57. [CrossRef]
58. Kambo, H.S.; Dutta, A. Strength, storage, and combustion characteristics of densified lignocellulosic biomass produced via torrefaction and hydrothermal carbonization. *Appl. Energy* **2014**, *135*, 182–191. [CrossRef]
59. Carone, M.T.; Pantaleo, A.; Pellerano, A. Influence of process parameters and biomass characteristics on the durability of pellets from the pruning residues of Olea europaea L. *Biomass Bioenergy* **2011**, *35*, 402–410. [CrossRef]
60. Jiang, L.; Liang, J.; Yuan, X.; Li, H.; Li, C.; Xiao, Z.; Huang, H.; Wang, H.; Zeng, G. Co-pelletization of sewage sludge and biomass: The density and hardness of pellet. *Bioresour. Technol.* **2014**, *166*, 435–443. [CrossRef] [PubMed]
61. Theerarattananoon, K.; Xu, F.; Wilson, J.; Ballard, R.; Mckinney, L.; Staggenborg, S.; Vadlani, P.; Pei, Z.J.; Wang, D. Physical properties of pellets made from sorghum stalk, corn stover, wheat straw, and big bluestem. *Ind. Crops Prod.* **2011**, *33*, 325–332. [CrossRef]
62. Monedero, E.; Portero, H.; Lapuerta, M. Pellet blends of poplar and pine sawdust: Effects of material composition, additive, moisture content and compression die on pellet quality. *Fuel Process. Technol.* **2015**, *132*, 15–23. [CrossRef]
63. Lisowski, A.; Olendzki, D.; Świętochowski, A.; Dąbrowska, M.; Mieszkalski, L.; Ostrowska-Ligęza, E.; Stasiak, M.; Klonowski, J.; Piątek, M. Spent coffee grounds compaction process: Its effects on the strength properties of biofuel pellets. *Renew. Energy* **2019**, *142*, 173–183. [CrossRef]

© 2020 by the authors. Licensee MDPI, Basel, Switzerland. This article is an open access article distributed under the terms and conditions of the Creative Commons Attribution (CC BY) license (http://creativecommons.org/licenses/by/4.0/).

Article

The Effect of the Addition of a Fat Emulsifier on the Amount and Quality of the Obtained Biogas

Jakub Sikora [1,*], Marcin Niemiec [2], Anna Szeląg-Sikora [1], Zofia Gródek-Szostak [3], Maciej Kuboń [1] and Monika Komorowska [4]

1. Faculty of Production and Power Engineering, University of Agriculture in Krakow, ul. Balicka 116B, 30-149 Kraków, Poland; Anna.Szelag-Sikora@ur.krakow.pl (A.S.-S.); Maciej.Kubon@ur.krakow.pl (M.K.)
2. Faculty of Faculty of Agriculture and Economics, University of Agriculture in Krakow, ul. Mickiewicza 21, 31-121 Kraków, Poland; Marcin.Niemiec@ur.krakow.pl
3. Department of Economics and Enterprise Organization, Cracow University of Economics, ul. Rakowicka 27, 31-510 Kraków, Poland; grodekz@uek.krakow.pl
4. Faculty of Biotechnology and Horticulture, University of Agriculture in Krakow, al. Mickiewicza 21, 31-121 Kraków, Poland; m.komorowska@ogr.ur.krakow.pl
* Correspondence: Jakub.Sikora@ur.krakow.pl

Received: 3 March 2020; Accepted: 7 April 2020; Published: 9 April 2020

Abstract: Slaughterhouse waste management is an important technological, economic, and environmental challenge. Recently, more and more attention has been paid to the possibility of obtaining biogas from waste generated by slaughterhouses. The aim of the paper was to examine the effect of an emulsifier addition in the form of a carboxymethyl cellulose solution to create animal waste fermentation media based on the quantity and quality of the generated biogas. The adopted research goal was achieved based on a laboratory experiment of methane fermenting poultry processing waste. The waste was divided into two fractions: soft (tissue) and hard (bone). A fat emulsifier in a concentration of 1%, 2.5%, 5%, and 10% of fresh weight of the substrate was added to each substrate sample made from the above fractions. The emulsifier used was a 55% carboxymethyl cellulose solution, since this emulsifier is most commonly used in food production. The experiment was conducted in order to determine how the addition of an emulsifier (55% carboxymethylcellulose solution) affects the hydration of fats during methane fermentation, as demonstrated on poultry slaughterhouse waste. The samples were subjected to static methane fermentation, according to the methodology of DIM DIN 38414(DIN Deutches Institut für Normung). The experiment lasted 30 days. The total amount of biogas obtained after fermentation was 398 mL·g^{-1} for the soft fraction and 402 mL·g^{-1} for the hard fraction. In the case of the soft waste fraction, the addition of carboxymethylcellulose at 1% of the mass to the biogas process increased the amount of obtained biogas by 16%. In the case of the hard fraction, no effect of the addition of emulsifier on the total amount of biogas obtained was identified. In each case, the biogas from substrates with added emulsifier contained less methane and slightly more carbon. The emulsifier added to the soft fraction of slaughterhouse waste from poultry processing allowed cutting the process of methanogenesis by over 50% while maintaining the efficiency of biogas production. In the case of biogasification of bone tissue, no unambiguous effect of the addition of emulsifier on the improvement of process efficiency was identified.

Keywords: biogas; renewable energy; poultry slaughterhouse waste; management

1. Introduction

Production of energy from biomass or organic waste materials is becoming increasingly important in developed countries. The policy on reducing greenhouse gas emissions, as well as the need to diversify energy sources, has become the foundation for the development of biomass fuel production technology [1]. One method of obtaining energy from biomass is to transform it in the process of anaerobic methane fermentation [2]. This allows obtaining energy in the form of gas, which can be used as a heat source or as a substrate for the production of electricity [3,4]. In most modern biogas plants, the input material is primarily various mixtures of manure and plants specially grown for this purpose, with the largest share of corn silage. However, the production of energy plants associated with the need to use large amounts of energy for agrotechnical treatments, to produce and use fertilizers and plant protection products, and to harvest and prepare the substrate. Therefore, the use of intentionally cultivated energy crops may be characterized by low energy efficiency and significant greenhouse gas emissions per unit of energy obtained. Therefore, the focus has recently shifted to the energy use of animal waste [5–8]. Thus far, the most common way to utilize such materials is thermal processing. However, this process is not very efficient, from both an energy and environmental point of view. The aspect of production optimization in terms of energy efficiency and environmental impact is an integral element of all modern quality management systems in primary production [9,10]. Obtaining energy from waste is a strategic element of sustainable waste management. The ecological aspect of the transformation of waste biomass using methane fermentation processes relates not only to obtaining energy from renewable sources but is also associated with the rational utilization of this waste. Moreover, it contributes to the reduction of greenhouse gas emitted from stored slaughterhouse waste and balances the production of energy from conventional sources [11,12]. The byproduct of methane fermentation is the post-fermentation sludge, which can be a valuable source of elements for plants if introduced into the soil. The use of digestate for fertilizing increases the level of carbon sequestration in the soil and is a factor supporting effective management of soil fertility. Structural and organizational changes in agriculture have led to a reduction in the use of organic fertilizers. In research related to the use of food industry waste for fertilizing purposes, special attention is paid to phosphorus, whose global resources will be exhausted by the late 21st century [13]. The use of this waste for biogas production, followed by the use of the obtained digestate for fertilization, may constitute an important link in the circulation of elements in agroecosystems as part of the implementation of rational agricultural production methods [14–16]. Improving the properties of the soil and enriching it with macro- and microelements reduces the demand for mineral fertilizers, the production of which also involves the emission of greenhouse gases. As a result of the methane fermentation process, the resulting digestate is free of pathogens of Salmonella and Escherichia coli bacteria, viruses, fungi, and parasites. The rate and effect of the disappearance of pathogens are affected by the parameters, such as pH, temperature, time, and level of volatile fatty acids. The sanitary aspect is an important role when the digestate is to be used for fertilizing [17].

For aquafarming and marine animal products, there is an increased risk of excessive accumulation of trace elements [18]. An important instrument supporting the development of methods of obtaining energy from waste is the EC (European Commission) legal acts, among which the most important is the Landfill Directive (1991/31/EC) [19], which imposes reduction of the amount of biodegradable waste sent to landfills, and the Framework Directive on waste (2008/98/EC) [20].

Based on the Regulation of the European Parliament and Council No. 1069/2009/EC of 21 October 2009 (1069/2009/EC 2009) [21], waste of animal origin is considered an animal byproduct. The regulation distinguishes three categories of waste in terms of animal and public health risk. Waste classified in Category 1 must be subjected to thermal utilization. Categories 2 and 3 are waste that can be used for biogas production. Waste generated during slaughter of poultry intended for food purposes is classified in Category 3. By law, such waste can be used for biogas after pasteurization at 70 °C for 60 min. The Act passed on 14 December 2012 on waste [22] does not include the provisions of the said Regulation, with the exception of products that are "intended for the landfill, for thermal transformation, or for use in a biogas plant or composting plant, in accordance with the said Regulation." That is why waste of animal origin intended for utilization, e.g., in a biogas plant, is still considered waste under the Act.

The aim of the paper was to examine the effect of the addition in the form of a carboxymethyl cellulose solution to create animal waste fermentation media, based on the quantity and quality of biogas generated during methane fermentation.

2. Materials and Methods

The adopted research goal was achieved based on a laboratory experiment of methane fermenting slaughterhouse waste under certain technological conditions. The waste mass used in the experiment was a waste from poultry processing, code 02 02 03, in accordance with the Regulation of the Minister of the Environment on the waste catalog [23,24]. The waste used in the experiment belonged to category 3, animal byproducts, recognized as the category not posing a safety risk to people or the environment. The waste samples were divided into two groups: hard and soft tissue. The first included bone tissues (legs, heads) and cartilage and the second included stomach contents, fat, digestive tract, epithelial tissues, and muscles. The collected mass of waste was approximately 40 kg. The collected waste was homogenized, and its moisture content and dry organic matter content were determined. In soft tissues, the content of dry mass was 25.8% and in hard tissues, 31.55%. Organic dry mass in soft tissues was 83%, compared to dry mass, and in hard tissues, 79.5%, compared to dry mass. A fat emulsifier in a concentration of 1%, 2.5%, 5%, and 10% of fresh weight of feedstock was added to each substrate prepared based on these samples. The biomass was hydrated to 90% humidity, which is the optimal humidity for wet methane fermentation, and then placed in 2 dm^3 fermentors, along with the inoculum. The support material was digestate from an agricultural biogas plant. The content in each of the fermentors was identical, i.e., 10% of the dry matter. Tests with such a fermentor content applied ensured that each organic molecule was surrounded with water. Each batch prepared with the addition of 55% carboxymethylcellulose solution was placed in four fermentors. The fifth fermentor, including only the inoculum, acted as the reference fermentor, against which the output from each of the four fermentors (with a batch containing the emulsifier addition in the form of 55% carboxymethyl cellulose solution) was compared.

Thus, prepared fermentors were placed in a chamber with temperature control. Next, the samples were subjected to static methane fermentation, according to the methodology of (DIN Deutches Institut für Normung) DIM DIN 38414. It consisted of a single introduction of substrates into the fermentation chambers, carrying out the process until its completion. The pH in the fermentors was maintained at 5.8–6.2 due to the alkalinity of the fats present in the fermented biomass. The appropriate pH was maintained due to the inoculation additive. Such a pH resulted from the chosen input medium, i.e., a digestate from an agricultural biogas plant, because the use of batches made from slaughterhouse waste and the fat emulsifier addition would then be tested by the authors on an industrial scale. Humidity of the digestate was approximately 96%, and its pH was approximately 6. Generally speaking, the optimum temperature of hydrolysis is between 30–50 °C and its optimum pH ranges between 5 and 7, although there is no evidence of improved hydrolytic activity below a pH of 7 [17]. The methane fermentation process lasted 30 days. The gas resulting from methane fermentation was collected in variable volume tanks for each fermentor. The NANO SENS 60 m was used to measure the moisture content of the produced biogas and to determine its chemical composition. The results of the parameters of the process carried out were read daily at the same time and automatically saved to the computer disk using a measuring software.

The Properties of the Tested Raw Materials

The components used in the research contained large amounts of protein and fats. In the raw waste of the soft fraction, the fat and protein content was 38.44 and 22.45, respectively, calculated per dry matter (Table 1). The hard fraction contained 34.39% protein and 13.22% fat. The ratio of the C:N content in the methane fermentation process is very important from the point of view of the efficiency of the methanogenesis process. The overall C:N ratio in the input material used in the methanogenesis process was 9.96 for soft tissue and 9.32 for hard tissue; the optimal C:N ratio in the methane fermentation feedstock material. The most optimal carbon to nitrogen value in the methanogenesis process ranges from 10–30:1. Values of this parameter for feedstock used in our own research were slightly below the optimal.

Table 1. Properties of the biogas process feedstock in individual research objects [%].

Emulsifier Additive	C%	Fat%	N%	Protein%	C:N	C%	N%	Fat%	Protein%	C:N
	Soft Fraction (Tissue)					Hard Fraction (Bone)				
0	44.52	22.45	4.69	38.44	9.49	40.01	4.395	13.82	34.39	9.10
1	43.20	22.21	4.534	38.24	9.53	39.81	4.241	13.56	34.18	9.32
2.5	39.92	21.65	4.359	37.88	9.16	38.65	4.025	13.07	33.98	9.30
5	39.52	21.15	4.190	37.32	9.43	37.36	4.111	12.76	33.54	8.11
10	38.79	20.35	3.336	36.08	11.63	36.54	3.869	11.44	32.7	8.79

3. Results

The results of the conducted research indicate that the addition of a fat emulsifier had a positive effect on the total biogas yield in the methane fermentation process, both in the fermentation of soft tissue and hard tissue, containing bone fraction, with a greater impact observed in the fermentation of the former. All the results shown in the graphs are the average of the four identical batches compared against the standard batch fermentor, in which only the inoculum was fermented. The standard deviation for all tested batches was 5 mL·g^{-1} TS. Such a low standard deviation is due to the fact that all the repetitions were made at the same time and under the same environmental conditions and that the batch was prepared uniformly and divided into four fermenters, according to the DIM DIN 38414 methodology. ANOVA (ANalysis Of Variance) was applied to analyze the results. The significance of mean differences among the objects was tested with the multiple comparison procedure, and Tukey's range test was applied at a significance level of $\alpha = 0.05$. Analysis was performed using the statistical software package Statistica v. 12.0. The Tukey's test was used to identify samples with homogeneous biogas yields (StatSoft Inc., Tulsa, OK, USA).

In the case of soft tissue, 398 mL of biogas per gram of dry weight of waste was obtained during a 30-day fermentation process (Figure 1). The addition of an emulsifier at 1% of the waste mass increased the total biogas yield by 16%. In this object, 465 mL of biogas per gram was obtained. Increasing the amount of emulsifier to 2.5% of the weight of the feedstock did not improve the efficiency of the process, and with the largest amount of emulsifier added, the amount of biogas was lower than in the control object. In the case of bone tissue, a smaller effect of the carboxymethylcellucose addition on the total amount of biogas obtained during methanogenesis was found (Figure 2). The total amount of biogas obtained under different amounts of the carboxymethyl cellulose addition was at a similar level.

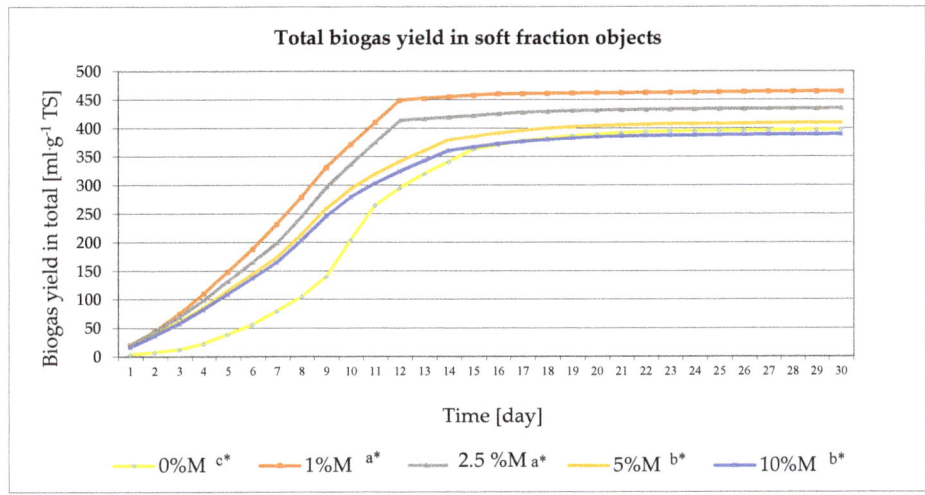

Figure 1. Total biogas yield in soft fraction objects. * Different letters mean statistically significant differences at the significance level $p = 0.05$.

A delay of fermentation was observed in both waste groups of objects without the added emulsifier (Figures 1 and 2). Only from the 6th day, the acceleration of the biogas production process was observed. This was probably due to the start of fat emulsification in the soft fraction objects. The use of carboxymethylcellulose as a technological additive to the biogasing of slaughterhouse waste cut the methanogenesis process. On day 12 of the process, a biogas yield of 449.1 mL·g^{-1} of dry waste was obtained in the object with the 1% emulsifier added. In this period, only 294.7 mL·g^{-1} dry waste of gas was obtained from the control object. In the following days, no increase in biogas production was observed in the object, in which the technological additive was used. It can be assumed that, in the object with the smallest addition of emulsifier, the end of the methanogenesis process occurred on the 12th day. In the case of a control object, biogas production was observed until day 23, when the total biogas yield was 395.3 mL·g^{-1} dry weight of the waste. The results of the conducted experiment indicate that the use of a technological additive, carboxymethyl cellulose, allowed to cut the methanogenesis process of slaughterhouse waste by over 50%. In the case of the hard fraction, the addition of the emulsifier did not have such a pronounced effect on the length of the methanogenesis process. At the initial stage of the process, the process was delayed in the object with no emulsifier added, but on the 13th day of the process, the biogas yield in individual research objects was at a comparable level, regardless of the amount of emulsifier added.

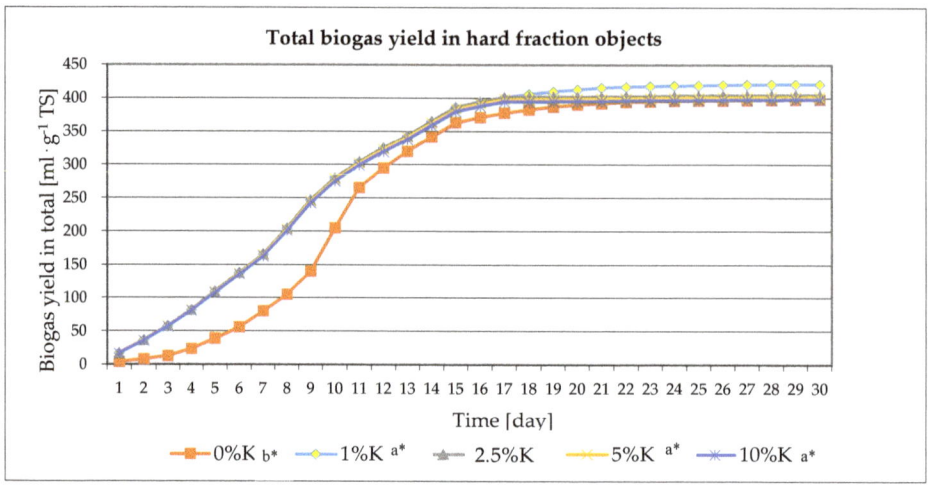

Figure 2. Total biogas yield in hard fraction objects. * Different letters mean statistically significant differences at the significance level $p = 0.05$.

The methane content is the most important parameter impacting the quality of produced biogas. The results of the conducted research indicate that the highest methane content was obtained in biogas produced from waste without the addition of an emulsifier. For both types of waste, the biogas content on the last day of the process was 75%. In the case of objects with emulsifier additives, the methane content was approximately 50%, regardless of the amount of emulsifier added (Figures 3 and 4). From the point of view of the quality of the obtained biogas, the best effects were observed on day 12 of the process. The methane content in the biogas obtained in the object without the added emulsifier was approximately. 81% at the time, while in the experimental object with 1% emulsifier added the value was approximately. 70% (Figures 3 and 4). With larger emulsifier additions, a lower methane content in biogas was found on the 12th day of the process, which was determined as the completion of the methanogenesis process in objects with the addition of carboxymethylcellulose.

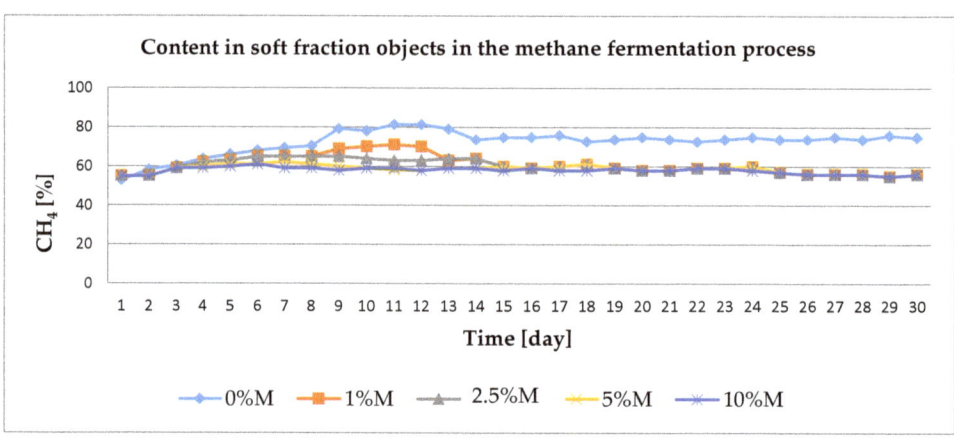

Figure 3. Methane content in soft fraction objects in the methane fermentation process.

14. Mangwandi, C.; Tao, L.J.; Albadarin, A.B.; Allen, S.J.; Walker, G.M. Alternative method for producing organic fertiliser from anaerobic digestion liquor and limestone powder: High Shear wet granulation. *Powder Technol.* **2013**, *233*, 245–254. [CrossRef]
15. Kuźnia, M.; Wojciech, J.; Łyko, P.; Sikora, J. Analysis of the combustion products of biogas produced from organic municipal waste. *J. Power Technol.* **2015**, *95*, 158–165.
16. Kasprzak, K.; Wojtunik-Kulesza, K.; Oniszczuk, T.; Kuboń, M.; Oniszczuk, A. Secondary Metabolites, Dietary Fiber and Conjugated Fatty Acids as Functional Food Ingredients against Overweight and Obesity. *Nat. Prod. Commun.* **2018**, *13*, 1073–1082. [CrossRef]
17. Weiland, P. Biomass digestion in agriculture: A successfull pathway for the energy production and waste treatment in Germany. *Eng. Life Sci.* **2006**, *6*, 302–309. [CrossRef]
18. Niemiec, M.; Komorowska, M.; Szeląg-Sikora, A.; Sikora, J.; Kuzminova, N. Content of Ba, B, Sr and As in water and fish larvae of the genus Atherinidae, L. sampled in three bays in the Sevastopol coastal area. *J. Elem.* **2018**, *23*, 1009–1020. [CrossRef]
19. Council of the European Union. Council Directive 1999/31/EC of 26 April 1999 on the landfill of waste. *Off. J. Eur. Commun.* **1999**, *182*, 1–19.
20. Directive 2008/98/EC of the European Parliament and of the Council of 19 November 2008 on waste and repealing certain Directives. *Off. J. Eur. Union L* **2008**, 312.
21. *Regulation (EC) No 1069/2009 of the European Parliament and of the Council of 21 October 2009, Laying Down Sanitary Provisions for Animal by-Products Not Intended for Human Consumption, and Repealing Regulation (EC) No 1774/2002 [Regulation on Animal by-Products]*; European Union: Brussels, Belgium, 2009.
22. The Act of 14 December 2012 on Waste. 2013. Available online: http://prawo.sejm.gov.pl/isap.nsf/DocDetails.xsp?id=wdu20130000021 (accessed on 7 April 2020).
23. Resolution of Minister of Environment Issued on December 9, 2014 on Waste Catalogue. 2014. Available online: http://isap.sejm.gov.pl/DetailsServlet?id=WDU20140001923 (accessed on 7 April 2020).
24. The Regulation of the Minister of the Environment of 20 January 2015 on the Process of R10 Recovery. 2015. Available online: http://prawo.sejm.gov.pl/isap.nsf/DocDetails.xsp?id=WDU20150000132 (accessed on 7 April 2020).
25. Cripa, F.B.; Arantes, M.K.; Sequinel, R.; Fiorini, A.; Rogério, F.; Rosado, F.R.; Alves, H.J. Poultry slaughterhouse anaerobic ponds as a source of inoculum for biohydrogen production. *J. Biosci. Bioeng.* **2020**, *129*, 77–85. [CrossRef] [PubMed]
26. Almomani, F.; Shawaqfah, M.; Bhosale, R.R.; Kumara, A.; Khraisheh, M.A.M. Intermediate ozonation to enhance biogas production in batch and continuous systems using animal dung and agricultural waste. *Int. Biodeterior. Biodegrad.* **2017**, *119*, 176–187. [CrossRef]
27. Gródek-Szostak, Z.; Malik, G.; Kajrunajtys, D.; Szeląg-Sikora, A.; Sikora, J.; Kuboń, M.; Niemiec, M.; Kapusta-Duch, J. Modeling the Dependency between Extreme Prices of Selected Agricultural Products on the Derivatives Market Using the Linkage Function. *Sustainability* **2019**, *11*, 4144. [CrossRef]
28. Salminen, E.; Rintala, J. Anaerobic digestion of organic solid poultry slaughterhouse waste- a review. *Bioresour. Technol.* **2002**, *83*, 17–18. [CrossRef]
29. Luostarinen, S.; Luste, S.; Sillanpää, M. Increased biogas production at wastewater treatment plants through co-digestion of sewage sludge with grease trap sludge from a meat processing plant. *Bioresour. Technol.* **2009**, *100*, 79–85. [CrossRef]
30. Latifi, P.; Karrabi, M.; Danesh, S. Anaerobic co-digestion of poultry slaughterhouse wastes with sewage sludge in batch-mode bioreactors (effect of inoculum-substrate ratio and total solids). *Renew. Sustain. Energy Rev.* **2019**, *107*, 288–296. [CrossRef]
31. Russo, V.; von Blottnitz, H. Potentialities of biogas installation in South African meat value chain for environmental impacts reduction. *J. Clean. Prod.* **2017**, *153*, 465–473. [CrossRef]
32. Gródek-Szostak, Z.; Szeląg-Sikora, A.; Sikora, J.; Korenko, M. Prerequisites for the cooperation between enterprises and business supportinstitutions for technological development. *Bus. Nonprofit Organ. Facing Increased Compet. Grow. Cust. Demand* **2017**, *16*, 427–439.
33. Fryda, L.; Panopoulos, K.; Vourliotis, P.; Pavlidou, E.; Kakaras, E. Experimental investigation of fluidised bed co-combustion of meat and bone meal with coals and olive bagasse. *Fuel* **2006**, *85*, 1685–1699. [CrossRef]

34. Vilvert, A.J.; Saldeira Junior, J.C.; Bautitz, I.R.; Zenatti, D.C.; Andrade, M.G.; Hermes, E. Minimization of energy demand in slaughterhouses: Estimated production of biogas generated from the effluent. *Renew. Sustain. Energy Rev.* **2020**, *120*, 109613. [CrossRef]
35. Azman, S. Anaerobic Digestion of Cellulose and Hemicellulose in the Presence of Humic Acids. Ph.D. Thesis, Wageningen University, Wageningen, The Netherlands, 2016.
36. Meegoda, J.N.; Li, B.; Patel, K.; Wang, L.B. A Review of the Processes, Parameters, and Optimization of Anaerobic Digestion. *Int. J. Environ. Res. Public Health* **2018**, *15*, 2224. [CrossRef]
37. Bücker, F.; Marder, M.; Peiter, M.R.; Lehn, D.N.; Esquerdo, V.M.; Luiz Antonio de Almeida Pinto, L.A.; Konrad, O. Fish waste: An efficient alternative to biogas and methane production in an anaerobic mono-digestion system. *Renew. Energy* **2020**, *147*, 798–805. [CrossRef]
38. Bulak, P.; Proc, K.; Pawłowska, M.; Kasprzycka, A.; Berus, W.; Bieganowski, A. Biogas generation from insects breeding post production wastes. *J. Clean. Prod.* **2020**, *244*, 118777. [CrossRef]
39. Wang, S.; Jena, U.; Das, K.C. Biomethane production potential of slaughterhouse waste in the United States. *Energy Convers. Manage.* **2018**, *173*, 143–157. [CrossRef]
40. Martí-Herrero, J.; Alvarez, R.; Floresd, T. Evaluation of the low technology tubular digesters in the production of biogas from slaughterhouse wastewater treatment. *J. Clean. Prod.* **2018**, *199*, 633–642. [CrossRef]
41. Park, S.; Yoon, Y.-M.; Han, S.K.; Kim, D.; Kim, H. Effect of hydrothermal pre-treatment (HTP) on poultry slaughterhouse waste (PSW) sludge for the enhancement of the solubilization, physical properties, and biogas production through anaerobic digestion. *Waste Manage.* **2017**, *64*, 327–332. [CrossRef]
42. Bustamante, M.A.; Restrepo, A.P.; Alburquerque, J.A.; Pérez-Murcia, M.D.; Paredes, C.; Moral, R.; Bernal, M.P. Recycling of anaerobic digestates by composting: Effect of the bulking agent used. *J. Clean. Prod.* **2013**, *47*, 61–69. [CrossRef]
43. Sikora, J.; Niemiec, M.; Szeląg-Sikora, A. Evaluation of the chemical composition of raw common duckweed (Lemna minor L.) and pulp after methane fermentation. *J. Elem.* **2018**, *23*, 685–695. [CrossRef]
44. Kapusta-Duch, J.; Szeląg-Sikora, A.; Sikora, J.; Niemiec, M.; Gródek-Szostak, Z.; Kuboń, M.; Leszczyńska, T.; Borczak, B. Health-Promoting Properties of Fresh and Processed Purple Cauliflower. *Sustainability* **2019**, *11*, 4008. [CrossRef]
45. Khalil, M.; Berawi, M.A.; Heryanto, R.; Rizalie, A. Waste to energy technology: The potential of sustainable biogas production from animal waste in Indonesia. *Renew. Sustain. Energy Rev.* **2019**, *105*, 323–331. [CrossRef]
46. Niemiec, M.; Chowaniak, M.; Sikora, J.; Szeląg-Sikora, A.; Gródek-Szostak, Z.; Komorowska, M. Selected Properties of Soils for Long-Term Use in Organic Farming. *Sustainability* **2020**, *12*, 2509. [CrossRef]
47. Sikora, J.; Niemiec, M.; Tabak, M.; Gródek-Szostak, Z.; Szeląg-Sikora, A.; Kuboń, M.; Komorowska, M. Assessment of the Efficiency of Nitrogen Slow-Release Fertilizers in Integrated Production of Carrot Depending on Fertilization Strategy. *Sustainability* **2020**, *12*, 1982. [CrossRef]
48. Kuboń, M.; Niemiec, M.; Tabak, M.; Komorowska, M.; Gródek-Szostak, Z. Ocena zasobności gleby w przyswajalne związki siarki z wykorzystaniem ekstrahentów o zróżnicowanej zdolności ekstrakcji. *Przem. Chem.* **2020**, *99*, 581–584. [CrossRef]

© 2020 by the authors. Licensee MDPI, Basel, Switzerland. This article is an open access article distributed under the terms and conditions of the Creative Commons Attribution (CC BY) license (http://creativecommons.org/licenses/by/4.0/).

Article

The Efficiency of Industrial and Laboratory Anaerobic Digesters of Organic Substrates: The Use of the Biochemical Methane Potential Correction Coefficient

Krzysztof Pilarski [1], Agnieszka A. Pilarska [2,*], Piotr Boniecki [1], Gniewko Niedbała [1], Karol Durczak [1], Kamil Witaszek [1], Natalia Mioduszewska [1] and Ireneusz Kowalik [1]

[1] Institute of Biosystems Engineering, Poznań University of Life Sciences, Wojska Polskiego 50, 60-637 Poznań, Poland; pilarski@up.poznan.pl (K.P.); bonie@up.poznan.pl (P.B.); gniewko.niedbala@up.poznan.pl (G.N.); kdurczak@up.poznan.pl (K.D.); kamil.witaszek@up.poznan.pl (K.W.); natalia.mioduszewska@up.poznan.pl (N.M.); ikowalik@up.poznan.pl (I.K.)
[2] Institute of Food Technology of Plant Origin, Poznań University of Life Sciences, Wojska Polskiego 31, 60-637 Poznań, Poland
* Correspondence: pilarska@up.poznan.pl; Tel.: +48-61-848-73-08

Received: 31 January 2020; Accepted: 9 March 2020; Published: 10 March 2020

Abstract: This study is an elaboration on the conference article written by the same authors, which presented the results of laboratory tests on the biogas efficiency of the following substrates: maize silage (MS), pig manure (PM), potato waste (PW), and sugar beet pulp (SB). This article presents methane yields from the same substrates, but also on a technical scale. Apart from that, it presents an original methodology of defining the Biochemical Methane Potential Correction Coefficient (BMPCC) based on the calculation of biomass conversion on an industrial scale and on a laboratory scale. The BMPCC was introduced as a tool to enable uncomplicated verification of the operation of a biogas plant to increase its efficiency and prevent undesirable losses. The estimated BMPCC values showed that the volume of methane produced in the laboratory was overestimated in comparison to the amount of methane obtained under technical conditions. There were differences observed for each substrate. They ranged from 4.7% to 17.19% for MS, from 1.14% to 23.58% for PM, from 9.5% to 13.69% for PW, and from 9.06% to 14.31% for SB. The BMPCC enables estimation of biomass under fermentation on an industrial scale, as compared with laboratory conditions.

Keywords: laboratory-scale efficiency; industrial-scale efficiency; biomass conversion; Biochemical Methane Potential Correction Coefficient; loss prevention

---

1. Introduction

The intensive development of agriculture causes an increase in the supply of organic waste. Waste matter disposal technologies are often based on the anaerobic digestion (AD) process, which takes place in biogas plants. In order to increase the efficiency of the installation, plants with high content of organic matter are added as a substrate [1,2]. Target crops are grown for this purpose—mainly maize and grass, which are used to produce silages [3]. Animal waste [4,5] and food waste [6,7] are also used as substrates. Biogas plants produce biogas, which is classified as a source of renewable energy because it contains methane. The resulting gas, which is an energy carrier, can be easily converted into electricity and heat.

The AD process is one of the most adequate and prospective methods of organic waste disposal and it is also a source of biofuel [8,9]. However, the process must be economically viable and stable. The main factors which affect AD-process efficiency are as follows: the chemical composition of the substrates, pH, temperature, the substrates mixing process, dry residue, the content of organic matter in the substrate, the digester load, and the concentration of inhibitors [10–12]. These parameters are controlled both in the laboratory-scale and industrial-scale process.

A slight change in the conditions of the methane digestion in a biogas plant may upset the process or halt it completely. Temperature is one of the factors that significantly affects the course of the digestion process. Small changes in temperature and the retention time affect bacterial activity [13]. As a consequence, biogas production is reduced. In Poland, the digestion process is usually carried out under mesophilic conditions at a temperature of about 39 °C. In Europe, mesophilic installations are predominant, but there are also thermophilic conditions, where more heat energy is used by the plant to sustain the process [14,15]. As mentioned before, pH is also important for methane fermentation. It should range from 6.8 to 7.5 because this guarantees optimal conditions for bacterial life and reproduction during all four phases of the methane fermentation process [16,17]. The biogas yield decreases when the pH value is higher than 7.5 or lower than 6.8.

When biomass is applied to digestion chambers with high-protein substrates, excessive amounts of ammoniacal nitrogen are usually formed. Excess ammonia significantly slows down the biogas production process. As temperature increases, so does the effect of ammonia, inhibiting methane fermentation [18,19]. Some substrates are rich in sulphur and they cause excessive amounts of this element in the digester. The element may occur in the form of ions dissolved during the liquid phase, or it may be found in hydrogen sulphide in a mixture of liquid and gas. Higher temperature increases the solubility of hydrogen sulphide, which results in its higher concentration during the liquid phase [20]. At the initial stage of operation of a biogas plant, it is very important to supply small portions of substrates with a homogeneous composition. This affects the overall normal digestion by individual bacteria at each stage of the fermentation process [21]. During the operation of a biogas plant, it is necessary to observe the retention time for individual organic substrates. This is the time a given substrate should spend in the digester until it achieves an appropriate level of degradation. These are the most common problems encountered in practice, which directly affect the actual production of biogas.

The pursuit of more efficient use of biomass to generate energy requires detailed verification of methane fermentation technologies. Detailed analysis and the selection of appropriate biogas production technologies gives investors a real opportunity to manage biogas plants effectively. Apart from that, more efficient and effective installations give a chance to reduce the financial and social costs of every kilowatt hour (kWh) of energy produced, as compared with the countries regarded as pioneers in this field. Poland is still a developing country, as it is attempting to catch up with the standard of Western European countries. Therefore, it should effectively control the expending of funds on solutions implemented in the field of renewable energy sources [1,12,22].

At the beginning of the development of biogas installations in Poland, i.e., between 2008 and 2009, investors were ready to expend money on large plants capable of generating a power of 2 MW or more [23]. However, the market very quickly verified these plans and showed that it was the wrong trend because it involved high costs of transport (it was necessary to supply thousands of tonnes of substrates). The dispersion of Polish agriculture was not taken into consideration either. After about 3 years, investors began to prefer biogas plants capable of generating a power of 0.5–1 MW. They were usually located on large farms [24] because it was necessary to provide a continuous supply of substrates and to solve the deodorisation problem. However, this method was also unsuccessful because the actual fermentation efficiency was much lower than the forecast. The energy consumption for fittings was higher than assumed, mostly due to the increased and unstable operation of the mixing systems in the installation [23]. These problems are still topical. Therefore, the proposed technologies should be analysed in detail and adjusted to national and local requirements. Scientists and biogas plant owners struggle with these problems not only in Poland but also in other countries [25–28].

So far, researchers have not prepared the assumptions that would clearly and easily verify the operation of a biogas plant. This verification system is more and more wanted by current and future investors. As seen in reference publications, most studies conducted by scientific institutions both in Poland and other countries usually refer to laboratory conditions [29–31]. The authors of most of these studies did not analyse the operation of biogas installations, which converted biomass into biogas/methane. However, these results directly translate into the economic and ecological effect of bioelectric plants. In general, scientific reports very rarely provide information about the efficiency of a particular technology proposed by researchers in terms of the relation between the amount of biogas generated in a laboratory and the potential efficiency of this technology applied on a technical scale. If any information is given, it is usually insufficient.

This study compares the efficiency of the anaerobic digestion technology applied on an industrial scale with its efficiency on a laboratory scale. The comparison was made for the same substrates. The study also verifies the level of significance of the determinants affecting the efficiency of operation of the biogas installation. The following substrates were used in the research: maize silage (MS); and agri-food waste, including pig manure (PM), potato waste (PW), and sugar beet pulp (SB). In view of the fact that numerous dependencies can be observed during the AD process in a biogas plant—and due to the fact that reference publications do not assess the efficiency of biomass conversion on an industrial scale—the Biochemical Methane Potential Correction Coefficient (BMPCC) was defined [23]. The coefficient provided information about the efficiency of organic matter decomposition into methane in a biogas plant and enabled comparison of these results with the AD process conducted in a laboratory.

2. Materials and Methods

2.1. Materials

Green substrates—maize silage (MS) and agri-food waste, pig manure (PM), potato waste (PW), and sugar beet pulp (SB) were used in the study. The waste materials were acquired from a farm and a sugar factory in the Wielkopolska region, Poland.

Maize silage was stored in silos, where it was ensilaged. Pig manure was supplied directly from a pig farm. Potato waste was also supplied directly after being customised. Sugar beet pulp was stored in silage sleeves. The degree of maize compaction in the ensilage process, the weather conditions, and storage time were the factors affecting the physicochemical properties after storage.

2.2. Physicochemical Analysis of Materials

The following standards were used in physicochemical analyses of the substrates (used organic materials) and samples (harvested fermentation mixture): pH—the potentiometric method, PN-EN 12176: 2004; dry residue—the weight method, PN-EN 12880: 2004; roasting losses (roasting residue) —the weight method, PN-EN 12879: 2004; sampling for chemical and physical tests, PN-EN ISO 5667-13: 2011; carbon, EN ISO 16948: 2015; hydrogen, EN ISO 16948: 2015; nitrogen, EN ISO 16948: 2015; oxygen, based on calculations; sulphur, PN-EN ISO 11885: 2009.

2.3. Laboratory-Scale Biogas Production

The anaerobic digestion process was conducted in a periodic mode of operation of digesters, under mesophilic conditions. The authors of this study presented a detailed diagram and described the construction and operation of biodigesters in their previous publications [32–34].

2.4. The Construction and Operation of an Industrial Installation

The biomass conversion tests were conducted for 6 months in an agricultural biogas plant. The facility was fed with the substrates listed above. Samples of the substrates were collected at monthly intervals because their physicochemical properties may have changed during storage.

The biogas plant consisted of two main digesters (F1 and F2) and a third tank, where digestate pulp was stored. The first tank (F1) was used for primary digestion, and the second tank (F2) was used for secondary digestion. The installation was also equipped with a primary tank (PT), into which waste potatoes were fed because they may have been contaminated with soil. Pig manure, which was the agent liquefying the entire fermenting mass, was also fed into this tank. The annual production capacity was about 3.5 million m^3 of biogas. The installation was equipped with a cogeneration system, whose power was 1 MW$_{el}$.

2.5. Collection of Samples for Tests

Samples for tests were collected once a month. Six samples of each substrate were tested during the six-month experiment. The following aspects were taken into account when collecting the samples: access to the sampling point, the possibility to safely interrupt the stream of material when samples were collected manually, and the type of construction of the fermentation chamber—due to the stratification of the material collected for tests. The safest and most practical station for manual sample collection was selected. The practicality of this location was analysed in terms of the representativeness of the material collected for tests. Each time, at least 3 samples were collected to increase confidence in the representativeness of the material collected.

2.6. Qualitative and Quantitative Analysis of Biogas

The quality and quantity of the biogas produced was analysed once a day. In order to make effective measurements of the biogas quality, a minimum quantity of 0.4 L had to be produced daily. When a smaller amount of biogas was produced, the chemical composition was not analysed. The measurement methodology was adopted from the German standard DIN 38414/S8, which was modified by the author in order to reduce measurement errors [35]. The qualitative composition of biogas was analysed by measuring the content of CH_4, CO_2, NH_3, and H_2S.

The quality and quantity of the biogas produced in the installation under real conditions was based on the measuring systems in place. These systems were permanently installed and met the German standard DIN 38414/S8. There were no breakdowns during the entire period under study. When estimating the uncertainty of measurement in this article, the procedures followed Polish standards and German standard [35–37].

2.7. Biochemical Methane Potential Correction Coefficient (BMPCC)—Calculation Methodology Based on

First, the yield of biogas from the substrate was measured (m^3·Mg^{-1} fresh matter (FM)) in a laboratory. Simultaneously, the composition of biogas was analysed by measuring the content of CH_4 and CO_2 (the concentrations of NH_3 and H_2S were omitted). Then, the biogas composition was used to calculate the volume of methane. At the next stage, the mass of methane contained in the biogas was measured under laboratory conditions (the mass of methane in the biogas obtained from the fresh matter of the substrate under laboratory conditions, MMB-L). The third stage involved analysing the substrate for dry residue, roasting losses, and the content of carbon, hydrogen, oxygen, nitrogen, and sulphur. Then, knowing the content of C, H, O, N, and S, the amount of methane that could theoretically be obtained was calculated (theoretical methane mass, TMM) according to the principle of mass conservation.

The fourth stage involved calculation of the conversion of organic matter contained in the biomass under laboratory conditions (conversion of organic matter under laboratory conditions—the laboratory degree of biomass conversion, COM-L). The following Equation (1) was used:

$$COM - L = \frac{MMB - L}{TMM}. \tag{1}$$

The fifth stage involved calculation of the conversion of organic matter contained in the biomass under the operating conditions of the installation (conversion of organic matter in the installation—the industrial degree of biomass conversion, COM-I). The following Equation (2) was used:

$$COM - I = \frac{MMB - I}{TMM}. \tag{2}$$

At the last stage, the BMPCC of each substrate was calculated as the ratio between the mass of methane produced in the installation and the mass of methane produced under laboratory conditions. The following Equation (3) was used:

$$\text{BMPCC} = 100 - \frac{\text{COM} - \text{I}}{\text{COM} - \text{L}} \times 100. \tag{3}$$

3. Results

3.1. Physicochemical Parameters of Substrates—Laboratory-Scale Measurements

3.1.1. pH of Substrates

The pH of maize silage used for the tests ranged from 4.21 (1MS) to 4.39 (3MS). The pH values of pig manure were similar during the entire test and ranged from 7.22 (5PM) to 7.56 (1PM). They were similar to data provided in reference publications [38,39]. The concentration of hydrogen ions in waste potatoes ranged from 7.44 (3PW) to 7.78 (4PW). As seen in results from the research on methane fermentation of waste potatoes conducted by [40], the pH of the substrate was 7. The pH of beet pulp ranged from 5.01 (1SB) to 5.18 (3SB). Table 1 shows the pH values of the substrates.

Table 1. Physicochemical properties and the laboratory-scale biogas efficiency of the substrates with the uncertainty of results, based on [1].

Substrate	pH (-)	TS (%)	VS (%)	Biogas (m³·Mg⁻¹ FM)	Biogas (m³·Mg⁻¹ TS)	Biogas (m³·Mg⁻¹ VS)	CH₄ (%)
1MS	4.21 ± 0.06	32.68 ± 0.51	95.15 ± 1.76	188 ± 3.6	575 ± 12.6	605 ± 13.4	51.2 ± 1.29
2MS	4.28 ± 0.06	32.21 ± 0.50	94.61 ± 1.75	183 ± 3.5	568 ± 12.4	601 ± 13.4	52.3 ± 1.31
3MS	4.39 ± 0.06	32.11 ± 0.50	94.21 ± 1.75	181 ± 3.5	564 ± 12.3	598 ± 13.3	50.4 ± 1.27
4MS	4.31 ± 0.06	31.86 ± 0.50	93.65 ± 1.74	178 ± 3.4	559 ± 12.2	595 ± 13.2	50.9 ± 1.28
5MS	4.35 ± 0.06	31.45 ± 0.49	94.83 ± 1.76	180 ± 3.5	572 ± 12.5	604 ± 13.4	51.6 ± 1.30
6MS	4.28 ± 0.06	31.06 ± 0.48	93.88 ± 1.74	184 ± 3.5	592 ± 12.9	631 ± 14.0	50.4 ± 1.27
1PM	7.56 ± 0.10	4.86 ± 0.08	76.16 ± 1.41	17 ± 0.3	350 ± 7.6	459 ± 10.2	52.6 ± 1.32
2PM	7.44 ± 0.10	4.32 ± 0.07	76.88 ± 1.43	19 ± 0.4	440 ± 9.6	572 ± 12.7	51.1 ± 1.28
3PM	7.31 ± 0.10	4.94 ± 0.08	78.49 ± 1.46	20 ± 0.4	405 ± 8.9	516 ± 11.5	51.8 ± 1.30
4PM	7.28 ± 0.10	4.65 ± 0.07	81.32 ± 1.51	22 ± 0.4	473 ± 10.3	582 ± 12.9	51.4 ± 1.29
5PM	7.22 ± 0.10	5.06 ± 0.08	80.11 ± 1.49	21 ± 0.4	415 ± 9.1	518 ± 11.5	51.3 ± 1.29
6PM	7.36 ± 0.10	5.01 ± 0.08	79.84 ± 1.48	18 ± 0.3	359 ± 7.8	450 ± 10.0	50.8 ± 1.28
1PW	7.36 ± 0.10	21.31 ± 0.33	94.87 ± 1.76	68 ± 1.3	319 ± 7.0	336 ± 7.5	51.6 ± 1.30
2PW	7.41 ± 0.10	21.33 ± 0.33	94.61 ± 1.75	69 ± 1.3	323 ± 7.1	342 ± 7.6	50.7 ± 1.27
3PW	7.44 ± 0.10	21.45 ± 0.33	94.83 ± 1.76	67 ± 1.3	312 ± 6.8	329 ± 7.3	51.4 ± 1.29
4PW	7.78 ± 0.11	21.85 ± 0.34	95.01 ± 1.76	70 ± 1.3	320 ± 7.0	337 ± 7.5	51.1 ± 1.28
5PW	7.65 ± 0.10	21.78 ± 0.34	95.12 ± 1.76	71 ± 1.4	326 ± 7.1	343 ± 7.6	52.2 ± 1.31
6PW	7.71 ± 0.10	21.86 ± 0.34	95.92 ± 1.78	70 ± 1.3	320 ± 7.0	334 ± 7.4	51.8 ± 1.30
1SB	5.01 ± 0.07	23.88 ± 0.37	94.16 ± 1.75	99 ± 1.9	415 ± 9.1	440 ± 9.8	50.2 ± 1.26
2SB	5.08 ± 0.07	23.44 ± 0.36	94.02 ± 1.74	97 ± 1.9	414 ± 9.0	441 ± 9.8	50.8 ± 1.28
3SB	5.18 ± 0.07	23.58 ± 0.37	93.88 ± 1.74	96 ± 1.8	407 ± 8.9	434 ± 9.6	51.4 ± 1.29
4SB	5.16 ± 0.07	23.41 ± 0.36	93.32 ± 1.73	93 ± 1.8	397 ± 8.7	426 ± 9.5	52.1 ± 1.31
5SB	5.11 ± 0.07	23.67 ± 0.37	93.46 ± 1.73	94 ± 1.8	398 ± 8.7	425 ± 9.4	50.7 ± 1.27
6SB	5.09 ± 0.07	23.33 ± 0.36	94.12 ± 1.74	95 ± 1.8	407 ± 8.9	433 ± 9.6	51.5 ± 1.29

3.1.2. Total Solids in Substrates

The content of total solids (TS) in the maize silage used in the installations ranged from 32.21% (2MS) to 31.06% (6MS). The TS content in the pig manure ranged from 4.32% (2PM) to 5.06% (6PM) and was consistent with the data provided in reference publications [41]. The TS content in the potato waste ranged from 21.33% (2PW) to 21.86% (6PW). The TS content in the beet pulp ranged from 23.33% (6SB) to 23.88% (1SB). The results were comparable with the data published in the study by [42]. Table 1 shows the TS content in the substrates.

3.1.3. Volatile Solids in Substrates

The content of volatile solids (VS), i.e., roasting losses, was another parameter under analysis. Substrates with high content of organic matter are a valuable raw material for biogas installations. There are three basic groups of organic matter in substrates: carbohydrates, protein, and fats. The materials used in this study mainly contained sugars and protein. The roasting losses for maize silage ranged from 93.88% (6MS) to 95.15% (1MS). The VS content in the pig manure ranged from 76.88% (2PM) to 81.32% (4PM). These values were similar to the data reported in reference publications [43,44]. The quality of liquid manure depends on the animals it comes from, their diet, and the degree of dilution with water. The VS content in the potato waste ranged from 94.61% (1PW) to 95.92% (6PW). Due to the content of carbohydrates in sugar beet pulp, this material positively influences the methane fermentation efficiency per digester volume unit. The VS content in the material used in the tests ranged from 93.32% (4SB) to 94.16% (1SB). Table 1 lists the results of this experiment.

3.2. Laboratory-Scale Biogas Efficiency of Samples

3.2.1. Volume of Biogas Obtained from Substrates per Fresh Matter

The yield of biogas obtained from maize silage amounted to 183 $m^3 \cdot Mg^{-1}$ FM. The biogas volumes obtained in the experiment were consistent with the data provided by other researchers [33]. The yield of biogas obtained from liquid manure amounted to 20.1 $m^3 \cdot Mg^{-1}$ FM. It was lower than the data provided in reference publications [45]. The volume of biogas obtained from waste potatoes amounted to 69.2 $m^3 \cdot Mg^{-1}$ FM. The volume of biogas obtained from the fresh matter of sugar beet pulp amounted to 96.4 $m^3 \cdot Mg^{-1}$ FM. There were similar values reported in reference publications [46,47]. Table 1 provides data on the volume of biogas obtained from the substrates.

3.2.2. Volume of Biogas Obtained from Substrates per Total Solids Content

Literature data provide the volume of biogas obtained in the AD process as the number of total solids contained in the samples so as to standardise the results without water. In this study, similar calculations were also performed. The volume of biogas obtained from maize silage at individual sample collection terms ranged from 559 (4MS) to 592 (6MS) $m^3 \cdot Mg^{-1}$ TS. The yield of biogas obtained from pig manure per TS in individual samples ranged from 350 (3PM) to 473 (4PM) $m^3 \cdot Mg^{-1}$ TS. In reference publications, there are big differences in the data on the biochemical methanogenic potential of pig manure obtained per TS [48]. The yield of biogas obtained from waste potatoes ranged from 323 (3PW) to 326 (5PW) $m^3 \cdot Mg^{-1}$ TS. The biochemical methanogenic potential of sugar beet pulp ranged from 397 (4SB and 5SB) to 415 (1SB) $m^3 \cdot Mg^{-1}$ TS. The volume of biogas obtained in this study was lower than in the study by [46]. The data on the volume of biogas per total solids content are shown in Table 1.

3.2.3. Volume of Biogas Obtained from Substrates per Volatile Solids Content

The yield of biogas obtained from maize silage per VS ranged from 595 (4MS) to 631 (6MS) $m^3 \cdot Mg^{-1}$ VS. The results of this experiment were in line with the data presented by [44]. The yield of biogas obtained from pig manure ranged from 450 (6PM) to 582 (3PM) $m^3 \cdot Mg^{-1}$ VS. The biogas volume was comparable to the results presented by [45]. The volume of biogas obtained from waste potatoes ranged from 329 (3PW) to 342 (2PW). The volume of biogas obtained from beet pulp ranged from 425 (5SB) to 440 (1SB and 2SB) $m^3 \cdot Mg^{-1}$ VS. The values reported in reference publications were lower than the results presented by [46], where the biochemical methanogenic potential of sugar beet pulp amounted to 504 $m^3 \cdot Mg^{-1}$ VS. The range of the results is given in Table 1.

The volume concentration of methane in the biogas obtained from the maize silage ranged from 50.4% (3MS) to 52.3% (2MS), see Table 1. The concentration noted in the presented experiment was identical with the data reported by [44]. The methane content in the biogas obtained from pig manure ranged from 50.8% (2PM) to 52.6% (5PM). According to the authors of scientific studies, the methane fraction in the biogas obtained as a result of the methane fermentation of pig manure was above 53% [48]. The concentration of methane in the biogas obtained from waste potatoes ranged from 50.7% (2PW) to 52.2% (5PW). These values were similar to the data presented by [49]. The volume concentration of methane in the biogas obtained from beet pulp ranged from 50.2% (1SB) to 52.8% (4SB). According to [50], the concentration is usually about 52%.

The pH range corresponding to the maize silage proved that the ensilage process was successful. During storage, the total solids as well as volatile solids of this substrate were reduced. It is most likely that it was caused by microorganisms responsible for the conservation of MS. However, despite the decrease in the MS organic matter, the amount of biogas produced in the following months was not reduced. The physicochemical properties of pig manure depend on the way animals are fed and their age. This significantly affects the biomethane efficiency of this material. It is similar with potato waste, which is a waste material but is not cultivated for energy purposes. However, as the table shows, this did not affect the values of the parameters corresponding to this material. Slight differences in the biogas yields of the sugar beet pulp were caused by the fact that there were various sugar beet species in this substrate.

3.3. pH—Industrial-Scale Measurements

In the PT tank, waste potatoes were comminuted and solid contaminants were removed. They were mixed and liquefied with pig manure (see Section 2.4). During the entire experiment, the pH in the PT tank ranged from 6.91 (3PT) to 7.34 (6PT). Samples were continuously fed from the PT tank to F1 and F2 (primary and secondary digestion), where all the substrates used in the experiment were mixed and underwent primary (F1) and secondary digestion (F2). There were the following pH values in the digesters: ranging from 7.02 (6F1) to 7.23 (1F1) in the first tank and from 7.19 (5F2) to 7.21 (6F2) in the secondary tank, as shown in Figure 1.

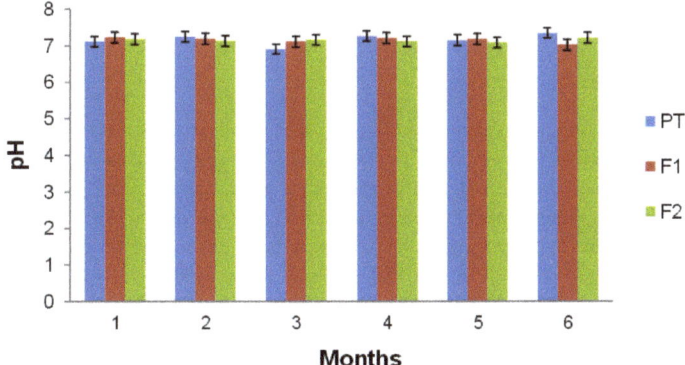

Figure 1. Industrial-scale pH measurements in each month of the experiment.

3.4. Temperature—Industrial-Scale Measurements

The large volume of liquid manure fed into the predigester significantly reduced the temperature, which was the lowest in this tank. This tank was not intended for the methane fermentation process. The temperature ranged from 12.4 °C to 18.2 °C for preliminary tank. The temperature in the first tank ranged from 38.4 °C to 40.3 °C, whereas in the secondary tank it ranged from 39.4 °C to 40.4 °C.

3.5. The Actual Amount of Substrate Fed each Month

The amount of substrates to be fed into the installation was planned on the basis of the results of laboratory tests. The amount was sufficient to generate a power of 1 MW$_{el}$. Depending on the month, the mass of maize silage fed into the installation ranged from 1668 Mg to 1910 Mg, pig manure—from 4540 Mg to 5160 Mg, waste potatoes—from 62 Mg to 186 Mg, and beet pulp—from 165 Mg to 217 Mg.

3.6. The Mass of Methane Produced from Each Substrate Each Month

Methane is one of the most important components of the biogas mixture as it is responsible for its calorific value.

Figure 2 shows the mass of methane obtained from each substrate in each month of the study.

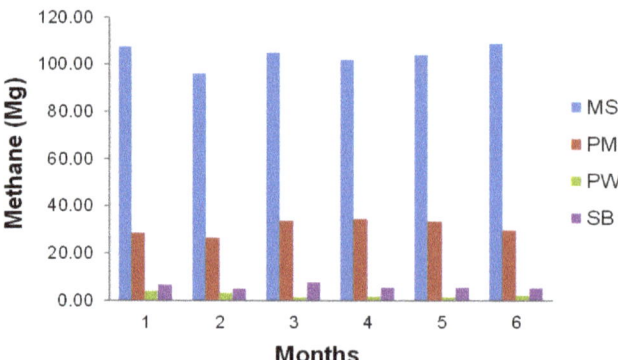

Figure 2. The mass of methane produced from the substrates in each month of the experiment.

The installation was equipped with devices that measured the volume of biogas produced and its percentage (volumetric) composition. The biogas volume was converted into mass. The resulting mass of methane refers to each month of the experiment. The average methane content in biogas ranged from 50.4% to 52.1%. These are standard values for substrates used in biogas installations.

3.7. Biomass Conversion in Laboratory

The degree of biomass conversion into biogas (methane and carbon dioxide) is important for the efficiency of the process. The content of methane in biogas is important because this compound determines its calorific value. The higher the methane volume concentration is, the less biogas is needed to produce the same amount of energy. It also reduces the demand for substrates because a smaller amount of inert gas (carbon dioxide) is necessary for the combustion process. Therefore, it is important to optimise the process in terms of the methane content in biogas. By testing the biochemical methanogenic potential of a particular organic material, it is possible to select the right amount of substrates for installations capable of generating a specific power.

Figure 3 shows the degree of biomass conversion into biogas (calculations based on Equation (1)). The maize silage was characterised by the best values of its conversion into biogas in the laboratory. However, the distribution of this biomass varied in individual periods of analysis and ranged from 78.8% (2) to 84.2% (6). The highest data amplitude was noted for pig manure, i.e., from 59.8% (6) to 76.9% (4). This indicates that the pig manure used in the test was not homogeneous and contained chemical compounds inhibiting the methane fermentation process, e.g., antibiotics and heavy metals from the feed provided to animals. The lowest degree of biomass conversion into biogas was noted for waste potatoes, i.e., from 43.54% (3) to 45.49% (2). Such poor results indicate that the potatoes may have contained residues from plant protection products, which inhibited biodegradation processes. Moreover, the potatoes were stored in open ground, which caused their decomposition. The degree of the conversion of beet pulp into biogas ranged from 55.90% (4) to 58.86% (2). This percentage of conversion was caused by the composition of organic matter contained in the pulp.

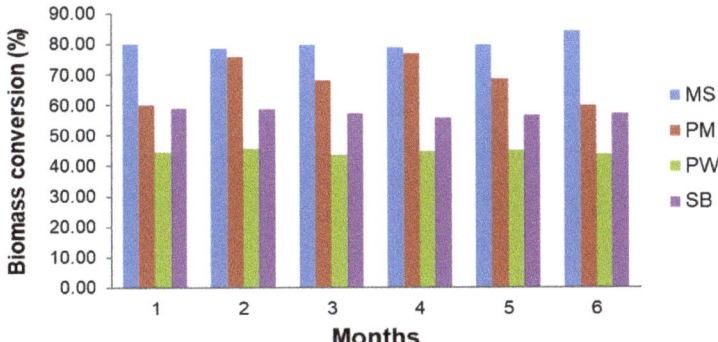

Figure 3. Biomass conversion in a laboratory.

3.8. Biomass Conversion under Industrial Conditions

In order to verify the degree of biomass conversion under the operating conditions of the installation, the amount of biogas obtained from each substrate was measured. It is easy to calculate the amount of methane from the biogas composition (Figure 4).

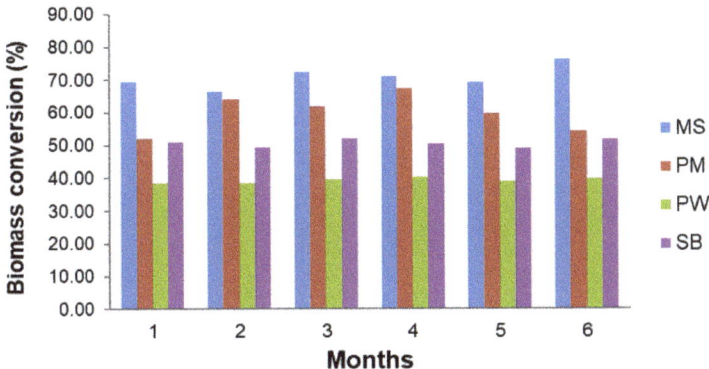

Figure 4. Biomass conversion under industrial conditions.

The values of biomass conversion conducted on a technical scale (Equation (2)) were lower than the results noted in the laboratory. This situation was caused by the fact that model tests are conducted in laboratories. However, by analysing the results of biomass conversion on an industrial scale, it is possible to verify the operation of the installation and implement a recovery plan to improve its efficiency. This procedure protects the owners of biogas plants from financial loss because each tonne of biomass which has not been converted into biogas/methane generates additional handling costs. The industrial-scale conversion of maize silage into biogas ranged from 66.44 (2) to 76.46 (6). For pig manure, it ranged from 52.10 (1) to 67.40 (4). Similar to the laboratory tests, the lowest conversion was obtained for waste potatoes, i.e., from 38.41 (2) to 40.27 (4). The conversion of beet pulp into biogas ranged from 49.17 (5) to 51.88 (6). Figure 5 shows the values of biomass conversion into biogas during the operation of the installation.

As resulted from the BMPCC values (Equation (3)), in comparison with the amount of methane obtained under technical conditions, the volume of methane produced in the laboratory was overestimated. There were differences noted for each substrate (see Figure 5). They ranged from 4.7% (6) to 17.19% (2) for maize silage, from 1.14% (4) to 23.58% (1) for pig manure, from 9.5% (4) to 13.69% (2) for waste potatoes, and from 9.06% (3) to 14.31% (5) for beet pulp.

To sum up, the biochemical methanogenic potential of the substrates used in the laboratory investigations was comparable to the data provided in reference publications [51–54]. The values of biogas and methane obtained in the industrial-scale production were lower than in the laboratory tests due to the disturbances that occurred in real installations. However, biogas plant owners should attempt to achieve a comparable efficiency of their facilities to the amounts obtained in laboratory tests. Model tests are used to estimate the amount of batch for the installation and the economic result of the project. Excessive losses of industrial-scale production cause owners to bear additional costs due to the need to purchase additional substrates.

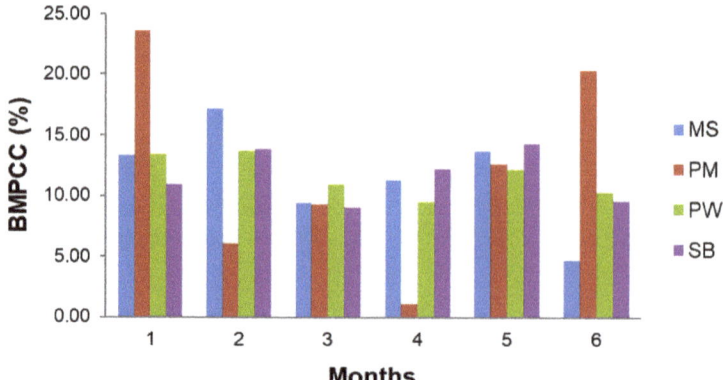

Figure 5. Values of the Biochemical Methane Potential Correction Coefficient for the substrates tested.

4. Conclusions

The BMPCC enables estimation of the industrial scale under fermentation of biomass compared with the laboratory conditions. Thus, it is possible to analyse the operation of the installation more accurately and eliminate the substrates that may inhibit the methane fermentation process. Waste potatoes seemed to be such a substrate in the installation analysed in this study because they were poorly converted in the laboratory. They may have been contaminated with the substances that weakened the methane fermentation process. When the substrate was fed into the digestion chambers, it may have reduced the process efficiency. In consequence, the biomass conversion in the entire bioreactor was affected, as was proved by the BMPCC.

The Biochemical Methane Potential Correction Coefficient is a solution that can be used as a system diagnosing a methane fermentation plant in a specific (selected) time interval. It can also be used to verify the biochemical methanogenic potential of individual substrates.

Author Contributions: Conceptualization, K.P.; methodology, K.P. and A.A.P.; software, P.B. and G.N.; validation, G.N. and P.B.; formal analysis, A.A.P. and K.D.; investigation, K.P., K.W., and N.M.; resources, K.P. and A.A.P.; data curation, K.P., A.A.P., and I.K.; writing—original draft preparation, K.P.; writing—review and editing, A.A.P. and P.B.; visualization, G.N. and P.B.; supervision, K.D.; project administration, K.P. and A.A.P.; funding acquisition, K.P. and K.D. All authors have read and agreed to the published version of the manuscript.

Funding: This research was co-funded by the company Biolab-Energy A&P, Poznan', Poland.

Conflicts of Interest: The authors declare no conflict of interest.

References

1. Pilarski, K.; Pilarska, A.A. Production efficiency of Poland farm-scale biogas plants: A case study. In Proceedings of the Renewable Energy Sources—6th International Conference, ICoRES 2019, Krynica, Poland, 12–14 June 2019. E3S Web of Conferences, Volume 154, 02002, 2020.
2. Stürmer, B. Feedstock change at biogas plants—Impact on production costs. *Biomass Bioenergy* **2017**, *98*, 228–235.
3. Samarappuli, D.; Berti, M.T. Intercropping forage sorghum with maize is a promising alternative to maize silage for biogas production. *J. Clean. Prod.* **2018**, *194*, 515–524.
4. Carmona, P.C.R.; Witaszek, K.; Janczak, D.; Czekała, W.; Lewicki, A.; Dach, J.; Pilarski, K.; Mazur, R. Methane fermentation of the poultry manure as an alternative and environmentally friendly technology of its management. *Arch. Waste Manag. Environ. Prot.* **2014**, *16*, 21–26.
5. Müller, F.P.C.; Maack, G.C.; Buescher, W. Effects of biogas substrate recirculation on methane yield and efficiency of a liquid-manure-based biogas plant. *Energies* **2017**, *10*, 325.
6. Pilarska, A.A.; Pilarski, K.; Wolna-Maruwka, A.; Boniecki, P.; Zaborowicz, M. Use of confectionery waste in biogas production by the anaerobic digestion process. *Molecules* **2019**, *24*, 37.
7. Stoknes, K.; Scholwin, F.; Krzesiński, W.; Wojciechowska, E.; Jasińska, A. Efficiency of a novel "Food to waste to food" system including anaerobic digestion of food waste and cultivation of vegetables on digestate in a bubble-insulated greenhouse. *Waste Manag.* **2016**, *56*, 466–476.
8. Ardolino, F.; Parrillo, F.; Arena, U. Biowaste-to-biomethane or biowaste-to-energy? An LCA study on anaerobic digestion of organic waste. *J. Clean. Prod.* **2018**, *174*, 462–476.
9. Hagos, K.; Zong, J.; Li, D.; Liu, C.; Lu, X. Anaerobic co-digestion process for biogas production: Progress, challenges and perspectives. *Renew. Sustain. Energy Rev.* **2017**, *76*, 1485–1496.
10. Silvestre, G.; Illa, J.; Fernández, B.; Bonmatí, A. Thermophilic anaerobic co-digestion of sewage sludge with grease waste: Effect of long chain fatty acids in the methane yield and its dewatering properties. *Appl. Energ.* **2014**, *117*, 87–94.
11. Pilarska, A.A.; Wolna-Maruwka, A.; Pilarski, K. Kraft lignin grafted with polyvinylpyrrolidone as a novel microbial carrier in biogas production. *Energies* **2018**, *11*, 3246.
12. Piechocki, J.; Solowiej, P.; Neugebauer, M. The use of biomass for electric power production in polish power plants. *Hung. Agric. Eng.* **2015**, *28*, 19–22.
13. Abubaker, J.; Cederlund, H.; Arthurson, V.; Pell, M. Bacterial community structure and microbial activity in different soils amended with biogas residues and cattle slurry. *Appl. Soil Ecol.* **2013**, *72*, 171–180.
14. Ruiz, D.; San Miguel, G.; Corona, B.; Gaitero, A.; Domínguez, A. Environmental and economic analysis of power generation in a thermophilic biogas plant. *Sci. Total Environ.* **2018**, *633*, 1418–1428. [PubMed]
15. Schlegel, M.; Kanswohl, N.; Rössel, D.; Sakalauskas, A. Essential technical parameters for effective biogas production. *Agron. Res.* **2008**, *6*, 341–348.
16. Zhou, J.; Zhang, R.; Liu, F.; Yong, X.; Wu, X.; Zheng, T.; Jiang, M.; Jia, H. Biogas production and microbial community shift through neutral pH control during the anaerobic digestion of pig manure. *Bioresour. Technol.* **2016**, *217*, 44–49. [PubMed]
17. Yan, L.; Yea, J.; Zhang, P.; Xu, D.; Wu, Y.; Liu, J.; Zhang, H.; Fang, W.; Wang, B.; Zeng, G. Hydrogen sulfide formation control and microbial competition in batch anaerobic digestion of slaughterhouse wastewater sludge: Effect of initial sludge pH. *Bioresour. Technol.* **2018**, *259*, 67–74. [PubMed]
18. Nie, H.; Fabian, J.H.; Strach, K.; Xu, C.; Zhou, H.; Liebetrau, J. Mono-fermentation of chicken manure: Ammonia inhibition and recirculation of the digestate. *Bioresour. Technol.* **2015**, *178*, 238–246.
19. Bousek, J.; Scroccaro, D.; Sima, J.; Weissenbacher, N.; Fuchs, W. Influence of the gas composition on the efficiency of ammonia stripping of biogas digestate. *Bioresour. Technol.* **2016**, *203*, 259–266.
20. Sun, Z.Y.; Yamaji, S.; Cheng, Q.S.; Yang, L.; Yue-Qin, T.; Kida, K. Simultaneous decrease in ammonia and hydrogen sulfide inhibition during the thermophilic anaerobic digestion of protein-rich stillage by biogas recirculation and air supply at 60 °C. *Process Biochem.* **2014**, *49*, 2214–2219.
21. Goberna, M.; Gadermaier, M.; Franke-Whittle, I.H.; García, C.; Wett, B.; Insam, H. Start-up strategies in manure-fed biogas reactors: Process parameters and methanogenic communities. *Biomass Bioenergy* **2015**, *75*, 46–56.

22. Woźniak, E.; Twardowski, T. The bioeconomy in Poland within the context of the European Union. *New Biotechnol.* **2018**, *40*, 96–102.
23. Pilarski, K. *Wydajność Procesu Fermentacji Metanowej w Biogazowniach Rolniczych (In English: Efficiency of Anaerobic Digestion Process in Agricultural Biogas Plants)*; Poznań University of Life Sciences: Poznań, Poland, 2019.
24. Piwowar, A.; Dzikuć, M.; Adamczyk, J. Agricultural biogas plants in Poland—Selected technological, market and environmental aspects. *Renew. Sustain. Energy Rev.* **2016**, *58*, 69–74.
25. Bouallagui, H.; Marouani, L.; Hamdi, M. Performances comparison between laboratory and full-scale anaerobic digesters treating a mixture of primary and waste activated sludge. *Resour. Conserv. Recycl.* **2010**, *55*, 29–33.
26. Kowalczyk, A.; Schwede, S.; Gerber, M.; Roland Span, R. Scale up of laboratory scale to industrial scale biogas plants, in bioenergy technology. In Proceedings of the 2011 World Renewable Energy Congress, Linköping, Sweden, 8–13 May 2011.
27. Hamawand, I.; Baillie, C. Anaerobic digestion and biogas potential: Simulation of lab and industrial-scale processes. *Energies* **2015**, *8*, 454–474.
28. Theuerl, S.; Herrmann, C.; Heiermann, M.; Grundmann, P.; Landwehr, N.; Kreidenweis, U.; Prochnow, A. The future agricultural biogas plant in Germany: A vision. *Energies* **2019**, *12*, 396.
29. Zhai, N.; Zhang, T.; Yin, D.; Yang, G.; Wang, X.; Ren, G.; Feng, Y. Effect of initial pH on anaerobic co-digestion of kitchen waste and cow manure. *Waste Manag.* **2015**, *38*, 126–131.
30. Pilarska, A.A.; Pilarski, K.; Wolna-Maruwka, A. Cell immobilization on lignin–polyvinylpyrrolidone material used for anaerobic digestion of waste wafers and sewage sludge. *Environ. Eng. Sci.* **2019**, *36*, 478–490.
31. Carrere, H.; Antonopoulou, G.; Affes, R.; Passos, F.; Battimelli, A.; Lyberatos, G.; Ferrer, I. Review of feedstock pretreatment strategies for improved anaerobic digestion: From lab-scale research to full-scale application. *Bioresour. Technol.* **2016**, *199*, 386–397.
32. Pilarska, A.A.; Pilarski, K.; Witaszek, K.; Waliszewska, H.; Zborowska, M.; Waliszewska, B.; Kolasiński, M.; Szwarc-Rzepka, K. Treatment of dairy waste by anaerobic digestion with sewage sludge. *Ecol. Chem. Eng. S* **2016**, *23*, 99–115.
33. Pilarska, A.A.; Pilarski, K.; Ryniecki, A.; Tomaszyk, K.; Dach, J.; Wolna-Maruwka, A. Utilization of vegetable dumplings waste from industrial production by anaerobic digestion. *Int. Agrophys.* **2017**, *31*, 93–102.
34. Pilarska, A.A.; Wolna-Maruwka, A.; Pilarski, K.; Janczak, D.; Przybył, K.; Gawrysiak-Witulska, M. The use of lignin as a microbial carrier in the co-digestion of cheese and wafer waste. *Polymers* **2019**, *11*, 2073–2092.
35. DIN Guideline 38 414-S8. *Characterisation of the Substrate, Sampling, Collection of Material Data, Fermentation Tests*; German Institute for Standardization: Berlin, Germany, 1985.
36. ISO. *Guide to the Expression of Uncertainty in Measurement (GUM)*; ISO: Geneva, Switzerland, 1993.
37. Konieczka, P.; Namieśnik, J. *Evaluation and Quality Control of Analytical Measurement Results*; Scientific-Technical Publishing House WNT: Warsaw, Poland, 2007.
38. Kukier, E.; Kwiatek, K.; Grenda, T.; Goldsztejn, M. Mikroflora kiszonek. *Życie Wet.* **2014**, *89*, 1031–1036.
39. Marszałek, M.; Kowalski, Z.; Makara, A. Physicochemical and microbiological characteristics of pig slurry. *Tech. Trans.* **2014**, *111*, 81–91.
40. Jacob, S.; Chintagunta, A.D.; Banerjee, R. Selective digestion of industrial potato wastes for efficient biomethanation: A sustainable solution for safe environmental disposal. *Int. J. Environ. Sci. Technol.* **2016**, *13*, 2363–2374.
41. Arumuganainar, S.; Choi, H.L.; Kannan, N.; Rajagopal, K. Biochemical methane potentials and organic matter degradation of swine slurry under mesophilic anaerobic digestion. *Ind. J. Sci. Technol.* **2016**, *9*, 1–6.
42. Sadecka, Z.; Suchowska-Kisielewicz, M. The possibility of using organic substrates in the fermentation process. *Ann. Set Environ. Prot.* **2016**, *18*, 400–413.
43. Fugol, M.; Szlachta, J. The reason for using corn and fermented liquid manure ensilage for biogas production. *Agric. Eng.* **2010**, *1*, 119–174.
44. Oleszek, M.; Krzemińska, I. Enhancement of biogas production by co-digestion of maize silage with common goldenrod rich in biologically active compounds. *BioResources* **2017**, *12*, 704–714.
45. Pessuto, J.; Scopel, B.S.; Perondi, D.; Godinho, M.; Dettmer, A. Enhancement of biogas and methane production by anaerobic digestion of swine manure with addition of microorganisms isolated from sewage sludge. *Proc. Saf. Environ. Prot.* **2016**, *104*, 233–239.

46. Miroshnichenko, I.; Lindner, J.; Lemmer, A.; Oechsner, H.; Vasilenko, I. Anaerobic digestion of sugar beet pulp in Russia. *Landtechnik* **2016**, *71*, 175–185.
47. Ziemiński, K.; Kowalska-Wentel, M. Effect of different sugar beet pulp pretreatments on biogas production efficiency. *Appl. Biochem. Biotechnol.* **2017**, *181*, 1211–1227. [PubMed]
48. Matulaitis, R.; Juškienė, V.; Juška, R. Measurement of methane production from pig and cattle manure in Lithuania. *Zemdir. Agric.* **2015**, *102*, 103–110.
49. Fugol, M.; Szlachta, J. Use of the potato industry waste for anaerobic digestion. *Przem. Chem.* **2013**, *92*, 824–828.
50. Hartmann, S.; Helmut, D. The economics of sugar beets in biogas production. *Landtechnik* **2011**, *66*, 250–253.
51. Szymańska, M.; Sosulski, T.; Szara, E.; Pilarski, K. Conversion and properties of anaerobic digestates from biogas production. *Przem. Chem.* **2015**, *94*, 1419–1423.
52. Witaszek, K.; Pilarska, A.A.; Pilarski, K. Selected methods of vegetable raw material pre-treatment used in biogas production. *Ekon. Environ.* **2015**, *53*, 138–152.
53. Pilarski, K.; Pilarska, A.A.; Witaszek, K.; Dworecki, Z.; Żelaziński, T.; Ekielski, A.; Makowska, A.; Michniewicz, J. The impact of extrusion on the biogas and biomethane yield of plant substrates. *J. Ecol. Eng.* **2016**, *17*, 264–272.
54. Pilarska, A.A.; Pilarski, K.; Waliszewska, B.; Zborowska, M.; Witaszek, K.; Waliszewska, H.; Kolasiński, M.; Szwarc-Rzepka, K. Evaluation of bio-methane yields for high-energy organic waste and sewage sludge: A pilot-scale study for a wastewater treatment plant. *Environ. Eng. Manag. J.* **2019**, *18*, 2023–2034.

© 2020 by the authors. Licensee MDPI, Basel, Switzerland. This article is an open access article distributed under the terms and conditions of the Creative Commons Attribution (CC BY) license (http://creativecommons.org/licenses/by/4.0/).

Article

The Follow-up Photobioreactor Illumination System for the Cultivation of Photosynthetic Microorganisms

Beata Brzychczyk [1], Tomasz Hebda [1,*], Jakub Fitas [1] and Jan Giełżecki [2]

1. Department of Mechanical Engineering and Agrophysics, Faculty of Production and Power Engineering, University of Agriculture in Krakow, ul. Balicka 120, 30-149 Krakow, Poland; beata.brzychczyk@urk.edu.pl (B.B.); jakub.fitas@urk.edu.pl (J.F.)
2. Department of Production Organization and Applied Computer Science, Faculty of Production and Power Engineering, University of Agriculture in Krakow, ul. Balicka 120, 30-149 Krakow, Poland; jan.gielzecki@urk.edu.pl
* Correspondence: Tomasz.hebda@urk.edu.pl

Received: 29 January 2020; Accepted: 27 February 2020; Published: 3 March 2020

Abstract: The article presents the basic conceptual assumptions of a photobioreactor with a complementary lighting system. The cylindrical bioreactor has three independent, interconnected, and fully controlled lighting systems. A characteristic feature is the combination of the lighting system with the measurement of photosynthetically active PAR (photosynthetically active radiation) and the optical density of the culture medium. The entire lighting system is based on RGBW ("red, green, blue, white") LED and RBG ("red, green, blue") LEDs. The pilot study was conducted on a simplified prototype of a photobioreactor designed for the distribution and optimization of light in algae cultures designed for energy purposes. The study was carried out on microalgae *Chlorella Vulgaris* BA0002a from the collection of marine algae cultures.

Keywords: cylindrical LED light coat; tracking lighting; photosynthetic microorganisms; photobioreactor; algae

1. Introduction

By definition a renewable resource, biomass remains an important aspect of the search for sustainable energy sources [1]. This includes not only refuse and energetic wood plantations, but also specially grown algae [2]. Algae-derived biomass may be used for energy production, whether by simple firing or co-firing with refuse-derived fuels, other biomass, or non-renewable fuels, or through production of biofuels [3]. The algae-derived biofuels may take the form of pyrolytic solid fuels [4], burnable gases such as hydrogen and methane [5], or liquid hydrocarbons and biodiesel [6,7]. Other chemical components, such as ammonia, may as well be produced through algae cultivation [8]. As photosynthetic organisms consume carbon dioxide in the process of photosynthesis, algae breeding also presents a desirable side-effect of carbon sequestration [9,10].

Like other autotrophic organisms, algae grow best when certain optimal conditions are reached. This makes outdoor algae farming impractical in colder climates, as the conditions for growth occur only through a part of the year at best; in addition to this, the natural conditions are subject to modification to a highly limited extent. However, unlike most multicellular plants, algae do not require soil and have a limited need for space. As a result, algae can be grown in specially prepared vessels or photobioreactors. This arrangement not only allows for year-round breeding of algae, but also for the adjustment of conditions such as, e.g., light wavelength [11] or light intensity, or even vessel geometry [12], in search for the optimal growth rate. Moreover, industrial photobioreactors must also allow for practical removal of algal matter without interrupting their operation. Therefore, research into algal growth conditions and photobioreactor construction remain an important and promising segment of renewable energy research.

As mentioned above, light is one of the most important factors affecting the rate of growth of photosynthetic organisms. The influence of light energy varies for different processes and different species. Both excessive lighting and lack of light adversely affect the production of biomass. Within a photobioreactor, the conversion of available light depends on the concentration of biomass, and thus on the optimal distribution of radiant energy, so that all cells receive the appropriate amount. It is assumed that the optimal value of the density of the photon flux for algae is in the range of 100–600 $\mu mol \cdot m^{-2} \cdot s^{-1}$ and is characteristic not only for a given species, but even for a given type [13]. In addition to the above, the microalgae cells reduce the intensity of light within the photobioreactor. Located farther from the source of light, they will receive less radiation as a result of mutual occlusion, which results in a decrease in the rate of growth. There is a process of absorption and dispersion called the effect of shading [14]. As the cell density increases, the intensity of light in the reactor is suppressed. Excessive lighting also acts as a brake causing the so-called photoinhibition phenomenon, i.e., a situation in which a photosynthetic microorganism is exposed to a greater amount of radiant energy than the one it is able to process. Thus, the second-order photositemator is damaged [15–17]. The situation usually takes place in the early stages, just after the establishment of the breeding.

Numerous constructional solutions of photobioreactors, such as columnary photobioreactors or cylindrical bubble columns made of light-transparent materials, with aeration culture of aerated algae in different gas mixtures mounted in them [18,19], can be found in the literature. Such solutions due to the large area of exposure and the price are ideal for outdoor farming, used on an industrial scale for the production of algae [20]. Tubular photobioreactors combined in a series are also used in outdoor cultures. The bundles of pipes directed towards the sun maximize the effect of solar radiation [21]. In artificial lighting cultures for illuminating photobioreactors, LED diodes are most often used due to the emission of radiation with narrow wavelength ranges and the ability to easily select the color of light. The research suggests that the growth of algae is largely influenced by the spectrum of red light, favoring their faster multiplication and growth. On the other hand, the spectrum of blue light causes cell growth [22].

Commonly used types of PBRs (photobioreactors) are flat panels used outdoors in conditions of natural surroundings. Due to their design and location, they are characterized by a perpendicular arrangement of the longest side of the bioreactor in relation to lighting, or a constant incapable of changing the inclination angle in relation to the radiation incidence plane [23]. These reactors do not allow for optimal exposure conditions for the photoautotrophic biomass located in them and thus do not allow control of the intensity of light falling on the surface of the photobioreactor. In the case of excessive intensity of lighting, the known solutions are used to inhibit the growth of biomass, which directly translates into the result of productivity. In the above systems it is also difficult to control temperature or gas removal [24,25]. There is also an example of a Polish photobioreactor solution contained in patent specification PL 224183 regarding a flat panel intended for growing microorganisms in the natural environment. The invention relates to the construction of a system following natural light. In respect to the method of gas introduction and the shape of the vessel, the air lift type of PBRs has seen interest. The reactors are divided into the light and the dark zone, while the driving force is provided by the air-lift module responsible for the circulation of the biomass and therefore the uniform exposition of algae cells to the light [26,27].

Another solution are photobioreactors made of a transparent film forming cultured reservoirs permeable to light [28]. Devices of this type are economically advantageous due to the possibility of easy assembly in places where it is possible to use, for example, waste carbon dioxide [14,29,30].

Laboratory solutions for photobioreactors of various constructions, fully automated, performing the task of breeding, research and optimization in multiplication processes of photoautotrophic microorganisms, despite the fact that they meet the tasks set before them, show a number of imperfections, including those related to the distribution of light. A number of solutions implement the radiant energy in an artificial way, hence the light structures are arranged both inside photobioreactors and outside the units [13,24,31]. The above solutions, however, do not ensure that the breeding facilities are illuminated from below. They also do not allow symmetrical and homogeneous illumination of cultures in a cylindrical bioreactor and thus do not provide a steady and uniform intensity of photosynthetically active radiation (PAR) reaching the cylindrical reactor, which can provide a cylindrical light jacket. They do not measure the radiation inside the reactor during real-time cultivation and do not compensate for changes related to the uneven distribution of radiation inside the reactor or the phytotron chamber. The aim of the article is to present an innovative concept of a photobioreactor prototype for the cultivation of photosynthetic microorganisms, in particular, the integrated follow-up lighting system. The model has been extended with additional elements by installing a circulating pump forcing the circulation of cultures so that all cells receive the same amount of light energy that directly affects their growth.

Taking the above into account, as part of research on photosynthetic microorganisms, a simplified photobioreactor model for studying the influence of light and other parameters on microalgae cultures was designed and built at the Department of Mechanical Engineering and Agrophysics. The photobioreactor prototype with the follow-up lighting system was submitted to the Patent Office and is currently assigned a number (P.429507) [32]. The invention relates to a specialized device for cultivating microorganisms, including microalgae, both in a continuous and periodic system. The solution for continuous production is intended for industrial needs, whereas the solution for periodic farming is for the needs of scientific research and to pilot optimization of breeding.

In order to ensure uniform and sufficient distribution of light along with dynamically changing conditions during cultivation, a photobioreactor prototype was designed with a follow-up, or lagging, lighting system. The key role in the lagging lighting system is played by a fully controlled and programmed unit, coupled with a measurement of the productivity of the farm and PAR (photosynthetically active radiation, a range of wavelengths of light in the range of 400–700 nm) [33] in the medium, enabling change of the lighting parameters relative to the changing culture conditions during the process. The entire lighting system is implemented by using LED RBGW and LED RBG lamps, selected in terms of wavelengths so that they best match the natural needs of photosynthetic microorganisms. LED lighting during operation generates less heat and is energy-saving. The controllers enable independent, smooth control of the intensity of colored LED panels, in individual modules, in the full power range together with the measurement of PAR radiation. The use of this type of lighting allows for smooth adjustment of light intensity and tunability of color intensity. In addition, the unit allows you to set independent brightening or blanking speeds of individual lighting modules with different time increments. The intensity of radiation is coupled with PAR measurement inside the reactor.

Functionally, the system has been divided into three basic lighting modules. A characteristic feature of the lighting system is a cylindrical light jacket ensuring an even illumination of the medium around the perimeter of the reactor. In addition, the system includes lighting for the bottom of the tank and lid. Thanks to the multidirectional illumination of cylindrical reactors, a large, developed light surface for breeding is obtained.

2. Materials and Methods

2.1. Construction of the System

A glass cylindrical reaction vessel (photobioreactor), transparent to solar and artificial radiation, ensures the transmission of as much energy as possible to the absorbing surface. The reactor consists of three lighting modules with control and registration system, feeding, mixing and sampling. The glass cylinder of the photobioreactor is placed in the light jacket, covered from the top with an airtight cover together with holes for taking samples, through which the feeding, mixing, venting, and transporting systems of the culture medium pass. The photobioreactor's lighting consists of the following lighting modules:

- module I—the cylindrical jacket panel inside which the reactor is placed,
- module II—lower lighting panel, implemented by lighting from the bottom of the reaction tank —the light source is lighting,
- module III—the panel of the illuminated reactor tank cover.

All light modules are coupled together and allow for full regulation and control of both color and intensity, duration of lighting time, and the possibility of setting independent speeds of simulating sunrises and sunsets or applying pulsed radiation of the required color. The lighting system is intended for cylindrical units with various L/d modules (aspect; height divided by diameter). The lighting is controlled by a computer (MCU), which selects the optimal parameters based on the sensor of the photosynthetic active PAR radiation intensity and the time of day and night. Knowledge of the distribution of radiation energy density and degradation in the PBR and energy absorption by algal biomass is necessary for the analysis and optimization of cultures in photobioreactors with artificial lighting. The control parameters are selected empirically and saved to the MCU in the form of patterns. The sensor of the intensity of photosynthetic PAR radiation can be moved additionally along the radius of the cylinder, which allows the measurement of the intensity of PAR radiation in the reactor space.

The upper light panel, which at the same time constitutes the reactor cover, ensures the supply and drainage of media necessary for culturing. In addition, the entire photobioreactor unit is equipped with:

- a sensor for measuring PAR (linear, point),
- a real-time optical density meter for breeding,
- a diffuser for the CO_2 feeding system and feeding of other process gases, equipped with an inlet nozzle with a shut-off valve and a meter to measure the amount of gas fed,
- a biomass mixing and sampling system—a suction and delivery pump with a discharge connector for removing biomass with a shut-off valve,
- a temperature measurement sensor,
- a system for measuring filtration and UV disinfection and nutrient solution,
- an inoculum delivery and nutrient replenishment system, and
- a cooling and heating system—a thermostabilization system for breeding.

The photobioreactor according to the invention works in strictly controlled and programmed conditions, suitable for use at a semi-industrial and industrial scale. The scheme of the photobioreactor is shown in Figure 1.

The photobioreactor (1) is a glass, cylindrical reaction vessel with a $dm^3 \cdot d^{-1}$ module (0.3–2.0), transparent in the visible and invisible spectrum for photosynthetically active radiation and embedded inside a lighting system that includes a cylindrical light jacket (2), lower light panel (3), and upper light panel (4) (constituting the reactor cover). The photobioreactor system is equipped with a suction and force pump (5), quantum immersion sensor for measuring photosynthetic radiation PAR (6), temperature sensor (7), CO_2 feeding system, and the introduction of other process gases (8). Moreover, the entire photobioreactor system also includes a biomass pickup system (9), UV measurement and disinfection unit of the filtrate (10), nutrient solution and inoculum (11), and heating-cooling system (12).

All systems, meters, and sensors are connected to the microcomputer controller (13), to which the lighting controls of the three light panels are also connected.

A simplified prototype was made at the University of Agriculture in Krakow. During the experimentation of cultivation, the *Chlorella Vulgaris* strain BA0002a obtained from the Collection of Baltic Sea Cultures of the University of Gdańsk (CCBA) was used. The first experiments were carried out in a prototype glass photobioreactor with an active volume of 4 dm^3. Before the experiments were started, the photobioreactor was sterilized and inoculated with 2 ml of microalgae in the medium. The experiments were carried out at a temperature of 23–25 °C and at a pH of 6.5–7.0. The culture was fed with carbon dioxide fed to the interior of the photobioreactor by means of a diffuser. Circulation of the culture medium was ensured using a circulating pump with an average capacity of 80.0 dm$^3 \cdot$h^{-1}. The lighting was a follow-up lighting system using LED RBGW diodes, with the number of lighting points being not less than 1500 RGBW LEDs and RGB LED per m^2, in the form of a cylindrical light jacket and a lower panel lighting system and illuminated cover.

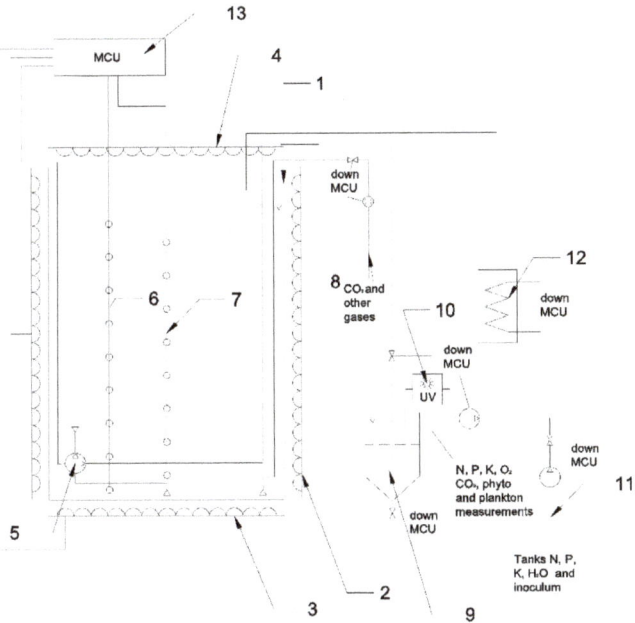

Figure 1. A photobioreactor scheme with a follow-up lighting system; source: own studies. Symbols described in text.

2.2. Methodology for Measuring the Intensity of Lighting

Preliminary pilot studies were carried out on the designed prototype of the photobioreactor. Measurement of the PAR radiation intensity inside the culture medium was carried out with the Quantum Meter MQ-100 Apogee Instruments measuring probe (Figure 2b) for measurements in an aqueous environment with a spectral range of 410–655 nm. The measuring device consisted of a digital meter with a quantum sensor. The tests were carried out during breeding, inside the culture medium with a directional probe set to the light source. Measurements were made along measurement path I marked on the diagram and on the axis of the photobioreactor (WWTP). The meter sensor was mounted on a tripod rail and moved vertically every 10 mm ± 1.0 mm along the marked measurement lines (Figure 2a). The measurements made will ultimately form the measurement grid of the PAR radiation intensity in the active volume of the reactors with a different lighting system.

The tests were carried out for the dual cylindrical light jacket switched on and the low light panel also switched on. LED directional lighting placed from the bottom contained 30 points of light of about 18 W, 1100 lm, 3000 K (warm white color), and 230 V, while the double light jacket consisted of: module I—150 LED RBG type LED lights with 36 W, 12 V, controller, white on; module II—300 points of light—18 W 950 lm, with a color temperature of 3100 K (warm white), 12 V power supply. The lighting worked in the day/night 14/10 system. The geometrical dimensions of the cylindrical light jacket were: diameter (24.5 ± 1.0) mm, height (19.5 ± 1.0) mm.

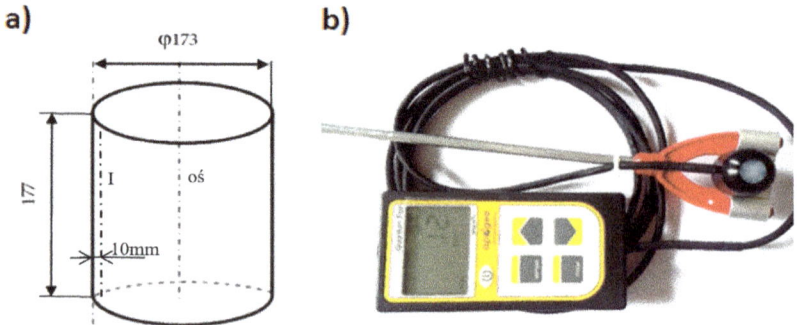

Figure 2. Diagram of PAR radiation measurement paths in the reactor (**a**) and Quantum Meter MQ-100 meter by Apogee Instruments (**b**); source: own work.

Growth of microalgae Chlorella vulgaris BA0002a from the collection of Baltic algae cultures at the CCBA (Culture Collection of Baltic Algae) was carried out in a sterilized cylindrical glass reactor with an $dm^3 \cdot d^{-1}$ module equal to 0.985 and 4.16 dm^3, including an active capacity of 4 dm^3 on nutrient medium about the composition of Table 1.

Table 1. Composition of culture medium.

Component	Concentration, mg·dm^{-3}
N	184.5
P	49.2
K	192.0

The medium of 4 dm^3 was inoculated with 2 ml of microalgae inoculum. Figure 3 presents graphs from pilot measurements of the intensity of photosynthetically active radiation, using a measuring probe whose head was directed towards the light jacket and was moved along measuring path I and on the axis of the reactor. Measurements of PAR radiation intensity were made for two different optical densities of the cultured 4.23 McF and 7.35 McF media, determined using a DEN-1B densitometer, in the McF degrees (0–180·10^4 cell·dm^{-3}).

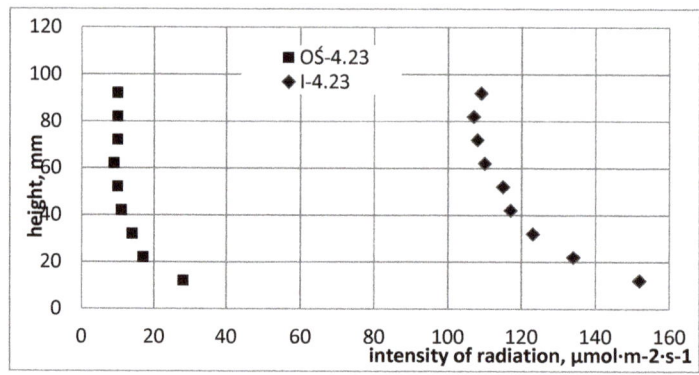

Figure 3. Characterization of photosynthetically active radiation distribution for optical density 4.23 McF ("OŚ" denotes a path along the reactor axis).

3. Results

Measurements of PAR intensity were made by means of a directional probe adjusted to one light source which was a double cylindrical light jacket. The probe was displaced every 10 mm along the reactor axis. It is easy to observe that for lower optical density the distribution of light inside the reactor has similar characteristics along measuring path I and along the PBR axis (Figure 3). The photon flux density in the reactor axis was the smallest and, about 25 mm from the bottom, decreased to values close to zero. In the case of the measurement along path I, the radiation reaching the reactor wall from about 50 mm from the bottom maintained a constant level of about 110 $\mu mol \cdot m^{-2} \cdot s^{-1}$. The richest zone with the highest radiation density in the set system was the bottom zone along the reactor wall.

Comparing the radiation distribution pattern inside the reactor at the optical density of 7.35 McF, we can notice a large uneven distribution of radiant energy in the reactor space (Figure 4). The radiation in the reactor axis is at zero, while immense unevenness and chaos of the photon flux prevails inside the culture medium in the measurement carried out according to path I. Both at the bottom and near the reactor cover there are larger values, because here the radiation from the cover module and the bottom panel is reaching the culture, while at the middle height of the tested system, the radiation reaching the culture is suppressed by 50%.

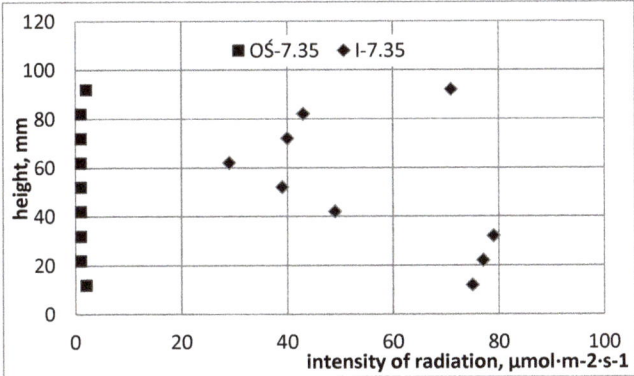

Figure 4. Characterization of photosynthetic PAR vector radiation distribution for optical density 7.35 McF.

4. Conclusions

The presented solution is innovative in terms of its comprehensiveness and multi-functionality and provides a fix for the problem of light distribution inside a PBR. The real-time light radiation intensity compensation using measured parameters can compensate for the deficits of uneven distribution of light and thus optimize the conducted breeding. The values set at the beginning of the cultivation during the production of biomass, after obtaining the feedback signal from on-line measurements, can be changed in real time, automatically as well as manually, during the process. There is a possibility of adjusting the amount of radiation to the increasing concentration of microorganisms based on PAR measurements obtained in real time from the inside the farm. None of the known solutions compensate for differences in light distribution in real time on the basis of photon flux density measurements reaching the breeding medium, which in this case is an innovation in the technical solution. The same applies to the administration of nutrients in periodic cultures. Pilot study results showed significant differences in the distribution of available light energy in the reactor space, which varies with the increase in biomass.

Author Contributions: Conceptualization, B.B.; Data curation, B.B. and T.H.; Formal analysis, T.H.; Methodology, J.G.; Project administration, B.B.; Resources, J.G.; Writing—original draft, B.B.; Writing—review & editing, T.H. and J.F. All authors have read and agreed to the published version of the manuscript.

Funding: This research was financed by the Ministry of Science and Higher Education of the Republic of Poland.

Conflicts of Interest: The authors declare no conflicts of interest.

References

1. Ahorsu, R.; Medina, F.; Constantí, M. Significance and Challenges of Biomass as a Suitable Feedstock for Bioenergy and Biochemical Production: A Review. *Energies* **2018**, *11*, 3366. [CrossRef]
2. Saad, M.G.; Dosoky, N.S.; Zoromba, M.S.; Shafik, H.S. Algal Biofuels: Current Status and Key Challenges. *Energies* **2019**, *12*, 1920. [CrossRef]
3. Amaro, H.M.; Macedo, A.C.; Malcata, F.X. Microalgae: An alternative as sustainable source of biofuels. *Energy* **2012**, *44*, 158–166. [CrossRef]
4. Lee, J.; Lee, K.; Sohn, D.; Kim, Y.M.; Park, K.Y. Hydrothermal carbonization of lipid extracted algae for hydrochar production and feasibility of using hydrochar as a solid fuel. *Energy* **2018**, *153*, 913–920. [CrossRef]
5. Ding, L.; Gutierrez, E.C.; Cheng, J.; Xia, A.; O'Shea, R.; Guneratnam, A.J.; Murphy, J.D. Assessment of continuous fermentative hydrogen and methane co-production using macro- and micro-algae with increasing organic loading rate. *Energy* **2018**, *151*, 760–770. [CrossRef]
6. Markou, G.; Angelidaki, I.; Nerantzis, E.; Georgakakis, G. Bioethanol Production by Carbohydrate-Enriched Biomass of Arthrospira (Spirulina) platensis. *Energies* **2013**, *6*, 3937–3950. [CrossRef]
7. Islam, M.A.; Magnusson, M.; Brown, R.J.; Ayoko, G.A.; Nabi, N.; Heimann, K. Microalgal Species Selection for Biodiesel Production Based on Fuel Properties Derived from Fatty Acid Profiles. *Energies* **2013**, *6*, 5676–5702. [CrossRef]
8. Wijayanta, A.T.; Aziz, M. Ammonia production from algae via integrated hydrothermal gasification, chemical looping, N2 production, and NH3 synthesis. *Energy* **2019**, *174*, 331–338. [CrossRef]
9. Ferreira, L.S.; Rodrigues, M.S.; Converti, A.; Sato, S.; Carvalho, L.C.M. Arthrospira (Spirulina) platensis cultivation in tubular photobioreactor: Use of no-cost CO_2 from ethanol fermentation. *Energy* **2012**, *92*, 379–385. [CrossRef]
10. Mohler, D.T.; Wilson, M.H.; Fan, Z.; Groppo, J.G.; Crocker, M. Beneficial Reuse of Industrial CO_2 Emissions Using a Microalgae Photobioreactor: Waste Heat Utilization Assessment. *Energies* **2019**, *12*, 2634. [CrossRef]
11. Zhang, Y.; Bao, K.; Wang, J.; Chao, Y.; Hu, C. Performance of mixed LED light wavelengths on nutrient removal and biogas upgrading by different microalgal-based treatment technologies. *Energy* **2017**, *130*, 392–401. [CrossRef]
12. Hosseini, N.S.; Shang, H.; Scott, J.A. Optimization of microalgae-sourced lipids production for biodiesel in a top-lit gas-lift bioreactor using response surface methodology. *Energy* **2018**, *146*, 47–56. [CrossRef]
13. Kumar, A.; Ergas, S.; Yuan, X.; Sahu, A.; Zhang, Q.; Dewulf, J.; Malcata, F.X.; van Langenhove, H. Enhanced CO_2 fixation and biofuel production via microalgae: Recent developments and future directions. *Trends Biotechnol.* **2010**, *28*, 371–380. [CrossRef]
14. Chisti, Y. Biodiesel from microalgae. *Biotechnol. Adv.* **2007**, *25*, 294–306. [CrossRef]
15. Carvalho, A.P.; Meireles, L.A.; Malcata, F.X. Microalgal reactors: A review of enclosed system design and performances. *Biotechnol. Prog.* **2006**, *22*, 1490–1506. [CrossRef]
16. Vass, I.; Cser, K.; Cheregi, O. Molecular mechanisms of light stress of photosynthesis. *Ann. N. Y. Acad. Sci.* **2007**, *1113*, 114–122. [CrossRef]
17. Pupo, O.R.; García, S.; Valencia, G.; Bula, A. Conceptual Design of Photobioreactor for Algae Cultivation. In Proceedings of the ASME 2011, International Mechanical Engineering Congress and Exposition, Denver, CO, USA, 11–17 November 2011.
18. Kumar, K.; Dasgupta, C.N.; Nayak, B.; Lindblad, P.; Das, D. Development of suitable photobioreactors for CO_2 sequestration addressing global warming using green algae and cyanobacteria. *Bioresour. Technol.* **2011**, *102*, 4945–4953. [CrossRef]
19. Singh, R.N.; Sharma, S. Development of suitable photobioreactor for algae production. *Renew Sustain. Energy Rev.* **2012**, *16*, 2347–2353. [CrossRef]

20. Richmond, A.; Hu, Q. *Handbook of Microalgal Culture*; Blackwell Publishers: Hoboken, NJ, USA, 2004; pp. 57–82.
21. Norsker, N.H.; Barbosa, M.J.; Vermuë, M.H.; Wijffels, R.H. Microalgal production—A close look at the economics. *Biotechnol. Adv.* **2011**, *29*, 24–27. [CrossRef]
22. Chen, P.; Min, M.; Chen, Y.; Wang, L.; Li, Y.; Chen, Q. Review of the biological and engineering aspects of algae to fuels approach. *Int. J. Agric. Biol. Eng.* **2009**, *2*, 1–30.
23. Frąc, M.; Jezierska-Tys, S.; Tys, J. Microalgae for biofuels production and environmental applications: A review. *Afr. J Biotechnol.* **2010**, *9*, 9227–9236.
24. Ugwu, C.U.; Aoyagi, H.; Uchiyama, H. Photobioreactors for mass cultivation of algae. *Bioresour. Technol.* **2008**, *99*, 4021–4028. [CrossRef]
25. Pierre, G.; Delattre, C.; Dubessay, P.; Jubeau, S.; Vialleix, C.; Cadoret, J.-P.; Probert, I.; Michaud, P. What Is in Store for EPS Microalgae in the Next Decade? *Molecules* **2019**, *24*, 4296. [CrossRef]
26. De Carvalho, L.R.; Fujü, M.T.; Roque, N.F.; Lago, J.H.G. Aldingenin derivatives from the red alga Laurencia aldingensis. *Phytochemistry* **2006**, *67*, 1331–1335. [CrossRef]
27. Posten, C. Design principles of photo-bioreactors for cultivation of microalgae. *Eng. Life Sci.* **2009**, *3*, 165–177. [CrossRef]
28. Kim, E.H.; Sim, S.J. Photobioreactor Made of a Transparent Film. Patent WO2011090330, 20 January 2011.
29. Borowitzka, M.A. Commercial production of microalgae: Ponds, tanks, tubes and fermenters. *J. Biotechnol.* **1999**, *70*, 313–321. [CrossRef]
30. Carlozzi, P.; Torzillo, G. Productivity of Spirulina in a strongly curved outdoor tubular photobioreactor. *Appl. Microbiol. Biot.* **1996**, *45*, 18–23. [CrossRef]
31. Kozieł, W.; Włodarczyk, T. Algae—biomass production. *Acta Agrophys.* **2011**, *17*, 105–116.
32. Talbierz, S.; Kukulska, N. Photobioreactor. Patent PL224183B1, 6 February 2012.
33. Suh, I.S.; Lee, S.B. A light distribution model for an internally radiating photobioreactor. *Biotechnol. Bioeng.* **2003**, *82*, 180–189. [CrossRef]

© 2020 by the authors. Licensee MDPI, Basel, Switzerland. This article is an open access article distributed under the terms and conditions of the Creative Commons Attribution (CC BY) license (http://creativecommons.org/licenses/by/4.0/).

Article

The Impact of a Controlled-Release Fertilizer on Greenhouse Gas Emissions and the Efficiency of the Production of Chinese Cabbage

Jakub Sikora [1,*], Marcin Niemiec [2,*], Anna Szeląg-Sikora [1,*], Zofia Gródek-Szostak [3,*], Maciej Kuboń [1,*] and Monika Komorowska [4,*]

1. Faculty of Production and Power Engineering, University of Agriculture in Krakow, 30-149 Kraków, Poland
2. Faculty of Agriculture and Economics, University of Agriculture in Krakow, 31-121 Kraków, Poland
3. Department of Economics and Enterprise Organization, Cracow University of Economics, 31-510 Krakow, Poland
4. Faculty of Biotechnology and Horticulture, University of Agriculture in Krakow, 31-121 Kraków, Poland
* Correspondence: Jakub.Sikora@ur.krakow.pl (J.S.); Marcin.Niemiec@ur.krakow.pl (M.N.); Anna.Szelag-Sikora@ur.krakow.pl (A.S.-S.); grodekz@uek.krakow.pl (Z.G.-S.); Maciej.Kubon@ur.krakow.pl (M.K.); m.komorowska@ogr.ur.krakow.pl (M.K.)

Received: 30 January 2020; Accepted: 17 April 2020; Published: 21 April 2020

Abstract: Optimization of plant fertilization is an important element of all quality systems in primary production, such as Integrated Production, GLOBAL G.A.P. (Good Agriculture Practice) or SAI (Sustainable Agriculture Initiative). Fertilization is the most important element of agricultural treatments, affecting the quantity and quality of crops. The aim of the study was to assess greenhouse gas (GHG) emissions in the cultivation of Chinese cabbage, depending on the technological variant. The factor modifying the production technology was the use of fertilizers with a slow release of nutrients. One tonne of marketable Chinese cabbage crop was selected as the functional unit. To achieve the research goal, a strict field experiment was carried out. Calculation of the total amount of GHG emitted from the crop was made in accordance with ISO 14040 and ISO 14044. The system boundaries included the production and use of fertilizers and pesticides, energy consumption for agricultural practices and the emission of gases from soil resources and harvesting residue. The use of slow-release fertilizers resulted in a greater marketable yield of cabbage compared to conventional fertilizers. The results of the research indicate a significant potential for the use of slow-release fertilizers in reducing agricultural emissions. From the environmental and production point of view, the most favourable variant is the one with 108 kg N·ha^{-1} slow-release fertilizers. At a higher dose of this element, no increase in crop yield was observed. At this nitrogen dose, a 30% reduction in total GHG emissions and a 50% reduction in fertilizer emissions from the use of per product functional unit were observed. The reference object was fertilization in accordance with production practice in the test area.

Keywords: greenhouse gases; efficiency; agriculture; slow release fertilizers; Chinese cabbage

1. Introduction

The environmental impact of plant production is mainly associated with the consumption of fossil fuels and the use of fertilizers and plant protection products. Greenhouse gas (GHG) emissions are also associated with energy consumption for the production, transport and application of fertilizers and plant protection products, irrigation, as well as the logistics process of food products [1]. An important source of agricultural pollution is the emission of GHG from the soil as a result of mineralization of dead organic matter and humus compounds [2]. Some of the elements introduced into the soil with organic and mineral fertilizers are dispersed in the environment; to the atmosphere, in the form of GHG, and to the underground and surface waters, causing intensified eutrophication processes and acidification of individual elements of the environment. At all its stages, food production is associated with the intensive use of natural resources such as soil, water, space and energy resources [3]. Intensive agricultural production, often regarded as conventional, has developed in response to the demand for cheap food production. It is characterized by maximizing production with the simultaneous increase in consumption of the means and energy resources [4]. The intensified production very quickly led to environmental degradation of large areas, especially in developed countries, with the most intensive agriculture. The transformation of natural ecosystems into monocultural agrocenosis is also associated with landscape changes that have a multidirectional, negative impact on the human environment [5]. The implementation and development of quality management systems in food production were somewhat the consumer market's reaction to the presence of poor quality products on the market [6,7].

Economic development of developed countries, and the related increase in the wealth of societies, has shaped consumer awareness regarding the environmental and health effects of overexploitation related to food production [8,9]. An important aspect of optimizing agriculture is reducing its environmental impact, especially in the context of emissions. Research results presented in the scientific literature report great potential for binding organic carbon in soil organic matter, which allows reducing the negative impact of agriculture on the environment, and the greenhouse effect [10–12]. Global GHG from fossil fuels used in agriculture is estimated at 0.4–0.6 Gt of CO_2 equivalent. Annually, total agricultural emissions amount to 4.6 Gt of CO_2 equivalent per year [13]. In the United States, agriculture accounts for 9% of the total emissions of these compounds [14]. There are primary and secondary sources of energy consumption and emission in plant production. Primary emission sources are the result of agrotechnical practices on the farm (e.g., tillage, sowing, fertilization, harvesting and transport, irrigation). Secondary (indirect) emission sources include emissions from the production of fertilizers, pesticides, production and maintenance of equipment, etc. What is important from the point of view of emissions is the emission of nitrogen oxides from soils intensively supplemented with nitrogen fertilizers. The amount of nitrogen compounds released to the atmosphere is related not only to the level of fertilization, but above all, to the efficiency of using the elements contained in fertilizers and the type and intensity of cultivation treatments. An important source of environmental impact of agricultural nitric oxide is its emission resulting from decomposition of harvesting residue. Total N_2O fertilizer emissions increased from 0.07 Gt of CO_2 equivalent per year in 1961 to 0.68 Gt CO_2 equivalent in 2010 [13]. Nitric oxide (N_2O) is a key compound responsible for the greenhouse effect. It retains 292 times more infrared radiation than carbon dioxide [15]. Assessing the environmental and economic efficiency of quality systems implemented in primary production is a very important element in the evaluation of the actual impact of producer regulations on the quality of produce and the degree of environmental impact [16–18]. One of the methods of comprehensive and multifaceted assessment of the quality system is to prepare a product life cycle taking into account the use of energy and means of production, as well as renewable and non-renewable environmental resources [19,20]. However, developing a reliable and universal method is very difficult because farms operate in a specific economic, social and climatic framework, which significantly affects the assessment and is very difficult to interpret [21]. That is why agricultural systems are often assessed based on a selected fragment of activity.

The aim of the study was to quantify and compare GHG emissions in the different technological variants of the cultivation of Chinese cabbage. The research variable was the fertilisation technology with conventional fertilisers and slow release ferilizers. One tonne of marketablemarketable Chinese cabbage crop was selected as the functional unit.

2. Materials and Methods

In terms of the scope of research, the number of sites and their geographical location, the experiment was planned based on the assumed qualitative and quantitative objectives, as well as technical possibilities. [22]. The choice of the plant variety, the individual fertilization options, the range of agrotechnical treatments, the level of irrigation and the time frame of the experiment was based on a risk analysis carried out in accordance with ISO 31000:2018 [23]. The risk defined was the impact of the above-mentioned risk factors on the reduction of plant yields. Different fertilization variants were applied in individual experiment objects, which resulted from the production practice applied in the research area, the requirements of quality management systems in primary production and economic aspects. The scope of the research and the selection of GHG emission sources were carried out based on literature study [15–21]. The sources of GHG emissions selected for the calculation of total emissions were selected based on their utility in the context of the determined research objectives.

In terms of the number of objects, the scope of research and geographical location, the experiment was planned based on the assumed qualitative and quantitative objectives and technical capabilities [22]. The choice of plant variety, individual fertilization variants, the range of agrotechnical treatments, the level of irrigation and the time frame of the experiment was based on the risk analysis carried out in accordance with ISO 31000:2018. The defined risk was the impact of the above-mentioned risk factors on the reduction of plant yields. The scope of the conducted research and the selection of GHG emission sources was based on a literature study. The scheme of the experimental design is as follows:

(1) Formulation of the strategic objective;
(2) Selection of the experience factors (a single-factor experience was selected);
(3) Selection of the plant (Chinese cabbage is a plant of high economic importance and high GHG emission potential due to high levels of fertilization and relatively low nitrogen use;
(4) Selection of the experiment site (geographical boundary of the system); the experiment was conducted in the area with large acreage of Chinese cabbage cultivations;
(5) Selection of agrotechnical treatments (except the experimental factor); the treatments were selected based on the recommendations of the integrated plant production methodology and based on production practices in the research area;
(6) Formulation of the experimental factor levels. The level of nitrogen fertilization and the forms of nitrogen applied in the subsequent experimental facilities were designed based on the following input data:

　(a) A control facility is necessary for the assessment of the site potential;
　(b) Objects fertilized with 400 and 500 kg of slow-release fertilizers. The amount of fertilization results from the manufacturer's recommendations in their advertising materials;
　(c) Objects fertilized with 600 and 800 kg of slow-release fertilizer and objects fertilized with 300 and 450 kg of ammonium nitrate·ha^{-1}. The level of fertilization was calculated based on the plants' fertilizing needs, at the estimated site productivity of 65 and 90 t·ha^{-1}. The two estimated levels result from the likelihood of adverse weather conditions during the vegetation period of the plants;
　(d) An object fertilized with 600 kg of ammonium nitrate·ha^{-1}. The level of fertilization results from production practices applied in the research area;

(7) Estimation of the system boundary in terms of GHG emission sources. The selection was based on the latest available literature and a risk analysis in the context of the assumed target, in accordance with ISO 31000:2018. These sources are:

　(a) GHG emissions related to the applied fertilizers;
　(b) GHG emissions related to the plant protection products used;
　(c) GHG emissions related to electricity consumption and combustion of fossil fuels;
　(d) GHG emissions related to decomposition of crop residues;
　(e) GHG emissions related to decomposition of soil organic matter.

The goal was achieved based on field experience which allowed calculating GHG emissions in individual technological variants. The experiment was established in the soil with a heavy clay grain size. The soil properties on which the experiment was established are included in Table 1. The forecrop for research plants was corn. The test plant was Chinese cabbage (*Brassica rapa* L.) of the Parkin F1 variety. The experiment was established on 15 May 2017. The plants were harvested on 22 July 2017. Plants were grown at a spacing of 50 × 30 cm. Cultivation and plant protection practices were carried out based on the integrated production methodology for the Chinese cabbage approved by the Main Inspector of Plant Health and Seed Inspection, based on art. 5 par. 3 pos. 2 of the Act of 18 December 2003 on Plant protection, consolidated text from 2008, no. 133, item 849 as amended. In terms of agricultural treatments, the following were applied: plowing, cultivating, planting, mechanical application of fertilizers (twice in facilities where conventional fertilizers were applied, and once in facilities where slow-release fertilizers were applied), as well as mechanical weeding (once). The harvesting was manual, and irrigation amounted to 80 $dm^3 \cdot m^{-2}$ from a deep water well, 36 m in depth.

Table 1. Selected soil properties at the beginning and at the close of the experiment. Used for experiments $mg \cdot kg^{-1}$ and after.

pH in H_2O	pH in KCl	[%]			$mg \cdot kg^{-1}$			
		N in Total	C Organic	N Mineral	P	K	Mg	Ca
7.01	6.65	0.16	1.41	56.65	147.8	359.5	199.4	850

Plants were irrigated to optimal water content to eliminate the possibility of water stress affecting the test results. Before establishing the experiment, the physicochemical and chemical properties of the soil were analyzed, and the following parameters were determined: pH, granulometric composition, the content of organic matter, mineral nitrogen and Kiejdahl nitrogen content, as well as the content of available forms of P, K, Mg and Ca.

Optimization of the agricultural system was based on the use of a multicomponent, slow-release fertilizer with NPK content (%) 18-05-10 + 4CaO + 2Mg (with a nitrogen release period of two months), ammonium nitrate (34% N), triple superphosphate (46% P_2O_5) and 60% potassium salt (60% K_2O). The experiment included seven fertilization levels and a control object (Table 2). The slow-release fertilizer was applied pointwise to each plant during planting, at a depth of 5–7 cm below the planting level. All phosphorus and potassium fertilizers were applied before sowing, while ammonium nitrate was applied in two doses: 60% before planting and 40% after planting. The date of late top dressing was selected based on the observation of meteorological conditions and monitoring condition of the plants. The experiment was conducted in four replicates, in randomized blocks.

Table 2. Experiment diagram.

Object Number	Slow-Release Fertilizer	Ammonium Nitrate	Triple Superphosphate	Potassium Salt	N	P_2O_5	K_2O
	kg of fertilizer·ha^{-1}				kg of component·ha^{-1}		
control	0	-	-	-	-	-	-
1	400	-	89	177	72	60	150
2	500	-	76	158	90	60	150
3	600	-	65	140	108	60	150
4	800	-	43	103	144	60	150
5	-	300	130	250	100	60	150
6	-	450	130	250	150	60	150
7	-	600	130	250	200	60	150

The system boundaries included:

(1) Production of fertilizers and agrochemicals used for growing plants;
(2) The farm's energy consumption for field work;
(3) Soil emissions (direct and indirect) related to fertilizer use;
(4) Emissions from harvesting residue management and from the mineralization of soil organic matter;
(5) Water consumption for irrigation.

In order to determine the environmental impact of the production of Chinese cabbage in various technological conditions, the following standards were applied: ISO 14040 "Environmental management—Life cycle assessment—Principles and framework" and ISO 14044 "Environmental management—Life cycle assessment—Requirements".

3. Results

The analysis was carried out according to the recommendations included in the document [24]. Product transport, packaging, agricultural tool wear and marketing were excluded from the research. The production and transport of seedlings were also excluded from the process due to the lack of data on this process. The adopted functional unit was 1 tonne of marketable product. The time frame of the system was one year. The potential for generating the greenhouse effect has been estimated based on the GHG emission calculated as per carbon dioxide equivalent (CO_2).

The input data for calculating the GHG emission value came from seven experimental objects and a control object where fertilization was not used (Table 2). Fertilization at object no. 7 was carried out in accordance with the production practices used in the area of testing. The GHG emission level for nitrogen production in ammonium nitrate was assumed at 7.99 kg $CO_2 \cdot kg^{-1}$ N. For triple superphosphate, this value was set at 0.36 kg $CO_2 \cdot kg^{-1}$ P_2O_5, while for potassium chloride it was 0.56 kg $CO_2 \cdot kg^{-1}$ K_2O [25]. Based on the data on the composition of the slow-release fertilizer used, its total carbon footprint was calculated at 8.2 kg $CO_2 \cdot kg^{-1}$ N, as per the mass of fertilizer [25]. Emission factors for harvesting residue were calculated based on the amount of waste generated during cabbage processing. Based on the results of the experiment, it was estimated that the ratio of marketable to collateral yield and harvesting residue in the cultivation of Chinese cabbage ranged from 19% to 28%. The adopted content of carbon fraction in dry matter of harvesting residue was 50%. The actual nitrogen content in harvesting residue was used for the calculations. The level of harvesting residue distribution was estimated at 25%. Carbon emissions from harvesting residue were calculated according to the methodology in the IPCC [26]. The value of nitric oxide emission from harvesting residue was adopted at 1.25% [27]. The value "N-N_2O emissions" was multiplied by 44/28 to convert it into N_2O. N_2O emission was used as a CO_2 equivalent by multiplying it by the global warming potential of 298 [15]. The adopted soil mineralization rate of organic matter was 2%. The assessment of the life cycle of Chinese cabbage includes emissions from the burning of fossil fuels used for agricultural procedures. Based on the data provided by EPA [28], the diesel emission from agricultural tractors was assumed at 3.864 kg $CO_2 \cdot dm^{-3}$ of fuel. Fuel consumption in particular agricultural procedures is presented in (Table 3). Owing to the small amount of nitrogen oxides and methane emitted the during diesel combustion in agricultural tractors, this source of GHG was omitted [28]. The amount of greenhouse gas emissions associated with the use of pesticides was estimated based on the data provided by Audsley [11] (Table 4). These authors report the total amount of GHG emissions per carbon dioxide equivalent at 25.5 kg CO_2 for 1 kg of active substance of pesticide. The amount of pesticides used in cabbage cultivation was 1950 g·ha^{-1} for all objects. For irrigation, deep well water pumped from 36 m was used using 40% electric pumps. The CO_2 emission factor from irrigation was calculated on the basis of the guidelines given by Wang [29] (Table 4). For the production of 1 kWh of electricity, CO_2 emissions were adopted at 0.9245 kg [29]. According to the methodology given by [30] the amount of nitrogen emitted as a result of mineral fertilization is 1% as direct emission and 0.27% as nitrogen dispersed in the environment. The amount of nitrogen oxides emitted from mineral fertilizer nitrogen was estimated at 0.75% of total nitrogen not used by plants during the growing season [30] (Tables 4 and 5). The N_2O to CO_2 conversion factor in the context of the greenhouse effect is 292 [15].

Table 3. Energy consumption and greenhouse gas (GHG) emissions associated with agricultural treatments.

Type of Agricultural Treatment	Diesel Use [dm^3]	Energy Use [MJ]	CO_2 Emission [kg]
Ploughing	57.0	2299.0	220.2
Mineral fertilization, one time	5.1	206.9	19.71
Cultivation with an aggregate	30.8	1241.5	119.0
Mechanical planting	154.0	6207.4	595.1
Application of plant protection products, four times	24.8	517.3	95.83
Mechanical weeding, one time	19.2	775.9	74.19
Total	1325.8	53,458.0	1124.0
Irrigation [KWh]		194.7	179.7
Total			1303.7

Table 4. CO_2 emissions from fertilizers [CO_2 equivalent·ha^{-1}).

Object Number	A	B	C	D	E	F
control					1323.4	353.1a *
2	592	72.29	664.29	48.75	1323.4	384.3a
3	740	64.00	804.00	48.75	1323.4	474.0b
4	888	56.34	944.34	48.75	1323.4	521.1bc
5	1184	40.70	1224.70	48.75	1323.4	563.9c
6		502.3	502.28	48.75	1303.7	412.8a
7		901.8	901.78	48.75	1303.7	476.9b
8		1301.3	1301.28	48.75	1303.7	458.1bc

* Different letters mean statistically significant differences at the significance level p = 0.05.

Table 5. Greenhouse gas (GHG) emissions.

Object Number	G	H	I	J	K
	equivalent of CO_2·ha^{-1}			equivalent of CO_2·ha^{-1}	
control					14.5
1	1.13	0.04	330.4	11.33	18.9
2	1.41	0.30	413.0	87.81	31.5
3	1.70	0.44	495.6	128.02	38.2
4	2.26	0.68	660.8	197.77	44.2
5	1.57	0.64	458.9	187.85	22.9
6	2.36	1.15	688.3	334.51	31.9
7	3.14	1.53	917.7	448.01	29.3

A—CO_2 equivalent from the slow-release fertilizer [25]; B—CO_2 equivalent from conventional fertilizers [25]; C—CO_2 equivalent from fertilizers in total; D—CO_2 equivalent from plant protection products [11]; E—CO_2 emissions from fuel combustion and electricity consumption [29]; F—CO_2 emissions from soil and harvesting residue and from the mineralization of organic matter^3 [26,27].

G—Direct N_2O emission from mineral fertilization [30]; H—Indirect N_2 emission related to mineral fertilization [30]; I—CO_2 equivalent from direct N_2 emission; J—CO_2 equivalent from the indirect emission of N_2 [15]; K—CO_2 equivalent from N_2 emission from the mineralization of harvesting residue^3 [26,27].

The obtained results were subjected to analysis of variance. The significance of differences in mean values was determined with the Tukey test ($\alpha \leq 0.05$) using the Statistica 13 (TIBCO Software Inc.) software to statistically process the results.

The presented work compares the total GHG emissions for various fertilization strategies within the assessment system adopted at taking inventory of their sources. The results of the calculations demonstrate a significant differentiation of the total GHG emissions in terms of CO_2 equivalent for the adopted system boundary. The adopted unit of account was 1 tonne of marketable product. The marketable yield in the non-fertilized object was 15.46 t. Fertilization in the amount of 72 kg N in the form of slow-release fertilizers and a full dose of phosphorus and potassium caused a double increase in the marketable yield of plants (Figure 1). When using traditional fertilizers, at 100 kg·ha^{-1} nitrogen, the marketable yield was approx. 56% higher compared to the unfertilized object. The maximum yield was obtained in an object fertilized with slow-release fertilizers, at 144 kg N, and it amounted to 44.02 t of marketable yield·ha^{-1} [31,32]. Niemiec [31], and Niemiec et al. [32] report a reduction in celery yield after applying slow-release fertilizers directly under the root. These authors stated that the reason for the decrease in yielding may be increased salinity of the soil solution directly under plant root zone. In this experiment, slow-release fertilizers were used below the planting level, which could limit the negative impact of soil solution salinity, especially at the initial stage of plant growth. To achieve the yield at the level observed when using 144 kg N in a slow-release fertilizer, 200 kg of nitrogen in conventional fertilizers was required. Due to the relatively low content of mineral forms of nitrogen in the soil, a strong reaction of plants to mineral fertilization was observed.

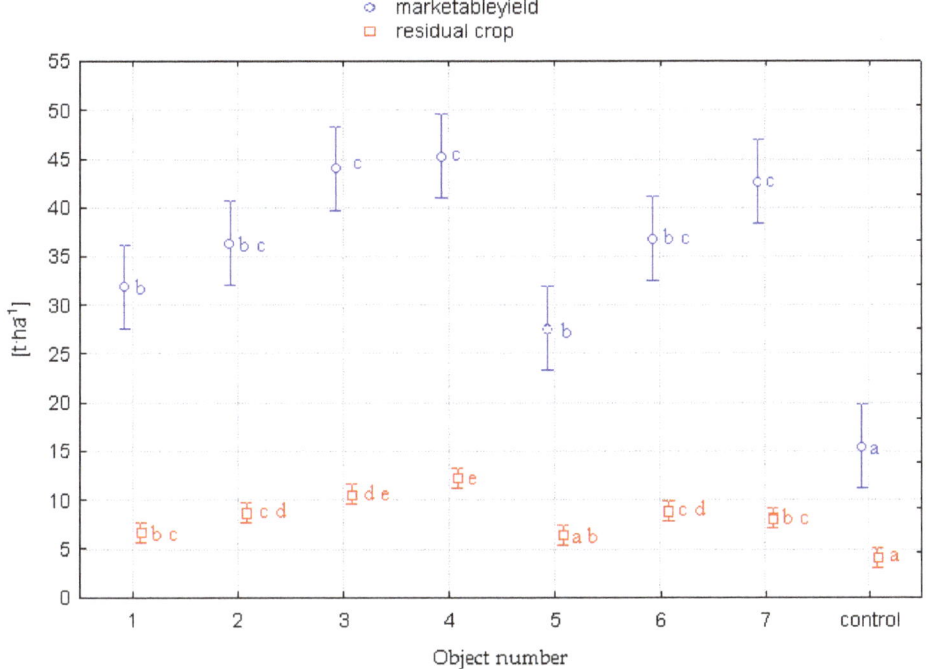

Figure 1. Amount of marketable and residual crop.

The experiment results indicate a significant impact of the proposed fertilization strategies on the development of GHG emissions, both from direct and indirect sources. Total GHG emissions, given as CO_2 equivalent ranged from 78.9 to 120.3 kg CO_2·t^{-1} fresh weight of marketable product (Figure 2. The lowest value of this parameter was obtained for an object fertilized with slow-release fertilizers at 108 kg N^{-1} fresh mass of the marketable product (option 3). For all facilities fertilized with conventional fertilizers, the emission factor was above 110 kg CO_2·t^{-1} fresh weight of marketable product. The value of GHG emissions for a nonfertilized object was 111.8 kg CO_2·t^{-1} product. There were no statistically significant differences between individual fertilization variants with the use of the slow-release fertilizer. In relation to the reference object, fertilized in accordance with production practice, a reduction of GHG emission level of 30% was achieved per functional unit of the product (Figure 2). From the point of view of environmental efficiency, this is an important technological achievement. In modern agricultural systems, improving efficiency by up to several percent can be problematic [33–35].

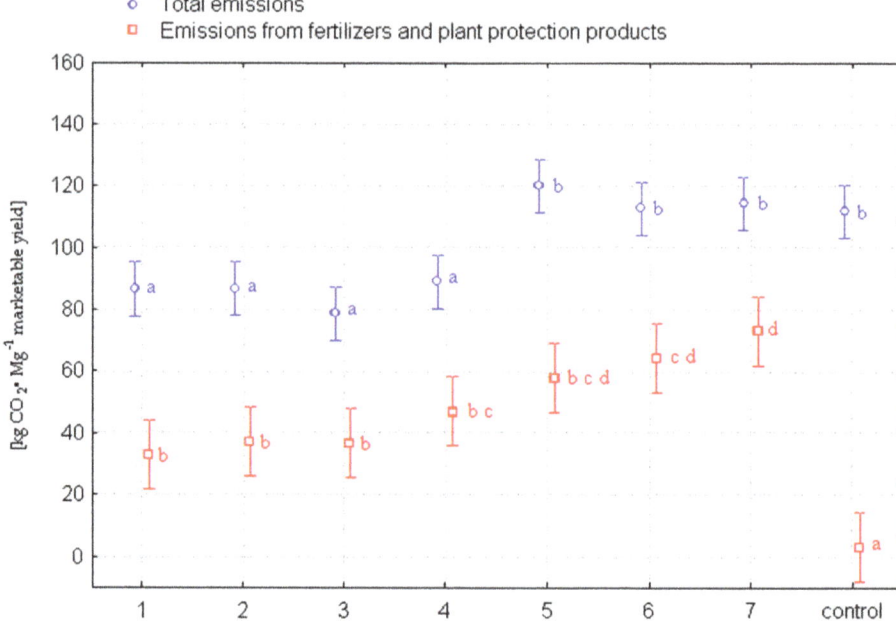

Figure 2. Total GHG emissions and emissions related to the use of fertilizers.

The share of mineral fertilizers in shaping GHG emissions is presented in Figure 3. In the object fertilized in accordance with production practice, the amount of GHG emissions associated with the use of mineral fertilizers and plant protection products was 73 kg of CO_2 equivalent·t^{-1}. When applying slow-release fertilizers, this value was 36.7 kg CO_2 equivalent·t^{-1}. The obtained research results indicate that the level of mineral fertilization efficiency is the most important element related to the optimization of plant production. This has been confirmed by studies of other authors [14,17,36]. Figure 3 presents the percentage share of individual GHG sources in the total emissions for the adopted system boundaries. The sources were divided into three groups:

1. Agricultural treatments and irrigation;
2. Production and use of fertilizers;
3. Greenhouse gas emissions from soil result from decomposition of harvesting residue and soil organic matter.

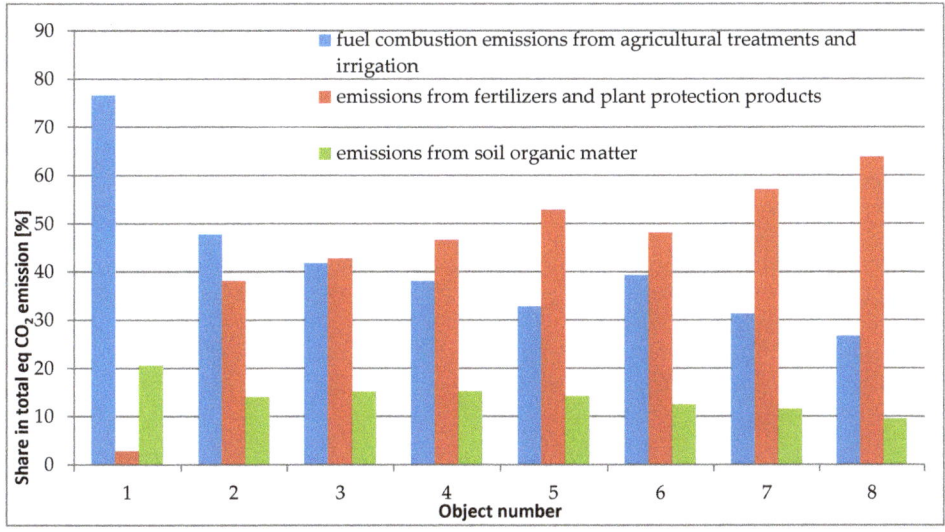

Figure 3. Structure of GHG emissions from individual sources [%].

Irrigation and agricultural treatments are another source of agricultural emissions included in the research. For the production technologies used, the level of GHG emissions associated with energy consumption was 1303.7 kg $CO_2 \cdot ha^{-1}$ (Table 4). In objects fertilized only with conventional fertilizers, this value was slightly higher, which results from the double application of the fertilizers (Table 4). According to the technology used, the application of slow-release fertilizers during planting did not cause additional GHG emissions. Energy consumption for irrigation and agricultural treatments was constant for all fertilization variants and is not correlated with production effects. However, it has a significant impact on the level of GHG emissions per unit of account. Very high values of this coefficient testify to the low efficiency of the use of production potential, while low values indicate unreasonable fertilization. In the conducted experiment, the share of emissions of GHG from fuel combustion and electricity consumption ranged from 26.70% to 76.60% (Figure 3). The lowest value of this parameter was found in a non-fertilized object, while the largest share of GHG emissions from energy consumption was found in the object fertilized with conventional fertilizers at the highest dose of 200 kg $N \cdot ha^{-1}$. In the experimental object fertilized with slow-release fertilizers, at 108 kg N, the agricultural and irrigation GHG emission rate was 38% (Figure 3). The amount of fertilizer GHG emissions is largely influenced by the total amount of fertilizers used, both organic and mineral, as well as the degree of nitrogen utilization by plants. Nitrogen not used by plants is dispersed in the natural environment and is a source of GHG. In the experiment, in the object where the lowest GHG emission rate was obtained, the GHG share associated with the use of fertilizers was 46%. Increasing the nitrogen dose increased the value of this parameter. In an object fertilized with conventional fertilizers, at 200 kg N, it was observed that over 60% of emitted GHG came from fertilizers and plant protection products (Figure 3). The factor that most affected the result of the GHG emission assessment was the unreasonable use of nitrogen from mineral fertilizers. The total nitrogen uptake of the cultivated Chinese cabbage in an object fertilized according to production practice amounted to approx. 30% of the nitrogen dose introduced with mineral fertilizers. In the case of an object with the slow-acting fertilizer at 108 kg$\cdot ha^{-1}$ this value was more than twice as high. The most important element in reducing the value of the carbon footprint for the adopted system boundary was the degree of utilization of nitrogen from mineral fertilizers. The object with the highest level of nitrogen fertilization was the reference for individual experiment variants. The share of GHG emissions from soil organic matter and the distribution of harvesting residue ranged from approx. 10% to over 20% of total GHG emissions. For individual fertilization variants, this value was slightly different and resulted from the amount of harvesting residue and its nitrogen content.

4. Discussion

The compared production technologies differed only by the use of an innovative nitrogen fertilizer. No modifications were made to other agricultural treatments. Fertilization is a very important element of agricultural production technology. It shapes the size of the crop and the efficiency of using other means of production, such as plant protection products, water used for irrigation and energy used for growing and harvesting products. The effectiveness of fertilization ultimately affects the efficiency of the use of production acreage and the energy efficiency of production. Unreasonable use of fertilizers results in a decrease in nutrient utilization efficiency. Plant nutrients not absorbed during the growing season significantly increase the impact of agriculture on GHG emissions. An important source of GHG is the soil environment. Optimizing the use of nitrogen introduced with fertilizers is one way to reduce GHG emissions. Liang [37] report that the use of biochar significantly increased the yield and nitrogen utilization in reed. Reduced losses of fertilizer elements as a result of the addition of biochar were also reported by Niemiec [38]. These authors stated that the fertilizers supplemented with biocarbon demonstrate the functional characteristics of slow-release fertilizers. Xiao [39] reported that the use of slow-release fertilizers in peach cultivation resulted in a reduction of the crop's impact on GHG emissions by 25%, while nitrogen emissions from fertilizers decreased by 50%. In turn, [36] report that the use of fertilizers with slow release of nutrients has allowed the reduction of nitrogen oxides directly from bamboo crops at approx. 20% to 80%. These values depended primarily on the pattern of weather conditions. In our own research, in an object optimally fertilized with slow-release fertilizers, the amount of nitrogen oxides emitted was 45% lower compared to an object fertilized in accordance with production practice in the area of research. In gaseous form, these elements are emitted from the soil surface in the form of carbon dioxide, nitrogen oxides and methane. Their amount results from the intensification of mineralization of organic matter and the level of nitrogen fertilization. Nitrogen compounds from organic and mineral fertilization are very important from the point of view of shaping the level of GHG emissions [24–26]. Each experiment was treated as a separate farm, functioning in a specific environment. Literature reports lots of data related to the calculation of the life cycle for agriculture in a broader perspective, both spatial and temporal [5,14,29].

The results presented by these authors relate to the general values of the impact of agriculture on the environment using specific production technologies. Life cycle assessment for strict experiments carried out is not a commonly used element of the assessment of the environmental impact of a production technology [39–45]. Vegetation experiments are conducted to optimize production technology or to assess the actual effectiveness of the use of a factor of production or production technique. Nowadays, for experiments using new fertilizers, the literature presents results related primarily to fertilization efficiency, the amount of ingredients dispersed in the environment or the quality of the crop. Fertilization is one of the factors impacting the level of GHG emissions and should be considered in a broader context, along with all elements of agricultural treatments. The paper attempts to scale the methodologies used in calculating GHG emissions for strict vegetation experiments. The compared production technologies differed only by the use of an innovative nitrogen fertilizer. Other elements and conditions remained unchanged. This approach allowed estimating the actual impact of using a fertilizer characterized by a slower release of nutrients. The level of GHG emissions is now a strategic element of agricultural characteristic and their emission from nitrogen fertilizers is significant. Agriculture is characterized by a very low level of efficiency in the use of nutrients derived from fertilizers, both mineral and organic. No modifications were made to other agricultural treatments. However, introducing modifications related to fertilization technologies is problematic. Reducing the total amount of elements introduced into the soil environment raises the risk of reducing yields, which agricultural producers fear [7]. Liang [37] has indicated the high effectiveness of subsidizing technological innovations in agricultural production in terms of limiting the negative impact on the environment [46–54]. Linking subsidies with pro-environmental measures in agriculture can bring positive effects in reducing GHG emissions.

5. Conclusions

The use of slow-release fertilizers resulted in a greater marketable yield of cabbage compared to conventional fertilizers. From the environmental and production point of view, the most favorable variant is the one with 108 kg N·ha^{-1} slow-release fertilizers. At a higher dose of this element, no increase in crop yield was observed. Fertilizing Chinese cabbage with slow-release fertilizers at 108 kg N·ha^{-1} resulted in a 30% reduction in total GHG emissions per functional unit, compared to fertilization in accordance with production practice. With slow-release fertilizers, fertilization GHG emissions were reduced by approx. 50% per functional unit of the produce. Efficiency of fertilizer nitrogen use is an important element shaping the level of agricultural GHG emissions.

The use of slow-release fertilizers should be promoted in agricultural production, since, if used rationally, they can reduce GHG emissions and climate change.

Author Contributions: Conceptualization, J.S., M.N., and M.K. (Monika Komorowska); methodology, M.K. (Maciej Kuboń), A.S.-S.; Z.G.-S.; software, J.S., M.N.; validation, A.S.-S. and Z.G.-S.; formal analysis, M.K. (Monika Komorowska), M.K. (Maciej Kuboń); resources, Z.G.-S., A.S.-S., M.K. (Maciej Kuboń); writing—original draft preparation: M.N., J.S., M.K. (Monika Komorowska) visualization, Z.G.-S., A.S.-S., M.K. (Maciej Kuboń). All authors have read and agreed to the published version of the manuscript.

Funding: This research received no external funding.

Conflicts of Interest: The authors declare no conflict of interest.

References

1. Cupiał, M.; Szeląg-Sikora, A.; Niemiec, M. Optimisation of the machinery park with the use of OTR-7 software in context of sustainable agriculture. *Agric. Agric. Sci. Procedia* **2015**, *7*, 64–69. [CrossRef]
2. Szeląg-Sikora, A.; Niemiec, M.; Sikora, J.; Chowaniak, M. Possibilities of Designating Swards of Grasses and Small-Seed Legumes From Selected Organic Farms in Poland for Feed. In Proceedings of the IX International Scientific Symposium Farm Machinery and Processes Management in Sustainable Agriculture, Lublin, Poland, 22–24 November 2017; pp. 365–370.
3. Li, L.; Wu, W.; Giller, P.; O'Halloran, J.; Liang, L.; Peng, P.; Zhao, G. Life Cycle Assessment of a Highly Diverse Vegetable. Multi-Cropping System in Fengqiu County, China. *Sustainability* **2018**, *10*, 983. [CrossRef]
4. Kocira, S.; Kuboń, M.; Sporysz, M. Impact of information on organic product packagings on the consumers decision concerning their purchase. In Proceedings of the 17th International Multidisciplinary Scientific GeoConference (SGEM 2017), Albena, Bulgaria, 29 June–5 July 2017; Volume 17, pp. 499–506.
5. Schmidt Rivera, X.C.; Bacenetti, J.; Fusi, A.; Niero, M. The influence of fertiliser and pesticide emissions model on life cycle assessment of agricultural products: The case of Danish and Italian barley. *Sci. Total Environ.* **2017**, *592*, 745–757. [CrossRef]
6. Szeląg-Sikora, A.; Sikora, J.; Niemiec, M.; Gródek-Szostak, Z.; Kapusta-Duch, J.; Kuboń, M.; Komorowska, M.; Karcz, J. Impact of Integrated and Conventional Plant Production on Selected Soil Parameters in Carrot Production. *Sustainability* **2019**, *11*, 5612. [CrossRef]
7. Niemiec, M.; Komorowska, M.; Szeląg-Sikora, A.; Sikora, J.; Kuboń, M.; Gródek-Szostak, Z.; Kapusta-Duch, J. Risk Assessment for Social Practices in Small Vegetable farms in Poland as a Tool for the Optimization of Quality Management Systems. *Sustainability* **2019**, *11*, 1913. [CrossRef]
8. Kasprzak, K.; Wojtunik-Kulesza, K.; Oniszczuk, T.; Kuboń, M.; Oniszczuk, A. Secondary Metabolites, Dietary Fiber and Conjugated Fatty Acids as Functional Food Ingredients against Overweight and Obesity. *Natl. Prod. Commun.* **2018**, *13*, 1073–1082. [CrossRef]
9. Gródek-Szostak, Z.; Szeląg-Sikora, A.; Sikora, J.; Korenko, M. Prerequisites for the cooperation between enterprises and business supportinstitutions for technological development. In *Business and Non-Profit Organizations Facing Increased Competition and Growing Customers' Demand*; Nowy Sacz Business School–National Louis University: Nowy Sącz, Poland, 2017; Volume 16, pp. 427–439.

10. Goglio, P.; Smith, W.N.; Grant, B.B.; Desjardins, R.L.; McConkey, B.G.; Campbell, C.A.; Nemecek, T. Accounting for soil carbon changes in agricultural life cycle assessment (LCA): A review. *J. Clean. Prod.* **2015**, *104*, 23–30. [CrossRef]
11. Audsley, E.; Stacey, K.; Parsons, D.J.; Williams, A.G. *Estimation of the Greenhouse Gas Emissions from Agricultural Pesticide Manufacture and Use*; Cranfield University: Cranfield, Bedford, 2009.
12. Eggleston, H.S.; Buendia, L.; Miwa, K.; Ngara, T.; Tanabe, K. *IPCC Guidelines for National Greenhouse Gas Inventories*; Institute for Global Environmental Strategies RIP: Hayama, Japan, 2009.
13. Tubiello, F.N.; Salvatore, M.; Rossi, S.; Ferrara, A.; Fitton, N.; Smith, P. The FAOSTAT database of greenhouse gas emissions from agriculture. *Environ. Res. Lett.* **2013**, *8*, 015009. [CrossRef]
14. Lan, K.; Yao, Y. Integrating Life Cycle Assessment and Agent-Based Modeling: A Dynamic Modeling Framework for Sustainable Agricultural Systems. *J. Clean. Prod.* **2019**, *238*, 117853. [CrossRef]
15. Forster, P.; Ramaswamy, V.; Artaxo, P.; Berntsen, T.; Betts, R.; Fahey, D.W.; Haywood, J.; Lean, J.; Lowe, D.C.; Myhre, G.; et al. Changes in Atmospheric Constituents and in Radiative Forcing. In *Climate Change 2007: The Physical Science Basis. Contribution of Working Group I to the Fourth Assessment Report of the Intergovernmental Panel on Climate Change*; Solomon, S., Qin, D., Manning, M., Chen, Z., Marquis, M., Averyt, K.B., Tignor, M., Miller, H.L., Eds.; Cambridge University Press: Cambridge, UK; New York, NY, USA, 2007; p. 212.
16. Kapusta-Duch, J.; Szeląg-Sikora, A.; Sikora, J.; Niemiec, M.; Gródek-Szostak, Z.; Kuboń, M.; Leszczyńska, T.; Borczak, B. Health-Promoting Properties of Fresh and Processed Purple Cauliflower. *Sustainability* **2019**, *11*, 4008. [CrossRef]
17. Bacenetti, J.; Lovarelli, D.; Fiala, M. Mechanisation of organic fertiliser spreading, choice of fertiliser and harvesting residue management as solutions for maize environmental impact mitigation. *Eur. J. Agron.* **2016**, *79*, 107–118. [CrossRef]
18. Liang, L.; Lal, R.; Ridoutt, B.G.; Du, Z.L.; Wang, D.P.; Wang, L.Y.; Wu, W.L.; Zhao, G.S. Life cycle assessment of China's agroecosystems. *Ecol. Indic.* **2018**, *88*, 341–350. [CrossRef]
19. Nemecek, T.; Dubois, D.; Huguenin-Elie, O.; Gaillard, G. Life cycle assessment of Swiss farming systems: I. Integrated and organic farming. *Agric. Syst.* **2011**, *104*, 217–232. [CrossRef]
20. De Luca, G.; Strkalj, N.; Manz, S. Nanoscale design of polarization in ultrathin ferroelectric heterostructures. *Nat. Commun.* **2017**, *8*, 1419. [CrossRef] [PubMed]
21. Devapriya, P.; Ferrell, W.; Geismar, N. Integrated production and distribution scheduling with a perishable product. *Eur. J. Oper. Res.* **2017**, *259*, 906–916. [CrossRef]
22. Montgomery, D.C. *Design and Analysis of Experiments*; John Wiley & Sons, Inc.: New York, NY, USA, 2001.
23. ISO 31000:2018. *Risk management*; ISO: Geneva, Switzerland, 2018.
24. ILCD. *General Giude for Lifr Cycle Assessment Detailed Guidance*; JRC European Commision, Publications Office of the European Union: Luxembourg, 2010; p. 394.
25. Kool, A.; Marinussen, M.; Blonk, H. *LCI Data for the Calculation Tool Feedprint for Greenhouse Gas Emissions of Feed Production and Utilization GHG Emissions of N, P and K Fertilizer Production*; Blonk Consultants: Gouda, The Netherlands, 2016. Available online: https://www.blonkconsultants.nl/wp-content/uploads/2016/06/fertilizer_production-D03.pdf (accessed on 26 March 2020).
26. IPCC. *Guidelines for National Greenhouse Gas Inventories, Volume 4: Agriculture, Forestry and Other Land Use*; Intergovernmental Panel on Climate Change: Geneva, Switzerland, 2006.
27. Novoa, R.; Tejeda, H. Evaluation of the N2O emissions from N in plant residues as affected by environmental and management factors. *Nutr. Cycl. Agroecosyst.* **2006**, *75*, 29–46. [CrossRef]
28. EPA Unated States Environmental Protection Agency. *Greenhouse Gas Inventory Guidance Direct Emissions from Mobile Combustion Sources*; U.S. EPA Center for Corporate Climate Leadership—GHG Inventory Guidance: Washington, DC, USA, 2016; p. 27.
29. Wang, J.; Rothausen, S.; Conway, D.; Zhang, L.; Xiong, W.; Holman, I.P.; Li, Y. China's water–energy nexus: Greenhouse-gas emissions from groundwater use for agriculture. *Environ. Res. Lett.* **2012**, *7*, 1–10. [CrossRef]
30. FAO. Global database of GHG emissions related to feed crops: Methodology. In *Version 1. Livestock Environmental Assessment and Performance Partnership*; FAO: Rome, Italy, 2017.
31. Niemiec, M. Efficiency of slow-acting fertilizer in the integrated cultivation of Chinese cabbage. *Ecol. Chem. Eng. A* **2014**, *21*, 333–346.
32. Niemiec, M.; Cupiał, M.; Szeląg-Sikora, A. Evaluation of the Efficiency of Celeriac Fertilization with the Use of Slow-acting Fertilizers. *Agric. Agric. Sci. Procedia* **2015**, *7*, 177–183. [CrossRef]

33. Gródek-Szostak, Z.; Malik, G.; Kajrunajtys, D.; Szeląg-Sikora, A.; Sikora, J.; Kuboń, M.; Niemiec, M.; Kapusta-Duch, J. Modeling the Dependency between Extreme Prices of Selected Agricultural Products on the Derivatives Market Using the Linkage Function. *Sustainability* **2019**, *11*, 4144. [CrossRef]
34. Cupiał, M.; Szeląg-Sikora, A.; Niemiec, M. Farm Machinery and Processes Management in Sustainable Agriculture Location: 7th International Scientific Symposium: Symposium proceedings, Gembloux, Belgium November, 25–27, 2015. *Agric. Agric. Sci. Procedia* **2015**, *7*, 64–69.
35. Sikora, J.; Niemiec, M.; Szeląg-Sikora, A.; Kuboń, M.; Olech, E.; Marczuk, A. Zgazowanie odpadów z przemysłowego przetwórstwa karpia. *Przem. Chem.* **2017**, *96*, 2275–2278. [CrossRef]
36. Zhang, J.; Jiang, J.; Tian, G. 2016. The potential of fertilizer management for reducing nitrous oxide emissions in the cleaner production of bamboo in China. *J. Clean. Prod.* **2016**, *112*, 2536–2544. [CrossRef]
37. Liang, J.-F.; An, J.; Gao, J.-Q.; Zhang, X.-Y.; Song, M.-H.; Yu, F.-H. Interactive effects of biochar and AMF on plant growth and greenhouse gas emissions from wetland microcosms. *Geoderma* **2019**, *346*, 11–17. [CrossRef]
38. Niemiec, M.; Komorowska, M.; Mudryk, K.; Jewiarz, M.; Sikora, J.; Szeląg-Sikora, A.; Rozkosz, A. *Evaluation of the Fertilizing Potential of Products Based on Torrefied Biomass and Valorized with Mineral Additives. Renewable Energy Sources: Engineering, Technology, Innovation*; Springer: Cham, Switherland, 2020; pp. 267–275.
39. Xiao, Y.; Peng, F.; Zhang, Y.; Wang, J.; Zhuge, Y.; Zhanga, S.; Gaoa, H. Effect of bag-controlled release fertilizer on nitrogen loss, greenhouse gas emissions, and nitrogen applied amount in peach production. *J. Clean. Prod.* **2019**, *234*, 258–274. [CrossRef]
40. Garg, A.; Kazunari, K.; Pulles, T. *IPCC Guidelines for National Greenhouse Gas Inventories*. Available online: https://www.ipcc-nggip.iges.or.jp/public/2006gl/pdf/2_Volume2/V2_1_Ch1_Introduction.pdf (accessed on 26 March 2020).
41. ISO 14040. *Environmental Management-Life Cycle Assessment-Principles and Framework*; ISO: Geneva, Switzerland, 2006.
42. Zhang, X.; Bol, R.; Rahn, C.; Xiao, G.; Meng, F.; Wu, W. Agricultural sustainable intensification improved nitrogen use efficiency and maintained high crop yield during 1980–2014 in Northern China. *Sci. Total Environ.* **2017**, *596–597*, 61–68. [CrossRef]
43. IFA. *Energy Efficiency and CO2 Emissions in Ammonia Production 2008–2009 Summary Report*; International Fertilizer Association: Paris, France, 2009.
44. Tongwane, M.; Mdlambuzi, T.; Moeletsi, M.; Tsubo, M.; Mliswa, V.; Grootboom, L. Greenhouse gas emissions from different crop production and management practices in South Africa. *Environ. Dev.* **2016**, *19*, 23–35. [CrossRef]
45. Jacxsens, L.; Luning, P.A.; Marcelis, W.J.; van Boekel, T.; Rovira, J.; Oses, S.; Kousta, M.; Drosinos, E.; Jasson, V.; Uyttendaele, M. Tools for the performance assessment and improvement of food safety management systems. *Trends Food Sci. Technol.* **2011**, *22* (Suppl. 1), 80–89. [CrossRef]
46. ISO. *TS-EN ISO 14067 Greenhouse Gases—Carbon Footprint of Products—Requirements and Guidelines for Quantification and Communication*; International Organization for Standardization: Geneva, Switzerland, 2013.
47. Wójcicki, Z. Methodology of examining energy consumption of agricultural production. *Problemy Inżynierii Rolniczej* **2015**, *23*, 17–29.
48. Niemiec, M.; Komorowska, M.; Szeląg-Sikora, A.; Sikora, J.; Kuzminova, N. Content of Ba, B, Sr and As in water and fish larvae of the genus Atherinidae, L. sampled in three bays in the Sevastopol coastal area. *J. Elem.* **2018**, *23*, 1009–1020. [CrossRef]
49. Goglio, P.; Williams, A.G.; Balta-Ozkan, N.; Harris, N.R.P.; Williamson, P.; Balta-Ozkan, N.; Huisingh, D.; Zhang, Z.; Tavoni, M. Advances and challenges of life cycle assessment (LCA) of greenhouse gas removal technologies to fight climate changes. *J. Clean. Prod.* **2020**, *244*, 118896. [CrossRef]
50. Kuźnia, M.; Wojciech, J.; Łyko, P.; Sikora, J. Analysis of the combustion products of biogas produced from organic municipal waste. *J. Power Technol.* **2015**, *95*, 158–165.
51. Niemiec, M.; Chowaniak, M.; Sikora, J.; Szeląg-Sikora, A.; Gródek-Szostak, Z.; Komorowska, M. Selected Properties of Soils for Long-Term Use in Organic Farming. *Sustainability* **2020**, *12*, 2509. [CrossRef]
52. Sikora, J.; Niemiec, M.; Tabak, M.; Gródek-Szostak, Z.; Szeląg-Sikora, A.; Kuboń, M.; Komorowska, M. Assessment of the Efficiency of Nitrogen Slow-Release Fertilizers in Integrated Production of Carrot Depending on Fertilization Strategy. *Sustainability* **2020**, *12*, 1982. [CrossRef]

53. Kuboń, M.; Niemiec, M.; Tabak, M.; Komorowska, M.; Gródek-Szostak, Z. Ocena zasobności gleby w przyswajalne związki siarki z wykorzystaniem ekstrahentów o zróżnicowanej zdolności ekstrakcji. *Przem. Chem.* **2020**, *99*, 581–584. [CrossRef]
54. Sikora, J.; Niemiec, M.; Szeląg-Sikora, A.; Gródek-Szostak, Z.; Kuboń, M.; Komorowska, M. The Effect of the Addition of a Fat Emulsifier on the Amount and Quality of the Obtained Biogas. *Energies* **2020**, *13*, 1825. [CrossRef]

© 2020 by the authors. Licensee MDPI, Basel, Switzerland. This article is an open access article distributed under the terms and conditions of the Creative Commons Attribution (CC BY) license (http://creativecommons.org/licenses/by/4.0/).

Article

A New Design for Wood Stoves Based on Numerical Analysis and Experimental Research [†]

Przemysław Motyl [1,*], Marcin Wikło [1,*], Julita Bukalska [2], Bartosz Piechnik [2] and Rafał Kalbarczyk [2]

1. Faculty of Mechanical Engineering, Kazimierz Pulaski University of Technology and Humanities in Radom, Stasieckiego 54, 26-600 Radom, Poland
2. Kratki.pl Marek Bal, W. Gombrowicza 4, 26-660 Wsola, Poland; j.bukalska@kratki.com (J.B.); b.piechnik@kratki.com (B.P.); r.kalbarczyk@kratki.pl (R.K.)
* Correspondence: p.motyl@uthrad.pl (P.M.); m.wiklo@uthrad.pl (M.W.)
† This paper is an extended version of our paper published in 6th International Conference Renewable Energy Sources engineering, technology, innovation, 12–14 June 2019, Krynica, Poland.

Received: 27 January 2020; Accepted: 20 February 2020; Published: 25 February 2020

Abstract: This work proposes a comprehensive approach to modifying the design of wood stoves with a heating power up to 20 kW, including design works, simulations, and experimental research. The work is carried out in two stages. In the first part, a numerical model is proposed of the fireplace insert including fluid flow, the chemical combustion reaction, and heat exchange (FLUENT software is applied to solve the problem). The results of the simulation were compared with the experiment carried out on the test bench. A comparison of the experimental and numerical results was made for the temperature distribution along with the concentration of CO, CO_2, and O_2. Construction changes were proposed in the second stage, together with numerical simulations whose goal was an increase in the efficiency of the heating devices. The results obtained show that the average temperature in the chimney flue, which has a low value that is a determinant of the higher efficiency of the heating devices, was reduced relative to the initial design of the fireplace intake by 11%–16% in all cases. The retrofit enhanced stable heat release from the wood stove, which increased the efficiency and reduced the harmful components of combustion.

Keywords: CFD; combustion simulation; stove; thermal measurements

1. Introduction

Renewable energy sources are an interesting alternative to fossil fuels, but their ecological effect is determined by the technology involved in their use in the production of heat or electricity. Resources of renewable energy sources, including biofuels, remain the subject of numerous analyses and scientific research works in many countries [1,2]. The combustion of wood remains one of the most popular sources of renewable energy in domestic appliances. Increasing environmental pollution and growing public awareness are among the main reasons for interest in renewable energy sources and the increase in scientists' efforts in the well-known energy conversion process. Burning biomass does not contribute to the greenhouse effect because burning wood releases the same amount of carbon dioxide into the atmosphere as plants absorb. The use of hearth-type fireplace inserts with manual fuel loading causes a high risk, and installations designed in accordance with the requirements of the relevant standards (given in the European document—Ecodesign) may not meet the environmental requirements. Tightening the environmental standards contributes to continuous work aimed at improving the performance of all devices admitted to trading in the EU. The requirements of the Ecodesign standard for fireplace inserts mainly relate to the value of CO and NO_x emissions and conducting the combustion process with high energy efficiency, as the high efficiency is strictly connected with low-emission combustion processes. Fireplace inserts are characterized by their simplicity of construction. Nevertheless, intensive research work in various areas connected with fireplaces has been observed.

Fireplaces are considered one of the parts of a micro energy system (MES), wherein multicriteria optimization is often utilized for the best management, i.e., the best environmental sustainability and lowest carbon emissions [3]. The fireplace as a source of heat can be included as an option when there is uncertainty about the MES, which is sometimes caused due to photovoltaic and wind power. Multicriteria optimization in the case of the bioeconomy is considered to support decision-making on residential heating alternatives. Taking into consideration the growth of the wood market, traditional log fireplaces are being replaced by pellets and wood stoves [4]. This article presents the fireplace as a heat source with an influence on the environment that is similar to the influence of wood stoves.

Research works have been carried out by both enterprises and research centers [5]. Some of the works are related to the modification of refractory materials used for the lining of combustion chambers, or the construction of external fireplace inserts [6]. The simplicity of design does not exclude work related to new air distribution solutions in the combustion chamber, which is crucial for ensuring minimum CO emissions [7]. The specificity of the product (fireplace inserts), as well as the increasingly popular design of fireplace inserts for individual consumer requirements, results in the need for continuous design work and rapid prototyping of new construction solutions. Detailed rules for the admission into the market of new constructions are defined by national and European standards. Numerical modeling is becoming an important element of the time limitation in the design of new constructions adhering to the tightened environmental standards. The process of numerical modeling of combustion processes and heat exchange in heating installations with manual fuel loading solves a number of problems related to the work cycle and the lack of automation of the process itself. The authors of [8–10] suggest a different method of numerical calculation for fireplaces up to 20 kW. Some authors [9] proposed only calculating the flow with the heat exchange where the wood logs are treated as a source of gas and heat, arguing that heat transfer is key to the final efficiency of the heating device. Utilizing only gas reactions in the fireplace is considered in [10]. This simplification can be correct with the assumption that the wood consists of 70–80% volatile parts. The authors of [8] proposed a more advanced simulation method of burning processes, considering solid and volatile components as well as the moisture naturally occurring in the wood.

The development of new construction solutions requires the use of both numerical models for fast prototyping and experimental research, which for this class of heating constructions should not be complicated or expensive. The research covered experimental and computational activities and the objective was to determine possible design improvements of a wood-open fireplace to minimize the total environmental impact. Two models of fireplaces with the same power will be introduced, with a difference in geometry.

2. Calculations and Experimental Research

The work was carried out in two stages. In the first part, a numerical model was proposed along with experimental tests of the fireplace insert, Case 0 (Figure 1), placed on the testing bench. The purpose of this stage was a preparation of a numerical tool to understand the physical-chemical process. A numerical model was prepared for a fireplace insert with manual fuel loading and a lockable furnace door with dimensions of 930 × 630 × 490 mm (Figure 1). Numerical simulations of the fluid flow, chemical combustion reaction, and heat exchange were performed by means of Fluent software of the Ansys INC (version 16.2) [11]. According to the construction data, the nominal power is 12 kW with a thermal efficiency of 70%. The outer casing was made of high-quality steel, the lining of the combustion chamber was chamotte, and the deflector was made of vermiculite. A numerical simulation was designed for the proposed fireplace insert to reflect the real operating conditions of the installation. From the simplified geometry of the fireplace insert, a solid was extracted as the computational domain and four cuboidal elements were added, being a model of wood logs (Figure 1b). For the prepared geometry, a tetrahedral numerical grid was prepared with a size of approx. 4 million elements. According to [5,8,10], for modeling combustion processes, the Finite-Rate/Eddy-Dissipation model [11] was used with the system of Equations (1)–(6): a two-variant turbulence model k–ε, where k is the turbulence kinetic energy, and ε is the dissipation rate [11], and radiation model DO [11]. A deciduous wood, hornbeam, was assumed as the fuel both during the experiment and in the simulation. Its composition is given in Table 1. In order to mimic the wood log in the model, it was assumed that it consists of the following:

- The outer part: this is the source of gas corresponding to the part of the volatile component of the wood. Volumetric reactions are represented by Equations (1) and (2):

$$C_{1.09} H_{2.1} O_{0.91} + 0.617 \cdot O_2 \rightarrow 1.097 \cdot CO + 1.054 \cdot H_2O \tag{1}$$

$$CO + 0.5 \cdot O_2 \rightarrow CO_2. \tag{2}$$

- The inner part corresponds to the part of the solid component of the wood, on the surface of which the reactions described by Equations (3)–(6) take place:

$$C(s) + 0.5 \cdot O_2 \rightarrow CO \tag{3}$$

$$C(s) + CO_2 \rightarrow 2CO \tag{4}$$

$$C(s) + H_2O \rightarrow H_2 + CO \tag{5}$$

$$H_2 + 0.5O_2 \rightarrow H_2O. \tag{6}$$

The stoichiometric coefficients for modeling the homogenous reaction (Equation (1)) have been calculated using a mass balance to each of the volatile element species, i.e., C, O, and H. Data on the assumed burning time, as well as the amount of wood (with the composition given in Table 1), were determined according to the experiment prepared earlier. When defining the model, the following were assumed: the vacuum at the outlet to the chimney was equal to −10 Pa; the fuel weight during the 66 min of the burning cycle was 3.92 kg, and the value of the throttle opening was 50%. The fuel mass assumed in the model was divided into 3960 s, which corresponds to 66 min of combustion in real conditions. An experiment was carried out following the norm [12] that determines that measurements should be taken throughout the entire burning process. The measurement was made on a laboratory test bench equipped with the following:

1. a flue gas analyzer recording the negative pressure, temperature, and flue gas composition (CO, CO_2, and O_2) in the chimney at a set distance from the outlet from the fireplace insert (Figures 2 and 3a);
2. a thermal camera registering the temperature on the surface of the walls and on the glass of the fireplace insert (Figure 3b); and,
3. a set of thermocouples recording temperature at the walls in the furnace chamber (Figures 3a and 4).

Table 1. Basic properties of wood.

Item	Value
Proximate analysis (wt %)	
Moisture	13.00
Volatile matter	73.08
Fixed carbon	13.92
Ultimate analysis (wt %) DAF	
Carbon	52.91
Hydrogen	5.95
Oxygen	41.14

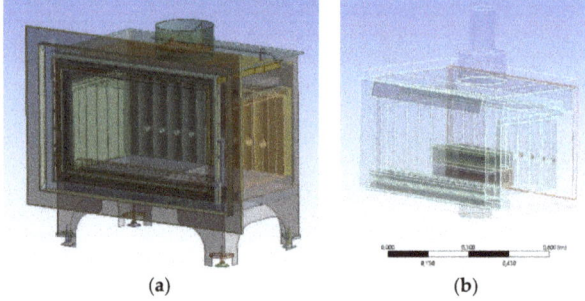

Figure 1. (a) View of the fireplace insert; (b) wood model (Case 0) [13].

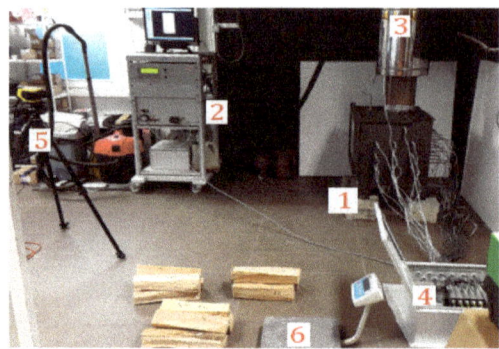

Figure 2. The facility of the testing bench: 1 fireplace insert with thermocouples, 2, 4 data acquisition units, 3 a vertical smoke conduit with mounted measuring section, 5 thermal camera, 6 electronic scale.

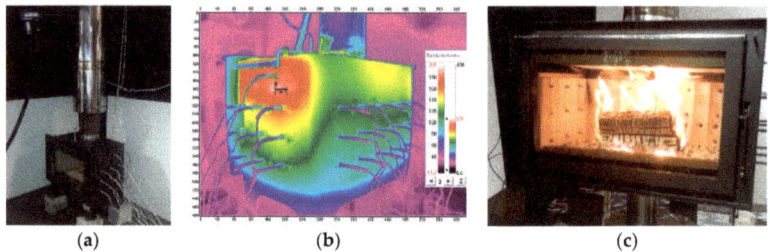

Figure 3. Testing bench facility, experiment: (a) fireplace with mounted thermocouples, (b) the view from the thermal imaging camera, (c) burning process [13].

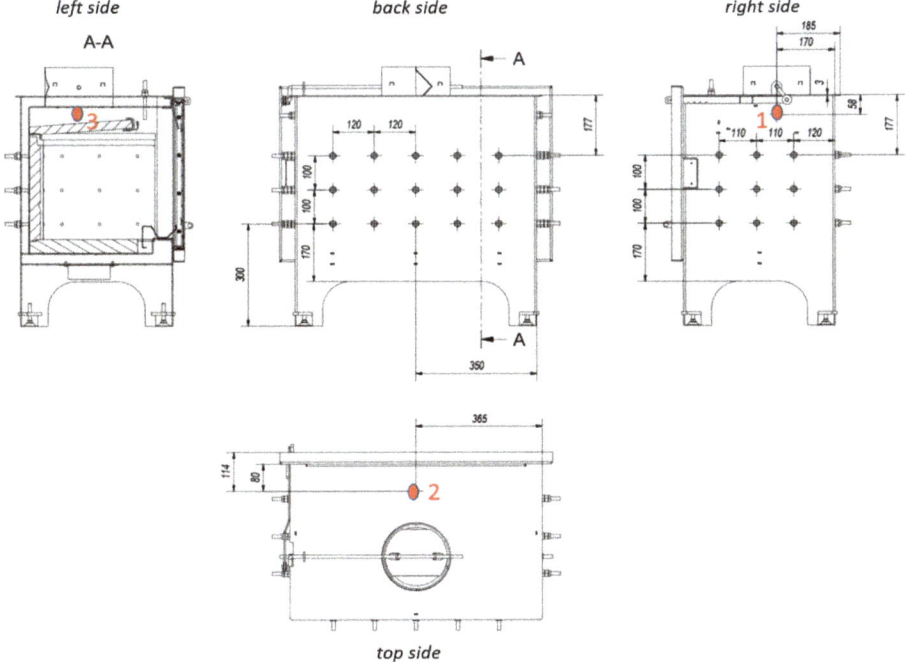

Figure 4. Diagram of three thermocouples' placement in the fireplace insert chosen for the comparative study (Case 0).

The next stage of the work was the introduction of structural changes leading to the higher efficiency of the device. For this purpose, a second numerical model was prepared covering the structural changes. It was assumed that the structural changes should be directed at increasing the heat exchange surface with the environment. Based on the measurement results of exhaust emissions, temperature, and simulation results obtained at the stage of preparation of the numerical model (Case 0), we decided to analyze several design options. The first of these (Case 1) consisted of extending the flue (Figure 5a), whose task was to extend the exhaust gas path and create an additional surface ensuring heat exchange with the environment. The second one (Case 2) assumed the addition of through channels (Figure 5b), whose purpose was to increase the heat exchange surface and disconnect the channels supplying additional air on the rear wall of the fireplace insert to simplify the design. The third (Case 3) assumed leaving the extended flue with through channels and leaving the channels supplying additional air on the rear wall of the fireplace insert (Figure 5c).

The chimney loss remains crucial in the final balance of thermal efficiency of the fireplace insert [12], expressed according to Equation (7):

$$\eta = 100 - (q_a + q_b), \quad (7)$$

where q_a is the relative flue gas loss and q_b is the relative loss of incomplete combustion.

Relative flue gas loss is directly proportional to the flue gas temperature in the chimney [12], which is why, in the analysis of fireplace insert design modification, it was decided to minimize the flue gas temperature in the chimney while maintaining CO emissions at the same level with the same fuel expenditure. The calculations were carried out using the numerical modeling technique from the first stage of the work (Case 0). The fuel output, vacuum in the chimney flue, and other boundary conditions were the same as in the first model (Case 0).

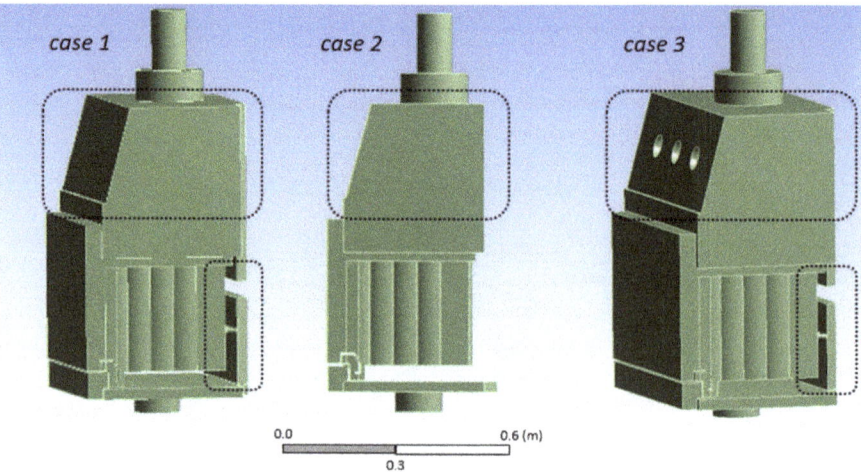

Figure 5. Three designs for a modified fireplace insert construction (Cases 1, 2, and 3) with marked structural elements that have been modified.

3. Simulation and Results

The degree of the simplification of the numerical model required conducting a parallel experiment at the laboratory stand for the first numerical model of the fireplace (Case 0). Two sets of data were analyzed, i.e., the composition of exhaust in the chimney, as well as the temperature values indicated by the thermocouples (Figures 6–8). Analyzing the results for the full work cycle of the fireplace insert, i.e., 66 min, it can be seen that, 15 min after the fireplace starts burning, there is a period between 15 and 45 min in which the operating conditions of the heating system are stabilized. The heating system reaches the highest flue gas temperature at the 45th min, after which the burning-off stage begins until the available fuel is consumed. Therefore, data from the numerical model (Case 0) were referred to for the period of stable operation (between 15 and 45 min). Table 2 presents the results of measurements and simulations in the chimney (Case 0). These are the results for the outlet plane to the chimney. In addition, the temperature values for three thermocouples, located as depicted in Figure 4, were compared (Figure 9). It is worth noting that the individual values are at a similar level and are subject to differences of several percentage points. Such differences may be acceptable considering the model simplifications that were made and the complex nature of the phenomena occurring in the said installation. The next stage of the work was to introduce construction changes leading to higher device efficiency. Based on the measurements taken and the results of simulations obtained at the stage of preparing the numerical model, a structural variant involving a change in the flue geometry was analyzed. Our task was to extend the route of the flue gas and create an additional surface to ensure the exchange of heat with the surroundings by swirling the exhaust stream.

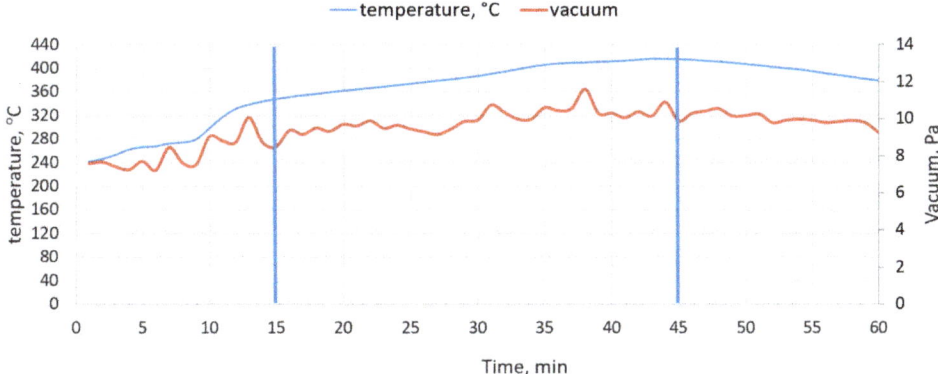

Figure 6. Vacuum in the combustion chamber and exhaust gas temperature during the full combustion cycle (Case 0).

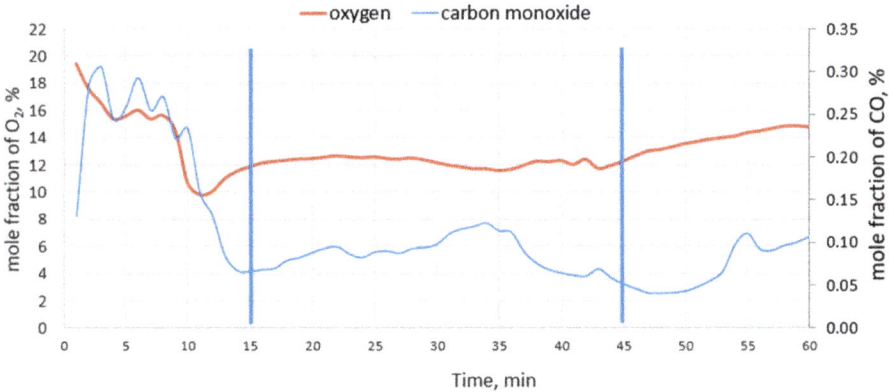

Figure 7. The concentration of O_2 and CO in the exhaust gas during the full combustion cycle (Case 0).

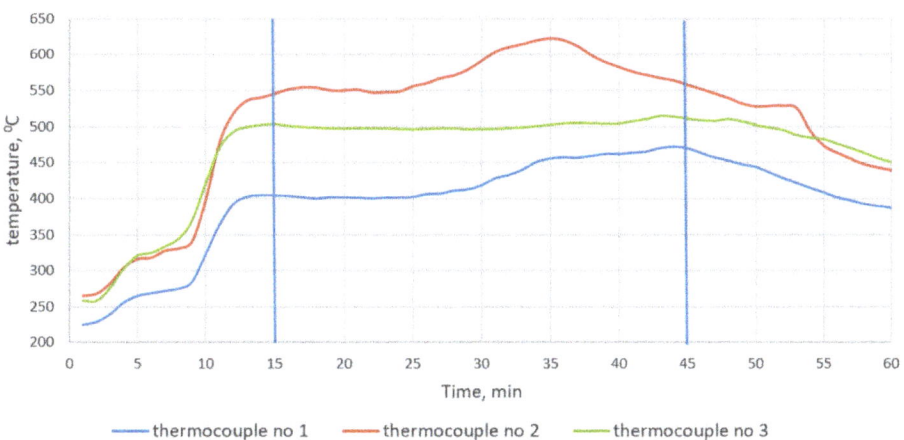

Figure 8. Change in temperature as a function of time for the three thermocouples (1, 2, and 3) shown in Figure 4, which register the exhaust gas temperature flowing around the deflector (Case 0).

Table 2. Measurement and simulation data acquired in the chimney.

Item	Experiment (Case 0) Minimum Value	Experiment (Case 0) Maximum Value	Simulation (Case 0)
Temperature, °C	348	416	447
Mole ratio O_2, %	11.61	12.69	11.1
Mole ratio, CO_2, %	8.1	8.87	8.47
Mole ratio, CO, %	0.1	0.12	0.05
Vacuum, Pa	9.9	11.63	10

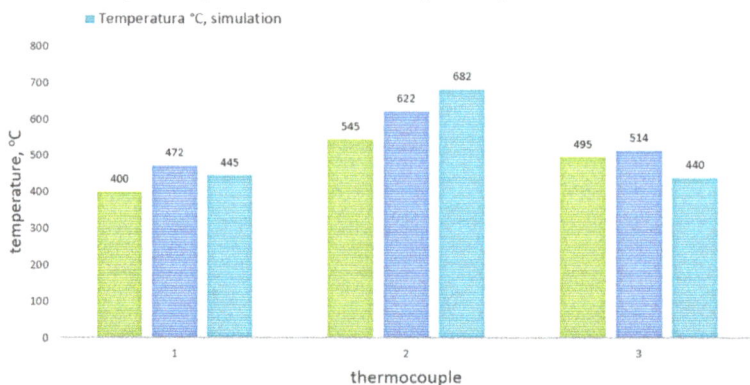

Figure 9. Temperature values (°C) comparison of the three thermocouples obtained from measurement and simulation (Case 0).

Figures 10–12 present the distribution of gas molecules along the temperature scale, as well as the predicted CO and O_2 concentrations in selected areas of the fireplace insert. The gas molecules in Figure 10a (Case 0) confirm the vertical flame propagation that could be observed during a bench experiment (Figure 3c). In the modified constructions, the redesigned flue gas chamber facilitates elongation of the exhaust gas path in the temperature zone, ensuring the combustion of the remaining combustible gases in the exhaust gas and minimal CO emission to the chimney. The obtained swirling of exhaust causes intense heat exchange with the environment and thus a reduction in the temperature of the flues in the chimney, which is directly related to the improvement of the efficiency of the modified fireplace insert. In both cases (Case 1 and Case 3), it should be emphasized that the holes in the back wall of the focal chamber fulfill their role of eliminating zones with a high concentration of CO. This treatment is necessary because the furnace chamber is characterized by large changes in temperature and the resulting lower temperature zone and high CO concentration can affect the emission standards obtained in the chimney. A comparison of the temperature in the chimney flue for all three variants of the modified structure (Figure 13) indicates that the most advantageous solution is the last design (Case 3). Figure 14 shows the final proposal for the fireplace insert (Case 3), together with a visualization of the swirl of exhaust gases that intensifies heat exchange with the environment heating installation. The recommended structural changes contribute to improving the efficiency of the final heating device.

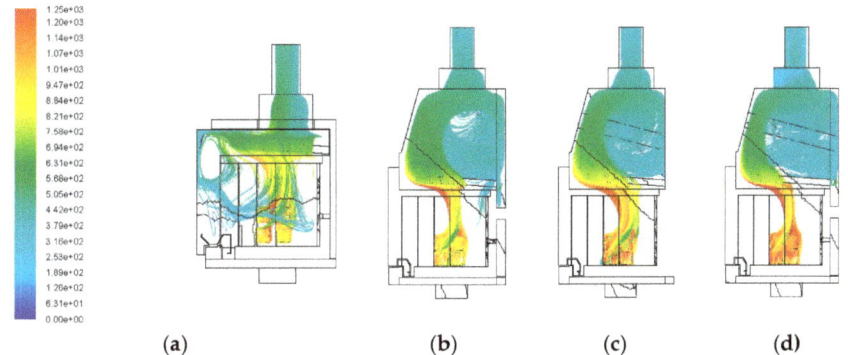

Figure 10. Path of the gas molecules and change of the temperature (°C) in the fireplace insert combustion chamber for: (**a**) Case 0, (**b**) Case 1, (**c**) Case 2, (**d**) Case 3.

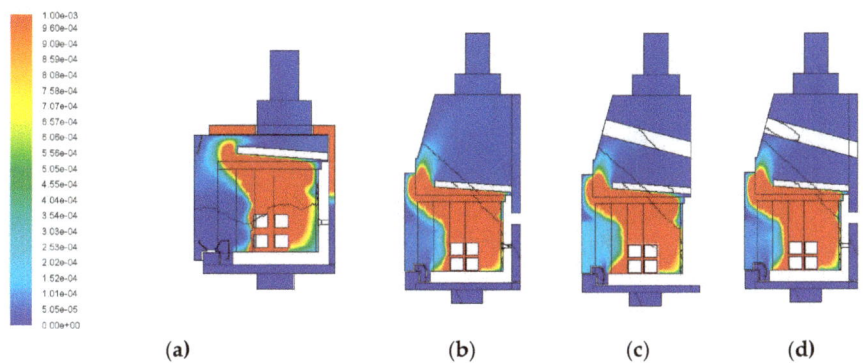

Figure 11. Concentration of CO (mole fraction) for (**a**) Case 0, (**b**) Case 1, (**c**) Case 2, (**d**) Case 3.

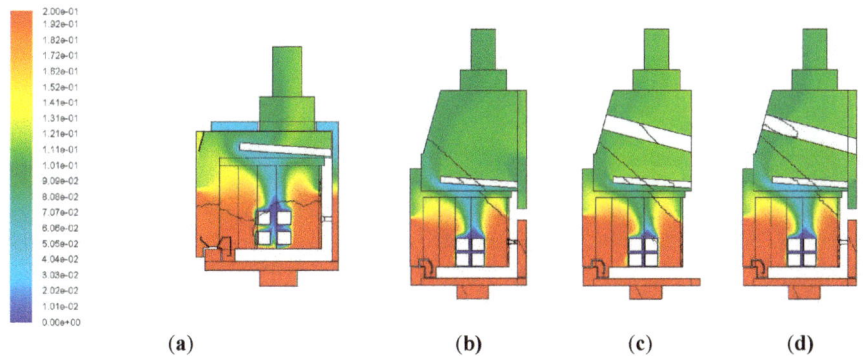

Figure 12. Concentration of O_2 (mole fraction) for (**a**) Case 0, (**b**) Case 1, (**c**) Case 2, (**d**) Case 3.

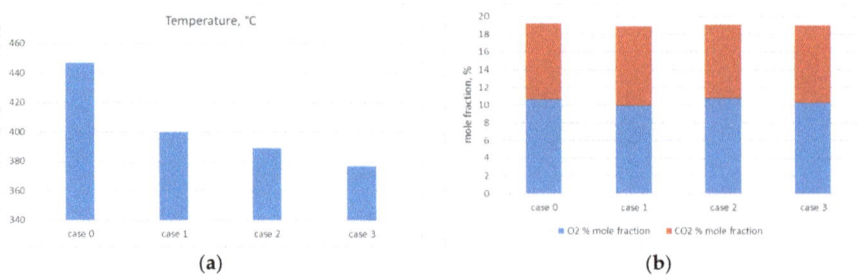

Figure 13. Comparison for the analysis structural variants of the fireplace insert (in the outlet plane of the fireplace for (**a**) temperatures; (**b**) O_2 and CO_2 emissions.

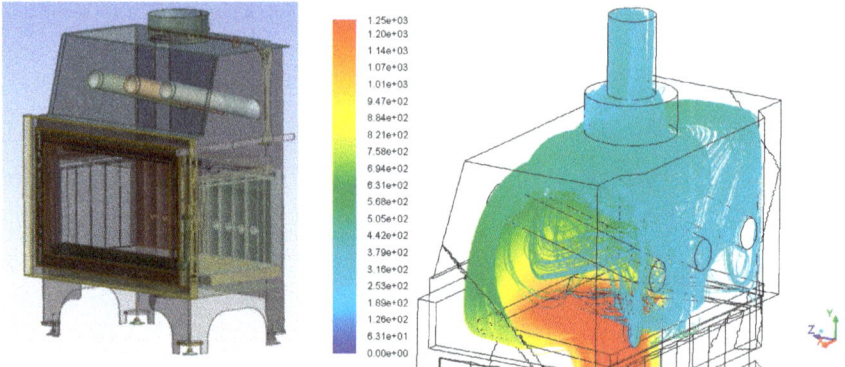

Figure 14. Gas particle paths in the fireplace (Case 3)—temperature scale (°C).

4. Conclusions

The need to implement many models of heating installations up to 20 kW with similar parameters under systematically tightened environmental standards means that numerical modeling of combustion processes and heat exchange is becoming an interesting method of rapid prototyping. The proposed computational technique, despite the introduction of model simplifications, allows for obtaining numerical data to lead to new construction designs within a few hours with the assumption of a numerical grid of 3–4 million elements. The obtained results show that the average temperature in the chimney flue was reduced relative to Case 0 by 11%–16% in all cases. Comparing the modified construction (Case 3) and the initial variant (Case 0), the design changes in the final design were recommended. The redesign of the smoke conduit to increase the heat exchange surface and flue gas turbulence, as well as modification of the air supply ducts to the rear wall of the combustion chamber, contributed to the lowering of the flue gas temperature, and thus the reduction of the chimney losses of the installation. Transient CFD simulations of log-fired stoves can be part of further research.

Author Contributions: P.M. was responsible for the preparation of the numerical burning process, numerical models, the conduction of the simulation, and a summary of the research. M.W. was responsible for the measurement methodology, the preparation of the testing equipment demands, and the analysis of the data. J.B. was responsible for supervision of the testing laboratory and the coordination of the tests. B.P. was responsible for conducting the measurements, the preparation of measuring equipment, and the preparation of the CAD model for the Case 0 simulation. R.K. was responsible for the preparation of the testing bench according to the codes, and the preparation and conducting of the thermal measurements. All authors have read and agreed to the published version of the manuscript.

Funding: This research was funded by National Centre for Research and Development, Poland, grant number POIG.01.04.00-14-351/13.

Conflicts of Interest: The authors declare no conflict of interest.

References

1. Lazaroiu, G.; Mihaescu, L.; Pisa, I.; Negreanu, G.P.; Berbece, V. The role of agricultural biomass in the production of ecological energy in Romania. In Proceedings of the International Multidisciplinary Scientific Geoconference, Albena, Bulgaria, 30 June–9 July 2018; Volume 18, pp. 721–726.
2. Lazaroiu, G.; Mihaescu, L.; Negreanu, G.; Pana, C.; Pisa, I.; Cernat, A.; Ciupageanu, D.A. Experimental investigations of innovative biomass energy harnessing solutions. *Energies* **2018**, *11*, 3469. [CrossRef]
3. Xing, T.; Lin, H.; Tan, Z.; Ju, L. Coordinated Energy Management for Micro Energy Systems Considering Carbon Emissions Using Multi-Objective Optimization. *Energies* **2019**, *12*, 4414. [CrossRef]
4. Martín-Gamboa, M.; Dias, L.C.; Quinteiro, P.; Freire, F.; Arroja, L.; Dias, A.C. Multi-Criteria and Life Cycle Assessment of Wood-Based Bioenergy Alternatives for Residential Heating: A Sustainability Analysis. *Energies* **2019**, *12*, 4391. [CrossRef]
5. Menghini, D.; Marra, F.S.; Allouis, C.; Beretta, F. Effect of excess air on the optimization of heating appliances for biomass combustion. *Exp. Therm. Fluid Sci.* **2008**, *32*, 1371–1380. [CrossRef]
6. Sornek, K.; Filipowicz, M.; Rzepka, K. Study of clean combustion of wood in a stove-fireplace with accumulation. *J. Energy Inst.* **2017**, *90*, 613–623. [CrossRef]
7. Carvalho, R.L.; Vicente, E.D.; Tarelho, L.A.C.; Jensen, O.M. Wood stove combustion air retrofits: A low cost way to increase energy savings in dwellings. *Energy Build.* **2018**, *164*, 140–152. [CrossRef]
8. Scharler, R.; Benesch, C.; Neudeck, A.; Obernberger, I. CFD based design and optimisation of wood log fired stoves. In Proceedings of the 17th European Biomass Conference and Exhibition, From Research to Industry and Markets, Hamburg, Germany, 29 June–3 July 2009; p. 7.
9. Menghini, D.; Marchione, T.; Martino, G.; Marra, F.S.; Allouis, C. Numerical and experimental investigations to lower environmental impact of an open fireplace. *Exp. Therm. Fluid Sci.* **2007**, *31*, 477–482. [CrossRef]
10. Bugge, M.; Skreiberg, Ø.; Haugen, N.E.L.; Carlsson, P.; Seljeskog, M. Predicting NOx Emissions from Wood Stoves using Detailed Chemistry and Computational Fluid Dynamics. *Energy Procedia* **2015**, *75*, 1740–1745. [CrossRef]
11. Ansys Fluent Theory Guide Release 16.2. Available online: http://www.pmt.usp.br/ACADEMIC/martoran/NotasModelosGrad/ANSYS%20Fluent%20Theory%20Guide%2015.pdf (accessed on 24 February 2020).
12. The Standard EN 13229:2001/A1:2003/A2:2004/AC:2007 Inset Application Including Open Fires Fired by Solid fuels—Requirements and Test Methods. Available online: https://standards.cen.eu/dyn/www/f?p=204:110:0::::FSP_PROJECT,FSP_ORG_ID:20787,6276&cs=18A8C4C833F81E864263EB7558C3F6684 (accessed on 24 February 2020).
13. Motyl, P.; Wikło, M.; Olejarczyk, K.; Kołodziejczyk, K.; Kalbarczyk Rafałand Piechnik, B.; Bukalska, J. Numerical Analysis of the Combustion Process in the Wood Stove. In *Renewable Energy Sources: Engineering, Technology, Innovation*; Wróbel, M., Jewiarz, M., Andrzej, S., Eds.; Springer International Publishing: Cham, Switzerland, 2020; pp. 229–237.

© 2020 by the authors. Licensee MDPI, Basel, Switzerland. This article is an open access article distributed under the terms and conditions of the Creative Commons Attribution (CC BY) license (http://creativecommons.org/licenses/by/4.0/).

MDPI
St. Alban-Anlage 66
4052 Basel
Switzerland
Tel. +41 61 683 77 34
Fax +41 61 302 89 18
www.mdpi.com

Energies Editorial Office
E-mail: energies@mdpi.com
www.mdpi.com/journal/energies

www.ingramcontent.com/pod-product-compliance
Lightning Source LLC
LaVergne TN
LVHW070428100526
838202LV00014B/1551